T0184824

Lectures on
Gravitation

Lectures on Gravitation

Ashok Das

University of Rochester, USA
Saha Institute of Nuclear Physics, India

 **World Scientific**

NEW JERSEY · LONDON · SINGAPORE · BEIJING · SHANGHAI · HONG KONG · TAIPEI · CHENNAI

Published by

World Scientific Publishing Co. Pte. Ltd.

5 Toh Tuck Link, Singapore 596224

USA office: 27 Warren Street, Suite 401-402, Hackensack, NJ 07601

UK office: 57 Shelton Street, Covent Garden, London WC2H 9HE

British Library Cataloguing-in-Publication Data
A catalogue record for this book is available from the British Library.

LECTURES ON GRAVITATION

Copyright © 2011 by World Scientific Publishing Co. Pte. Ltd.

All rights reserved. This book, or parts thereof, may not be reproduced in any form or by any means, electronic or mechanical, including photocopying, recording or any information storage and retrieval system now known or to be invented, without written permission from the Publisher.

For photocopying of material in this volume, please pay a copying fee through the Copyright Clearance Center, Inc., 222 Rosewood Drive, Danvers, MA 01923, USA. In this case permission to photocopy is not required from the publisher.

ISBN-13 978-981-4329-37-8
ISBN-10 981-4329-37-1
ISBN-13 978-981-4329-38-5 (pbk)
ISBN-10 981-4329-38-X (pbk)

Printed in Singapore by World Scientific Printers.

To
Cristiane
and
Débora and Ivana

Preface

Over the past several years, I have taught a one-semester graduate course on gravitation at the University of Rochester. The principles of general relativity were developed systematically in a traditional manner, and this book comprises my lectures of this course. The class was attended by both graduate as well as several undergraduate students. Correspondingly, the subject matter was developed in detail using tensor analysis (rather than the compact differential geometric concepts). The concepts and ideas of the material were supplemented by a lot of worked out examples dispersed through out the lectures, which the students found extremely useful.

Although it is the normal convention in a course on gravity to use a metric of signature $(-, +, +, +)$, I have used through out the lectures the conventional Bjorken-Drell metric signature $(+, -, -, -)$ for consistency with my other books. Since the book consists of my actual lectures in the course, it is informal with step by step derivation of every result which is quite helpful to students, particularly to undergraduate students.

There are many references to the material covered in this course and it is impractical to give all the references. Instead, let me note only the few following books on the subject which can be used as references and which contain further references on the subject.

1. *Gravitation and Cosmology*, S. Weinberg, John Wiley (1972), New York.

2. *Gravitation*, C. W. Misner, K. S. Thorne and J. A. Wheeler, W. H. Freeman (1973), San Francisco.

3. *Introduction to General Relativity*, R. Adler, M. Bazin and M. Schiffer, McGraw-Hill (1975), New York.

The figures in this book were drawn using Jaxodraw as well as PSTricks. I am grateful to the people who developed these extremely useful software. Finally, I would like to thank Fernando Méndez for his help with some of the figures and Dave Munson for sorting out various computer related problems.

Ashok Das
Rochester

Contents

CHAPTER 1

Basics of geometry and relativity

1.1 Two dimensional geometry

We have all heard of the statement that "Gravitation is the study of geometry". Let us see what this exactly means. We are, of course, familiar with Euclidean geometry. For simplicity we will restrict here only to two dimensions so that the study of geometry of the manifold corresponds to the study of surfaces and Euclidean geometry corresponds simply to the all familiar plane geometry.

A geometry, of course, has built in assumptions. For example, one has to define the basic elements (concepts) of geometry such as points. Furthermore, we know that given any two points we can draw many curves joining them. However, the straight line, as we know, defines the shortest path between two points in Euclidean geometry. Euclid's axioms, among other things, state that a straight line is determined by two points, it may have indefinite length, that any two right angles are equal, that only one line parallel to a given line can pass through a point outside the line in the same plane, that when equals (equal line segments) are added to equals, the sums are always equal and that the whole is greater than any of its parts. From these axioms Euclid deduced hundreds of theorems which tell us a lot about Euclidean geometry.

For example, let us consider a triangle ABC in plane geometry shown in Fig. 1.1. Through the vertex A let us draw a straight line DAE parallel to BC. From the properties of the parallel lines we immediately conclude that ($\widehat{DAB} = $ Angle DAB)

$$\text{Angle } DAB = \text{ Angle } ABC, \quad \text{or,} \quad \widehat{DAB} = \widehat{ABC}, \tag{1.1}$$

and

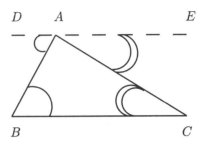

Figure 1.1: A triangle in plane geometry.

Angle $EAC =$ Angle ACB,   or, $\widehat{EAC} = \widehat{ACB}$.   (1.2)

On the other hand, we also know that

$$\widehat{DAB} + \widehat{BAC} + \widehat{EAC} = 180°. \qquad (1.3)$$

Thus, we conclude that

$$\widehat{DAB} + \widehat{BAC} + \widehat{EAC} = \widehat{ABC} + \widehat{BAC} + \widehat{ACB} = 180°. \qquad (1.4)$$

This, of course, proves the familiar result that the sum of the angles of a triangle on a plane equals 180°. Similarly many more interesting results can be obtained by using Euclid's axioms.

However, it is Euclid's axiom on the parallel lines that generated much interest. It says, "If a straight line falling on two straight lines makes the interior angles on the same side less than two right angles, the two straight lines produced indefinitely meet on that side on which the angles are less than two right angles", which is shown in Fig. 1.2. Since this axiom involves extending straight lines to the inexplorable region of spatial infinity, people tried hard to prove that this follows from the other axioms of Euclid and that this is not truly an independent axiom.

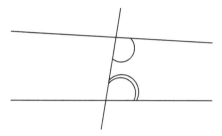

Figure 1.2: Two straight lines in plane geometry which are not parallel according to Euclid's axiom.

The study of this question, on the other hand, led to the birth of other, non-Euclidean geometries which satisfy all the axioms of Euclid except the one on parallel lines. These are geometries of curved surfaces where the definition of a straight line carries over to what we know as geodesics. (As we will see later, geodesics are paths along which the length is extremal.) One such geometry is the geometry of the surface of a sphere. On a sphere the geodesics ("straight lines") are parts of great circles joining the opposite pole points. It is clear that since any two great circles must intersect (at the poles), on a sphere we cannot draw any line parallel to a given line through a point outside of the line.

Another geometry which was developed independently by Gauss, Bolyai and Lobachevski is opposite in its properties to the previous one (namely, the geometry of a sphere). Therefore, it is also known as the geometry of the surface of a pseudosphere. It is rather hard to visualize. But a crude example of this geometry is the surface of a horse saddle. On such a surface, many geodesics can be drawn through a given point not intersecting a given line outside of this point and hence in this geometry, it is possible to draw many lines parallel to a given line through a point outside of the line. These are the three possible distinct geometries we can have.

In a geometry described by curved surfaces it is no longer true that the sum of all angles in a triangle equals 180°. In fact, one can simply show that for a triangle on a sphere the sum of angles is greater than 180° whereas for the geometry of Gauss, Bolyai and Lobachevski, it is less than 180°. Construction of these geometries showed that the fifth postulate of Euclid is indeed independent.

A question we can ask at this point is what is the geometry of the space we live in. Is it flat, or is it like a sphere or is it like a pseudosphere? If we were two dimensional creatures living in a two dimensional space, we can simply study the properties of geometrical figures and deduce the nature of the geometry ourselves. But this does not seem right. There ought to exist a general quantity which distinguishes one geometry from the others. It must be an inner property of the surface (or the manifold) and let us determine this.

First of all let us note that although a straight line is defined as the shortest path between two points, we still have not defined the concept of a distance. In plane geometry if two points have coordinates (x_1, x_2) and (y_1, y_2), then we know that the shortest distance between the two is given by

$$d(x,y) = \left[(x_1 - y_1)^2 + (x_2 - y_2)^2 \right]^{\frac{1}{2}}. \qquad (1.5)$$

Furthermore, it was shown that for the geometry of Gauss, Bolyai and Lobachevski, a point with coordinates (x_1, x_2) satisfies (suitably scaled so that the coordinates are dimensionless)

$$x_1^2 + x_2^2 < 1, \qquad (1.6)$$

and the shortest distance between two points (x_1, x_2) and (y_1, y_2) is given by

$$\cosh \frac{d(x,y)}{a} = \frac{1 - x_1 y_1 - x_2 y_2}{\left(1 - x_1^2 - x_2^2\right)^{\frac{1}{2}} \left(1 - y_1^2 - y_2^2\right)^{\frac{1}{2}}}. \qquad (1.7)$$

Here "a" is a length scale set by the problem under study.

Gauss realized that the distance $d(x,y)$ which gives the shortest path between two points determines the inner properties of a space. Thus for example, a cylinder or a cone has the same inner properties as a plane since a plane can be rolled into a cylinder or a cone without stretching or tearing. (The formula for the distance on a cylinder or a cone is the same as on a plane.) However, as we all know, a sphere cannot be unrolled into a plane without tearing it and correspondingly the formula for the distance on a sphere is different.

Gauss also recognized that any space can be locally identified with a Euclidean space. Hence locally we can set up a Euclidean coordinate system $\xi_i, i = 1, 2$, to describe the space so that the infinitesimal distance between the points (ξ_1, ξ_2) and $(\xi_1 + d\xi_1, \xi_2 + d\xi_2)$ can be written as

$$ds^2 = d\xi_1^2 + d\xi_2^2. \tag{1.8}$$

Of course, if the surface is curved, this coordinate system cannot cover any finite space still satisfying the laws of plane geometry. Let us suppose that there exists a coordinate system (x_1, x_2) which covers the entire surface. Then we can write the distance between two points with the coordinates (x_1, x_2) and $(x_1 + dx_1, x_2 + dx_2)$ (which locally coincides with (ξ_1, ξ_2) and $(\xi_1 + d\xi_1, \xi_2 + d\xi_2)$) as

$$
\begin{aligned}
ds^2 &= d\xi_1^2 + d\xi_2^2 \\
&= \left(\frac{\partial \xi_1}{\partial x_1} dx_1 + \frac{\partial \xi_1}{\partial x_2} dx_2 \right)^2 + \left(\frac{\partial \xi_2}{\partial x_1} dx_1 + \frac{\partial \xi_2}{\partial x_2} dx_2 \right)^2 \\
&= \left[\left(\frac{\partial \xi_1}{\partial x_1} \right)^2 + \left(\frac{\partial \xi_2}{\partial x_1} \right)^2 \right] dx_1^2 \\
&\quad + 2 \left[\frac{\partial \xi_1}{\partial x_1} \frac{\partial \xi_1}{\partial x_2} + \frac{\partial \xi_2}{\partial x_1} \frac{\partial \xi_2}{\partial x_2} \right] dx_1 dx_2 \\
&\quad + \left[\left(\frac{\partial \xi_1}{\partial x_2} \right)^2 + \left(\frac{\partial \xi_2}{\partial x_2} \right)^2 \right] dx_2^2 \\
&= g_{11}(x) dx_1^2 + 2g_{12}(x) dx_1 dx_2 + g_{22}(x) dx_2^2 \\
&= g_{ij} dx_i dx_j, \tag{1.9}
\end{aligned}
$$

where we have identified

$$g_{11}(x) = \left(\frac{\partial\xi_1}{\partial x_1}\right)^2 + \left(\frac{\partial\xi_2}{\partial x_1}\right)^2,$$

$$g_{12}(x) = \frac{\partial\xi_1}{\partial x_1}\frac{\partial\xi_1}{\partial x_2} + \frac{\partial\xi_2}{\partial x_1}\frac{\partial\xi_2}{\partial x_2} = g_{21}(x),$$

$$g_{22}(x) = \left(\frac{\partial\xi_1}{\partial x_2}\right)^2 + \left(\frac{\partial\xi_2}{\partial x_2}\right)^2. \tag{1.10}$$

The form of ds^2 in (1.9) is the sign of a metric space and $g_{ij}, i, j = 1, 2$ represents the metric tensor of the two dimensional space. Thus, for example, the metric for a plane where

$$ds^2 = dx_1^2 + dx_2^2, \tag{1.11}$$

is given by

$$g_{11} = g_{22} = 1, \quad g_{12} = 0. \tag{1.12}$$

For a sphere of fixed radius r, on the other hand,

$$ds^2 = r^2 \left(d\theta^2 + \sin^2\theta d\phi^2\right). \tag{1.13}$$

Thus, in this case, the metric takes the form

$$g_{\theta\theta} = r^2, \quad g_{\theta\phi} = 0, \quad g_{\phi\phi} = r^2\sin^2\theta. \tag{1.14}$$

It is this dependence of the metric on the coordinate θ that gives the sphere different inner properties from that of a plane. We can also show that for the geometry of Gauss, Bolyai and Lobachevski, the metric is given by

$$g_{11}(x) = \frac{a^2\left(1 - x_2^2\right)}{\left(1 - x_1^2 - x_2^2\right)^2},$$

$$g_{12}(x) = \frac{a^2 x_1 x_2}{\left(1 - x_1^2 - x_2^2\right)^2},$$

$$g_{22}(x) = \frac{a^2\left(1 - x_1^2\right)}{\left(1 - x_1^2 - x_2^2\right)^2}. \tag{1.15}$$

Once we know the metric functions, the length of a path can be obtained by integrating ds along that path. Thus the metric functions determine all the inner properties of a space. Unfortunately, however, they are also coordinate dependent. Note that if we parameterize a plane with Cartesian coordinates as in (1.12), then we have

$$g_{11} = g_{22} = 1, \quad \text{and} \quad g_{12} = 0. \tag{1.16}$$

However, if we parameterize the same surface with polar coordinates, then we can write

$$ds^2 = dr^2 + r^2 d\theta^2, \tag{1.17}$$

so that

$$g_{rr} = 1, \qquad g_{\theta\theta} = r^2, \qquad \text{and} \qquad g_{r\theta} = 0. \tag{1.18}$$

This does not look like an Euclidean space (the metric component $g_{\theta\theta}$ is a function of coordinates), but it really is the same space. This example shows that although the metric functions determine all the inner properties of a space, since their forms depend on the particular coordinate system used, they are not very useful in classifying spaces. We should look for an object constructed from g_{ij} that does not depend on the choice of coordinate system.

In fact, Gauss had found a unique function (in two dimensions) that does this. It is called the Gaussian curvature and is defined (for two dimensional spaces) as

$$
\begin{aligned}
K(x_1, x_2) = \frac{1}{2g} &\left[2 \frac{\partial^2 g_{12}}{\partial x_1 \partial x_2} - \frac{\partial^2 g_{11}}{\partial x_2^2} - \frac{\partial^2 g_{22}}{\partial x_1^2} \right] \\
&- \frac{g_{11}}{4g^2} \left[\left(\frac{\partial g_{22}}{\partial x_2} \right) \left(2 \frac{\partial g_{12}}{\partial x_1} - \frac{\partial g_{11}}{\partial x_2} \right) - \left(\frac{\partial g_{22}}{\partial x_1} \right)^2 \right] \\
&- \frac{g_{22}}{4g^2} \left[\left(\frac{\partial g_{11}}{\partial x_1} \right) \left(2 \frac{\partial g_{12}}{\partial x_2} - \frac{\partial g_{22}}{\partial x_1} \right) - \left(\frac{\partial g_{11}}{\partial x_2} \right)^2 \right]
\end{aligned}
$$

$$+ \frac{g_{12}}{4g^2} \left[\left(\frac{\partial g_{11}}{\partial x_1} \right) \left(\frac{\partial g_{22}}{\partial x_2} \right) - 2 \left(\frac{\partial g_{11}}{\partial x_2} \right) \left(\frac{\partial g_{22}}{\partial x_1} \right) \right.$$

$$\left. + \left(2 \frac{\partial g_{12}}{\partial x_1} - \frac{\partial g_{11}}{\partial x_2} \right) \left(2 \frac{\partial g_{12}}{\partial x_2} - \frac{\partial g_{22}}{\partial x_1} \right) \right].$$

$$(1.19)$$

Here g represents the determinant of the two dimensional metric g_{ij}, namely,

$$g = \det g_{ij} = g_{11}g_{22} - g_{12}^2. \tag{1.20}$$

It is clear from the expressions in (1.19) as well as (1.16) that the Gaussian curvature vanishes for a plane surface (geometry),

$$K = 0. \tag{1.21}$$

For a sphere, on the other hand, we note from (1.14) that (we can identify $x_1 = \theta, x_2 = \phi, g_{11} = g_{\theta\theta}, g_{22} = g_{\phi\phi}, g_{12} = g_{\theta\phi}$)

$$g = r^4 \sin^2 \theta,$$

$$\frac{\partial g_{\theta\theta}}{\partial \theta} = \frac{\partial g_{\theta\theta}}{\partial \phi} = \frac{\partial g_{\phi\phi}}{\partial \phi} = 0,$$

$$\frac{\partial g_{\phi\phi}}{\partial \theta} = 2r^2 \sin \theta \cos \theta = r^2 \sin 2\theta,$$

$$\frac{\partial^2 g_{\phi\phi}}{\partial \theta^2} = 2r^2 \cos 2\theta. \tag{1.22}$$

Thus, for a sphere we obtain

$$\begin{aligned} K &= \frac{1}{2g} \left(-\frac{\partial^2 g_{\phi\phi}}{\partial \theta^2} \right) - \frac{g_{\theta\theta}}{4g^2} \left(-\left(\frac{\partial g_{\phi\phi}}{\partial \theta} \right)^2 \right) \\ &= \frac{1}{2r^4 \sin^2 \theta} \left(-2r^2 \cos 2\theta \right) - \frac{r^2}{4r^8 \sin^4 \theta} \left(-r^4 \sin^2 2\theta \right) \\ &= -\frac{1}{r^2 \sin^2 \theta} \left(1 - 2\sin^2 \theta \right) + \frac{1}{4r^2 \sin^4 \theta} \left(4\sin^2 \theta \cos^2 \theta \right) \end{aligned}$$

$$= -\frac{1}{r^2 \sin^2 \theta} \left(1 - 2\sin^2 \theta - \cos^2 \theta\right)$$

$$= -\frac{1}{r^2 \sin^2 \theta} \left(-\sin^2 \theta\right) = \frac{1}{r^2} > 0. \tag{1.23}$$

That is, the surface of a sphere is a space of constant (since r is a fixed constant) positive curvature. Similarly, one can show that the surface of a pseudosphere has a constant negative curvature,

$$K = -\frac{1}{a^2} < 0. \tag{1.24}$$

We can check that the Gaussian curvature is independent of the coordinate system used by simply calculating the curvature of a plane in polar coordinates. Let us note from (1.18) that, in this case, (we can identify $x_1 = r, x_2 = \theta, g_{11} = g_{rr}, g_{22} = g_{\theta\theta}, g_{12} = g_{r\theta}$)

$$g = r^2,$$

$$\frac{\partial g_{rr}}{\partial r} = \frac{\partial g_{rr}}{\partial \theta} = \frac{\partial g_{\theta\theta}}{\partial \theta} = 0,$$

$$\frac{\partial g_{\theta\theta}}{\partial r} = 2r,$$

$$\frac{\partial^2 g_{\theta\theta}}{\partial r^2} = 2. \tag{1.25}$$

Thus, in these coordinates

$$K(r, \theta) = \frac{1}{2g} \left[-\frac{\partial^2 g_{\theta\theta}}{\partial r^2} \right] - \frac{g_{rr}}{4g^2} \left[-\left(\frac{\partial g_{\theta\theta}}{\partial r}\right)^2 \right]$$

$$= \frac{1}{2r^2} [-2] - \frac{1}{4r^4} \left[-(2r)^2 \right]$$

$$= -\frac{1}{r^2} + \frac{1}{r^2} = 0. \tag{1.26}$$

Thus we see that if we can calculate the Gaussian curvature of a surface, we can distinguish between the three geometries independent of the choice of coordinate system.

So far we have been dealing with only two dimensions and hence there is only one quantity independent of the coordinate system used. This can be intuitively seen from the fact that since the metric is symmetric, in two dimensions, there can only be three independent metric components. There are, however, two coordinates to represent any system. Hence the number of quantities that can be formed which are independent of the coordinates used is exactly one (namely, $3 - 2 = 1$). In higher dimensions, say D dimensions, the number of independent metric components is

$$\frac{1}{2} D(D + 1). \tag{1.27}$$

However, since there are D coordinates to specify a system, the number of independent quantities that can be constructed which are independent of the coordinate system used is

$$\frac{1}{2} D(D + 1) - D = \frac{1}{2} D(D - 1). \tag{1.28}$$

As a result, the study of geometry becomes more complicated when $D > 2$. This general problem was solved by Riemann and was extended to physics by Einstein.

▶ **Example (Pseudosphere in two dimensions).** For the pseudosphere in two dimensions we have (see (1.7))

$$\cosh \frac{d(x, y)}{a} = \frac{1 - x_1 y_1 - x_2 y_2}{(1 - x_1^2 - x_2^2)^{\frac{1}{2}} (1 - y_1^2 - y_2^2)^{\frac{1}{2}}}, \tag{1.29}$$

where $d(x, y)$ denotes the distance between the points (x_1, x_2) and (y_1, y_2). We would like to determine the metric as well as the Gaussian curvature for this space.

For any two vectors A, B in two dimensions, let us introduce the compact notation

$$A \cdot B = A_i B_i = A_1 B_1 + A_2 B_2, \quad A^2 = A_i A_i = A_1 A_1 + A_2 A_2, \tag{1.30}$$

where $i = 1, 2$ and we use the summation convention over repeated indices. Then we can write (1.29) as

$$\cosh \frac{d(x, y)}{a} = \frac{1 - x \cdot y}{\sqrt{1 - x^2} \sqrt{1 - y^2}}. \tag{1.31}$$

To determine the metric tensor for this space, we need to consider only infinitesimal distances (see discussion following (1.7)) so that let us assume $y_i = x_i + dx_i, i = 1, 2$ and we can identify the infinitesimal distance with the proper length $d(x, y) = ds$. Expanding both sides of the equation (1.31) and keeping terms to quadratic order in the infinitesimal changes, we obtain

$$\cosh \frac{ds}{a} = 1 + \frac{ds^2}{2a^2} + \cdots$$

$$= \frac{(1 - x^2 - x \cdot dx)}{1 - x^2} \left[1 + \frac{1}{2} \frac{2x \cdot dx + dx^2}{1 - x^2} + \frac{3}{8} \left(\frac{2x \cdot dx + dx^2}{1 - x^2} \right)^2 + \cdots \right]$$

$$= \left(1 - \frac{x \cdot dx}{1 - x^2} \right) \left[1 + \frac{1}{2} \frac{2x \cdot dx + dx^2}{1 - x^2} + \frac{3}{8} \left(\frac{2x \cdot dx + dx^2}{1 - x^2} \right)^2 + \cdots \right]$$

$$= 1 + \frac{1}{2} \frac{dx^2}{1 - x^2} + \frac{1}{2} \frac{(x \cdot dx)^2}{(1 - x^2)^2}, \tag{1.32}$$

where $dx^2 = dx \cdot dx$. Note that the linear terms (in the differentials) on the right hand side of (1.32) in the expansion for the distance cancel and the quadratic terms lead to

$$\frac{ds^2}{2a^2} = \frac{1}{2} \frac{dx^2}{1 - x^2} + \frac{1}{2} \frac{(x \cdot dx)^2}{(1 - x^2)^2}$$

$$= \frac{1}{2(1 - x^2)} \left(\delta_{ij} + \frac{x_i x_j}{(1 - x^2)} \right) dx_i dx_j, \tag{1.33}$$

where $i, j = 1, 2$ and summation over repeated indices is understood. This is known as the line element for the space and determines the metric components for the pseudosphere to be

$$g_{11} = \frac{a^2(1 - x_2^2)}{(1 - x^2)^2}, \quad g_{12} = g_{21} = \frac{x_1 x_2 a^2}{(1 - x^2)^2}, \quad g_{22} = \frac{a^2(1 - x_1^2)}{(1 - x^2)^2}. \tag{1.34}$$

Starting with the components of the metric tensor in (1.34), we obtain the determinant of the metric to be (see (1.20))

$$g = g_{11}g_{22} - g_{12}^2$$

$$= \frac{a^4}{(1 - x^2)^4} [(1 - x_1^2)(1 - x_2^2) - x_1^2 x_2^2]$$

$$= \frac{a^4}{(1 - x^2)^3}. \tag{1.35}$$

Next let us evaluate the three expressions involved in the calculation of the Gaussian curvature (see (1.19)). For simplicity, let us define A, B_{12}, and C as

$$A = 2\frac{\partial^2 g_{12}}{\partial x_1 \partial x_2} - \frac{\partial^2 g_{11}}{\partial x_2^2} - \frac{\partial^2 g_{22}}{\partial x_1^2}$$

$$= \frac{2a^2}{(1-x^2)^2} + \frac{8a^2 x^2}{(1-x^2)^3} + \frac{48a^2 x_1^2 x_2^2}{(1-x^2)^4}$$

$$+ \frac{2a^2}{(1-x^2)^2} - \frac{4a^2(1-5x_2^2)}{(1-x^2)^3} - \frac{24a^2 x_2^2(1-x_2^2)}{(1-x^2)^4}$$

$$+ \frac{2a^2}{(1-x^2)^2} - \frac{4a^2(1-5x_1^2)}{(1-x^2)^3} - \frac{24a^2 x_1^2(1-x_1^2)}{(1-x^2)^4}$$

$$= \frac{2a^2}{(1-x^2)^2} - \frac{4a^2}{(1-x^2)^3} = -\frac{2a^2(1+x^2)}{(1-x^2)^3}, \tag{1.36}$$

$$B_{12} = \frac{\partial g_{22}}{\partial x_2}\left(2\frac{\partial g_{12}}{\partial x_1} - \frac{\partial g_{11}}{\partial x_2}\right) - \left(\frac{\partial g_{22}}{\partial x_1}\right)^2$$

$$= \frac{4a^2 x_2(1-x_1^2)}{(1-x^2)^3}\left[\frac{2a^2 x_2}{(1-x^2)^2} + \frac{8a^2 x_1^2 x_2}{(1-x^2)^3} + \frac{2a^2 x_2}{(1-x^2)^2}\right.$$

$$\left. - \frac{4a^2 x_2(1-x_2^2)}{(1-x^2)^3}\right] - \left[-\frac{2a^2 x_1}{(1-x^2)^2} + \frac{4a^2 x_1(1-x_1^2)}{(1-x^2)^3}\right]^2$$

$$= -\frac{4a^4 x_1^2}{(1-x^2)^4}, \tag{1.37}$$

$$C = \frac{\partial g_{11}}{\partial x_1}\frac{\partial g_{22}}{\partial x_2} - 2\frac{\partial g_{11}}{\partial x_2}\frac{\partial g_{22}}{\partial x_1} + \left(2\frac{\partial g_{12}}{\partial x_1} - \frac{\partial g_{11}}{\partial x_2}\right)\left(2\frac{\partial g_{12}}{\partial x_2} - \frac{\partial g_{22}}{\partial x_1}\right)$$

$$= \frac{4a^2 x_1(1-x_2^2)}{(1-x^2)^3}\frac{4a^2 x_2(1-x_1^2)}{(1-x^2)^3}$$

$$- 2\left[-\frac{2a^2 x_2}{(1-x^2)^2} + \frac{4a^2 x_2(1-x_2^2)}{(1-x^2)^3}\right] \times (1 \leftrightarrow 2)$$

$$+ \left[\frac{2a^2 x_2}{(1-x^2)^2} + \frac{8a^2 x_2^2 x_1^2}{(1-x^2)^3} + \frac{2a^2 x_2}{(1-x^2)^2} - \frac{4a^2 x_2(1-x_2^2)}{(1-x^2)^3}\right] \times (1 \leftrightarrow 2)$$

$$= \frac{8a^4 x_1 x_2}{(1-x^2)^4}. \tag{1.38}$$

Substituting (1.36)-(1.38) into (1.19), we obtain the Gaussian curvature K to be

$$K = \frac{1}{2g}A - \frac{g_{11}}{4g^2}B_{12} - \frac{g_{22}}{4g^2}B_{21} + \frac{g_{12}}{4g^2}C$$

$$= \frac{1}{2\frac{a^4}{(1-x^2)^3}}\left[-\frac{2a^2(1+x^2)}{(1-x^2)^3}\right]$$

$$- \frac{1}{4\left(\frac{a^4}{(1-x^2)^3}\right)^2} \left[-\frac{4a^6 x_1^2(1-x_2^2)}{(1-x^2)^6} - (1 \leftrightarrow 2) \right]$$

$$+ \frac{\frac{x_1 x_2 a^2}{(1-x^2)^2}}{4\left(\frac{a^4}{(1-x^2)^3}\right)^2} \frac{8a^4 x_1 x_2}{(1-x^2)^4}$$

$$= -\frac{1}{a^2}\left(1 + x^2 - x_1^2(1-x_2^2) - x_2^2(1-x_1^2) - 2x_1^2 x_2^2\right)$$

$$= -\frac{1}{a^2}\left(1 + x^2 - x^2\right) = -\frac{1}{a^2}, \tag{1.39}$$

which coincides with (1.24).

◀

1.2 Inertial and gravitational masses

Newton's second law of motion says that any force applied on a particle is proportional to the acceleration it produces. Hence we can write

$$\mathbf{F} = m_I \mathbf{a}. \tag{1.40}$$

This refers to the inertial properties of the system since the force can be any force in general. However, when a particle is subjected to a gravitational pull, it experiences a force given by

$$\mathbf{F} = m_g \mathbf{g}, \tag{1.41}$$

where $\mathbf{g}$ represents the acceleration due to gravity. Here m_I and m_g are constants of proportionality known respectively as the inertial and the gravitational mass of the particle and can, in principle, be different. However, if we combine both the equations (1.40) and (1.41), we deduce that the acceleration produced by a gravitational force has the form

$$\mathbf{a} = \left(\frac{m_g}{m_I}\right) \mathbf{g}. \tag{1.42}$$

If the inertial mass does not coincide with the gravitational mass ($m_I \neq m_g$) and $\frac{m_g}{m_I}$ is not the same for different particles, then clearly different objects can fall with different accelerations under the influence of gravity, the pendula of the same length but made of different material can have different time periods. On the other hand, if $\frac{m_g}{m_I}$ is the same for all materials, they would fall with the same acceleration. Newton tried to measure this ratio with pendula of different material and Fresnel improved on it. It has been checked by now that (in fact, it holds even up to 10^{-13})

$$\left| \frac{m_g}{m_I} - 1 \right| < 10^{-11}, \tag{1.43}$$

for different materials.

Let us examine the experimental set up which measures this ratio. The experiment was designed by Eötvös of Hungary and consists of suspending two weights of different material from a finely balanced beam as shown in Fig. 1.3. Here g' and g'_s represent the vertical and the horizontal components of the centrifugal acceleration due to the rotation of earth as shown in Fig. 1.4 (g'_s is along the surface).

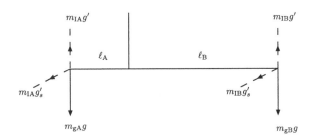

Figure 1.3: Experimental setup for measuring $\frac{m_g}{m_I}$.

The beam is in equilibrium so that we have

$$\ell_A \left(m_{gA}\, g - m_{IA}\, g' \right) = \ell_B \left(m_{gB}\, g - m_{IB} g' \right). \tag{1.44}$$

Therefore, we can solve for ℓ_B from (1.44) to obtain

$$\ell_B = \ell_A\, \frac{\left(m_{gA}\, g - m_{IA}\, g' \right)}{\left(m_{gB}\, g - m_{IB}\, g' \right)}. \tag{1.45}$$

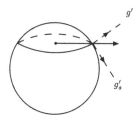

Figure 1.4: Vertical and horizontal components of the centrifugal acceleration.

On the other hand, the horizontal component of the centrifugal acceleration would lead to a torque

$$
\begin{aligned}
T &= \ell_A m_{IA}\, g'_s - \ell_B m_{IB} g'_s \\[2mm]
&= g'_s \left(\ell_A m_{IA} - \ell_A\, \frac{(m_{gA}\, g - m_{IA}\, g')}{m_{gB}\, g - m_{IB}\, g'}\; m_{IB} \right) \\[2mm]
&= \ell_A g'_s \left(\frac{m_{IA} m_{gB} - m_{IB} m_{gA}}{m_{gB}\, g - m_{IB}\, g'} \right) g,
\end{aligned}
\tag{1.46}
$$

where we have used (1.45). Furthermore, if we assume that the vertical component of the centrifugal acceleration is much smaller than the acceleration due to gravity ($g' \ll g$), we can write

$$
\begin{aligned}
T &= \ell_A g'_s \left(m_{IA} - \frac{m_{IB} m_{gA}}{m_{gB}} \right) \\[2mm]
&= m_{gA} \ell_A g'_s \left(\frac{m_{IA}}{m_{gA}} - \frac{m_{IB}}{m_{gB}} \right).
\end{aligned}
\tag{1.47}
$$

Any inequality in the ratio $\frac{m_I}{m_g}$ for the two weights would lead to a twist in the wire from which the beam is suspended. No twist was observed and Eötvös concluded from this that the difference of $\frac{m_I}{m_g}$ (or conversely $\frac{m_g}{m_I}$) from unity can be at most one part in 10^9. Dicke and his coworkers later improved the experiment and the limit we presently have is one part in 10^{11} (or even 10^{13}). This is, of course, an impressive result because *a priori* we do not know why the gravitational mass and the inertial mass should be equal to such a high precision. (We note that if $\frac{m_g}{m_I}$ is a constant for all materials, then this constant can be absorbed into the definition of Newton's constant and we can identify $\frac{m_g}{m_I} = 1$.)

1.3 Relativity

The Aristotelian view, the Galilean view and the modern view of relativity have one point in common. Each assumes that space-time is a collection of events and that each event is labelled by three spatial coordinates and one time coordinate. That is to say, the space-time manifold is a four dimensional continuum. However, beyond this the different views differ drastically from one another.

For example, in the Aristotelian view, events take place at absolute space and time coordinates and three different kinds of events can take place in such a manifold. An Aristotelian observer would observe an event A to precede event B or follow event B or happen simultaneously with event B. All events that happen simultaneously can happen anywhere in space and hence define a three dimensional hypersurface. Therefore, the causal structure of space-time is described as in Fig. 1.5.

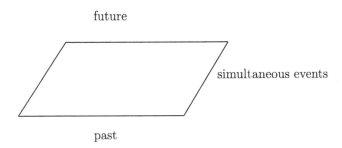

Figure 1.5: Causal structure of space-time in Aristotelian and Galilean relativity.

On the other hand, if we look at Newton's law of motion for a particle under the influence of gravitational forces, the equation has the form

$$m \frac{d^2 \mathbf{x}}{dt^2} = \sum_N \frac{m m_N (\mathbf{x}_N - \mathbf{x})}{|\mathbf{x}_N - \mathbf{x}|^3}, \tag{1.48}$$

where we are assuming that the gravitational force is due to a distribution of point masses in space with $\mathbf{x}_N$ representing the location of the mass m_N. We see that Newton's law selects out a family of inertial frames of reference. For example, let us consider the transformation

$$\mathbf{x} \rightarrow R\mathbf{x} + \mathbf{v}t + \mathbf{a}, \qquad \det R = 1,$$

$$t \rightarrow t + T, \tag{1.49}$$

where R is a rotation matrix (constant, independent of $\mathbf{x}, t$), $\mathbf{v}$, $\mathbf{a}$ and T are constants. Clearly, then, under such a transformation,

$$\frac{d^2 \mathbf{x}}{dt^2} \rightarrow R \frac{d^2 \mathbf{x}}{dt^2},$$

$$\frac{(\mathbf{x}_N - \mathbf{x})}{|\mathbf{x}_N - \mathbf{x}|^3} \rightarrow \frac{R (\mathbf{x}_N - \mathbf{x})}{|\mathbf{x}_N - \mathbf{x}|^3}. \tag{1.50}$$

Therefore, Newton's equation (1.48) remains invariant under the transformation (1.49) which is known as the Galilean transformation. It is a ten parameter family (group) of transformations (three parameters of rotation, three velocity parameters $\mathbf{v}$, three parameters of translation $\mathbf{a}$ for the coordinates and one time translation parameter T) and Newton's equations are invariant under Galilean transformations. We note that Newton's equation would not be invariant if we go to a rotating or linearly accelerating frame, i.e., if R and $\mathbf{v}$ are time dependent.

Galilean transformations select out a family of inertial frames in which Newton's equations (laws) are valid. An observer in such a frame would observe an event A to precede an event B or follow B or happen simultaneously with B. In this sense it is the same as the view of the Aristotelian observer and the causal structure of space-time still remains the same as in Fig. 1.5. However, unlike the Aristotelian observer, no two Galilean observers in two different inertial frames would agree on the exact position or the time of an event. The only thing that they can agree on is the time interval between two events and the space interval between them. Furthermore, we note that in the Galilean view neither the concept of a constant absolute speed nor the concept of a limiting speed exists (namely, frames can move with any arbitrary velocity $\mathbf{v}$).

According to Newton the view of space-time was the following. He believed that there is an absolute fixed space-time and the inertial forces are due to the frames of reference accelerating with respect to the absolute space-time. He tried to justify his view through the example of a rotating bucket of water. A great number of arguments followed and the one that stands out in particular is known as Mach's principle. This says that the acceleration experienced by a particle is influenced by the distribution of matter in space-time. That is, the presence of big stars would change the acceleration experienced by a particle. We can contrast this with the Newtonian view which would say that the presence of big stars does not matter to the acceleration of a particle. We will see later during the course of these lectures how Einstein's theory reconciles both these ideas.

The Newtonian theory along with Galilean relativity, namely that the relative separation of events in space and time are meaningful, met with a large number of success. The place where theoretical prediction disagreed with observation was in the precession of the perihelion of mercury. The theoretical calculations using Newton's equations yielded a value of 8"/century for the value of the precession in contrast to the observed value of 43"/century. (We will do this calculation later using Einstein's general relativity.)

We note here only two other phenomena that would not conform to the views of Galilean relativity (the first was observed much later). We experimentally observe that μ mesons decay at rest in the laboratory with a lifetime $\tau_\mu = 10^{-6}$ sec. On the other hand, μ mesons are also produced in cosmic ray showers. And it was measured that the cosmic ray muons have a lifetime about ten times longer than the measured lifetime in the laboratory. Various explanations were put forward to explain this, one of which is that the muon lifetime changes with the speed of the particle. This was indeed verified in the laboratory by producing muons at various speeds. However, this poses a serious problem to Galilean relativity since we assume that observers in different inertial frames have clocks which read the same time intervals and hence can be compared. The μ meson example, on the other hand, shows that if we regard the muon to be a clock (in the sense that it defines a time interval of 10^{-6} sec) then the time reading depends on whether the muon is at rest or is in motion.

The second problem comes from Maxwell's equations which, in vacuum (without sources), take the form

$$\nabla \cdot \mathbf{E} = 0,$$

$$\nabla \cdot \mathbf{B} = 0,$$

$$\nabla \times \mathbf{E} = -\frac{1}{c} \frac{\partial \mathbf{B}}{\partial t},$$

$$\nabla \times \mathbf{B} = \frac{1}{c} \frac{\partial \mathbf{E}}{\partial t}. \tag{1.51}$$

First of all it is straightforward to check that these equations are no longer invariant under the Galilean transformations (1.49). Secondly, equations (1.51) lead to

$$\nabla \times (\nabla \times \mathbf{E}) = -\frac{1}{c} \frac{\partial}{\partial t} (\nabla \times \mathbf{B}),$$

$$\text{or,} \quad \nabla (\nabla \cdot \mathbf{E}) - \nabla^2 \mathbf{E} = -\frac{1}{c} \frac{\partial}{\partial t} \left(\frac{1}{c} \frac{\partial \mathbf{E}}{\partial t} \right),$$

$$\text{or,} \quad \nabla^2 \mathbf{E} = \frac{1}{c^2} \frac{\partial^2 \mathbf{E}}{\partial t^2}. \tag{1.52}$$

This predicts that electromagnetic waves propagate with the constant speed of light c. A constant absolute speed is, of course, against the spirit of Galilean relativity and hence Maxwell postulated that electromagnetic waves propagate through a medium called ether. That is, one selects out a unique frame of reference to define the speed of light. However, experiments failed to detect ether and the Michelson-Morley experiment was decisive in establishing that there was no ether and that the speed of light nonetheless is an absolute constant.

These conflicts, among other issues, led people to reevaluate the Galilean view and Einstein proposed to replace the requirement of invariance of physical laws under Galilean transformations by that under Lorentz transformations. This goes under the name of special theory of relativity. The first step in this direction and a great one is to realize that space and time should be treated on equal footing. Thus one defines a four dimensional space-time manifold as well as a coordinate four vector (four dimensional vector) on this manifold of the form

$$x^\mu = (ct, \mathbf{x}), \qquad \mu = 0, 1, 2, 3. \tag{1.53}$$

The length of a vector on this manifold is defined as

$$x^2 = \eta_{\mu\nu} x^\mu x^\nu, \tag{1.54}$$

where $\eta_{\mu\nu}$ is the metric of the four dimensional space known as the Minkowski space. The metric is symmetric and diagonal with components

$$\begin{aligned}
\eta_{\mu\nu} &= 0, \quad \text{if} \quad \mu \neq \nu, \\
\eta_{00} &= -\eta_{ii} = 1, \qquad i = 1, 2, 3 \ , \ i \text{ not summed.}
\end{aligned} \tag{1.55}$$

We can define the inverse metric $\eta^{\mu\nu}$ from the relation

$$\eta^{\mu\nu} \eta_{\nu\gamma} = \delta^\mu_\gamma. \tag{1.56}$$

It is easy to see that the inverse metric also has the form

$$\begin{aligned}
\eta^{\mu\nu} &= 0, \quad \text{if} \quad \mu \neq \nu, \\
\eta^{00} &= -\eta^{ii} = 1, \qquad i \text{ not summed.}
\end{aligned} \tag{1.57}$$

Note that we can raise and lower the index of a given four vector with the help of the metric as

$$\begin{aligned}
x_\mu &= \eta_{\mu\nu} x^\nu, \\
x^\mu &= \eta^{\mu\nu} x_\nu,
\end{aligned} \tag{1.58}$$

where repeated indices are summed and this leads to

$$x_\mu = (ct, -\mathbf{x})\,. \tag{1.59}$$

The two four vectors x^μ, x_μ are known respectively as contravariant and covariant four vectors and the length of the coordinate vector (1.54) takes the form

$$x^2 = \eta^{\mu\nu} x_\mu x_\nu = \eta_{\mu\nu} x^\mu x^\nu = x^\mu x_\mu = c^2 t^2 - \mathbf{x}^2\,. \tag{1.60}$$

Clearly x^2 must be a four dimensional scalar since it does not carry any index. (We note that unlike the length of a vector in the Euclidean space, here the length need not be positive since it involves a difference of two terms.)

We can ask for the form of the four dimensional transformation which leaves this length invariant. It is clear that if

$$x^\mu \to x'^\mu = \Lambda^\mu{}_\nu\, x^\nu, \tag{1.61}$$

then the length remains invariant under this transformation only if

$$x'^2 = \eta_{\mu\nu} x'^\mu x'^\nu = x^2 = \eta_{\mu\nu} x^\mu x^\nu,$$

$$\text{or,} \quad \eta_{\mu\nu} \Lambda^\mu{}_\rho x^\rho \Lambda^\nu{}_\sigma x^\sigma = \eta_{\mu\nu} x^\mu x^\nu,$$

$$\text{or,} \quad \eta_{\mu\nu} \Lambda^\mu{}_\rho \Lambda^\nu{}_\sigma = \eta_{\rho\sigma}\,. \tag{1.62}$$

This has the same form as $\Lambda^{\mathrm{T}} \eta \Lambda = \eta$ in matrix notation (the superscript T denotes matrix transposition) which defines rotations in this manifold. Therefore, we conclude that rotations in the four dimensional space-time manifold which leave the length invariant also leave the metric invariant. These are known as homogeneous Lorentz transformations (they consist of three dimensional rotations and boosts).

However, in addition to four dimensional rotations if we also translate the coordinate axes by a constant amount, we obtain the inhomogeneous Lorentz transformations (also known as Poincaré transformations) under which

$$x^\mu \to x'^\mu = \Lambda^\mu{}_\nu\, x^\nu + a^\mu. \tag{1.63}$$

The infinitesimal length interval between two points (as well as the finite length interval) remains invariant under this transformation, namely,

$$
\begin{aligned}
\mathrm{d}s^2 = \eta_{\mu\nu}\mathrm{d}x^\mu\mathrm{d}x^\nu \quad &\to \quad \mathrm{d}s'^2 = \eta_{\mu\nu}\mathrm{d}x'^\mu\mathrm{d}x'^\nu \\
&= \quad \eta_{\mu\nu}\Lambda^\mu{}_\rho\mathrm{d}x^\rho\Lambda^\nu{}_\sigma\mathrm{d}x^\sigma \\
&= \quad \eta_{\mu\nu}\Lambda^\mu{}_\rho\Lambda^\nu{}_\sigma\mathrm{d}x^\rho\mathrm{d}x^\sigma \\
&= \quad \eta_{\rho\sigma}\mathrm{d}x^\rho\mathrm{d}x^\sigma = \mathrm{d}s^2 = c^2\mathrm{d}t^2 - \mathrm{d}\mathbf{x}^2. \tag{1.64}
\end{aligned}
$$

Here $\mathrm{d}s$ is known as the "proper length" of the interval. As we know, Lorentz transformations mix up space and time coordinates and hence neither a space interval nor a time interval remains invariant under these transformations. However, it is the particular combination of the space and time interval given by the proper length in (1.64) which is invariant under an inhomogeneous Lorentz transformation (Poincaré transformation). (We can contrast this with the Galilean view according to which different observers will agree on both space and time intervals.)

For a light wave, the wave velocity is given by $\left|\frac{\mathrm{d}\mathbf{x}}{\mathrm{d}t}\right| = c$ and hence the proper length for the path of a light wave (photon) follows from (1.64) to vanish

$$\mathrm{d}s = 0. \tag{1.65}$$

If we go to a different inertial frame the proper length does not change so that for a light wave we still have

$$\mathrm{d}s' = 0. \tag{1.66}$$

Hence in a different Lorentz frame the speed of light would still be equal to c. That is, Lorentz transformations can accommodate a constant absolute speed and the speed of light can be the same irrespective of the inertial frame of reference we choose. This is, of course, the content of the Michelson-Morley experiment. Clearly since this represents a departure from the Galilean view, the causal structure of space-time must be different. Indeed, in this framework, we see that space-time can be divided into four disconnected regions, as shown in Fig. 1.6, of the following three distinct types:

$$
\begin{aligned}
(x-y)^2 &= (x-y)_\mu (x-y)^\mu > 0, &&\text{time-like,}\\
(x-y)^2 &= 0, &&\text{light-like,}\\
(x-y)^2 &< 0, &&\text{space-like.}
\end{aligned}
$$

$$(1.67)$$

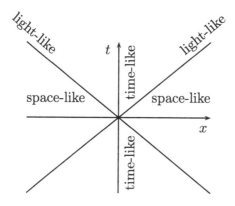

Figure 1.6: Different invariant regions of Minkowski space.

In the time-like regions, the propagation of any signal occurs with a speed less than the speed of light and hence such regions represent causal regions. The light-like regions correspond to regions where signals travel at the speed of light. Finally, in the space-like regions the speed of propagation of any signal exceeds the speed of light and hence, in such regions, no two points can be causally connected. These three types of regions cannot be mapped into one another through a Lorentz transformation. The Einstein view of space-time, therefore, is that at every point in the manifold there exists a cone which corresponds to a causal future cone where the time evolution of dynamical systems takes place. Furthermore, as is evident from the transformations themselves, the notion of simultaneity does not exist in this manifold. Namely, simultaneity is not a Lorentz invariant concept.

Note that the inhomogeneous Lorentz transformations (Poincaré transformations) also constitute a family (group) of ten parameter transformations (three rotations, three boosts and four space-time translations) – exactly like the Galilean transformations. Therefore, they do not represent a larger symmetry group. However, the nature of the transformations is different. In addition to three dimensional rotations Lorentz transformations also involve boosts. If we consider only the homogeneous Lorentz group we can further restrict ourselves to the proper Lorentz transformations by demanding that

$$\Lambda^0_0 \geq 1, \qquad \det \Lambda^\mu{}_\nu = 1, \tag{1.68}$$

representing orthochronous transformations.

Let us examine how, with these ideas, we can explain the decay of the μ mesons. Let us assume that an observer at rest observes the ticks between his clock to be with intervals $dt = \Delta t$. Therefore, he will calculate the proper length to be

$$ds = \left(c^2 dt^2 - d\mathbf{x}^2\right)^{\frac{1}{2}} = cdt \left(1 - \frac{\mathbf{v}^2}{c^2}\right)^{\frac{1}{2}} = cdt = c\Delta t, \tag{1.69}$$

where we have used $\mathbf{v} = 0$ for an observer at rest. However, if another observer traveling with a velocity $\mathbf{v}$ observes the ticks to be of interval $dt' = \Delta t'$, then he will calculate the proper length to be

$$
\begin{aligned}
ds' &= \left(c^2 dt'^2 - d\mathbf{x}'^2\right)^{\frac{1}{2}} = cdt' \left(1 - \frac{\mathbf{v}^2}{c^2}\right)^{\frac{1}{2}} \\
&= c\Delta t' \left(1 - \frac{\mathbf{v}^2}{c^2}\right)^{\frac{1}{2}}.
\end{aligned}
\tag{1.70}
$$

However, since the proper length is an invariant quantity we have

$$\mathrm{d}s' \; = \; \mathrm{d}s,$$

$$\text{or,} \quad \Delta t' \; = \; \Delta t \left(1 - \frac{\mathbf{v}^2}{c^2} \right)^{-\frac{1}{2}} = \gamma \Delta t, \tag{1.71}$$

where we have defined

$$\gamma = \left(1 - \frac{\mathbf{v}^2}{c^2} \right)^{-\frac{1}{2}}. \tag{1.72}$$

This shows that the faster an observer is moving, the longer would be the time interval between the ticks of his clock. (The time interval in a clock depends on the speed with which it is moving.) This explains why the μ mesons in cosmic rays have a longer lifetime. The constant γ is conventionally known as the Lorentz factor.

We show next that Maxwell's equations in vacuum without sources are covariant under Lorentz transformations (they are not invariant under Galilean transformations). But before doing that let us introduce the compact notation for four dimensional derivative four vectors (we are going to assume $c = 1$ from now on). The four dimensional covariant derivative is defined as

$$\partial_\mu = \frac{\partial}{\partial x^\mu} = \left(\frac{\partial}{\partial t} , \, \boldsymbol{\nabla} \right), \tag{1.73}$$

which leads to the contravariant derivative

$$\partial^\mu = \frac{\partial}{\partial x_\mu} = \eta^{\mu\nu} \partial_\nu = \left(\frac{\partial}{\partial t} , \, -\boldsymbol{\nabla} \right). \tag{1.74}$$

Furthermore, we note that

$$\partial_\mu \partial^\mu = \eta^{\mu\nu} \partial_\mu \partial_\nu = \frac{\partial^2}{\partial t^2} - \boldsymbol{\nabla}^2 = \Box, \tag{1.75}$$

which is known as the D'Alembertian, is invariant under Lorentz transformations.

To show that Maxwell's equations are Lorentz covariant, let us combine the electric and the magnetic fields into a Lorentz tensor in the following manner. Let us define a second rank anti-symmetric tensor $F_{\mu\nu} = -F_{\nu\mu}$ (it has six independent components in four dimensions) with

$$F_{0i} = E_i \equiv (\mathbf{E})_i, \tag{1.76}$$

and

$$
\begin{aligned}
F_{ij} &= -\epsilon_{ijk}B_k \equiv -\epsilon_{ijk}(\mathbf{B})_k, \\
\text{or,} \quad B_i &\equiv (\mathbf{B})_i = -\frac{1}{2}\,\epsilon_{ijk}F_{jk},
\end{aligned}
\tag{1.77}
$$

where ϵ_{ijk} denotes the three dimensional Levi-Civita tensor. With this identification, two of Maxwell's equations in vacuum without sources (see (1.51)) can be written as

$$\partial^\mu F_{\mu\nu} = 0. \tag{1.78}$$

To see this let us note that this actually reduces to two sets of equations. Namely, for $\nu = 0$, we have ($F_{00} = 0$ by anti-symmetry)

$$
\begin{aligned}
&\partial^\mu F_{\mu 0} = 0, \\
\text{or,} \quad &\partial^i F_{i0} = 0, \\
\text{or,} \quad &(-\boldsymbol{\nabla}) \cdot (-\mathbf{E}) = \boldsymbol{\nabla} \cdot \mathbf{E} = 0.
\end{aligned}
\tag{1.79}
$$

Similarly, for $\nu = j$, we obtain

$$
\begin{aligned}
&\partial^\mu F_{\mu j} = 0, \\
\text{or,} \quad &\partial^0 F_{0j} + \partial^i F_{ij} = 0, \\
\text{or,} \quad &\frac{\partial E_j}{\partial t} + \partial^i\left(-\epsilon_{ijk}B_k\right) = 0, \\
\text{or,} \quad &\frac{\partial \mathbf{E}}{\partial t} - \boldsymbol{\nabla} \times \mathbf{B} = 0, \\
\text{or,} \quad &\boldsymbol{\nabla} \times \mathbf{B} = \frac{\partial \mathbf{E}}{\partial t}.
\end{aligned}
\tag{1.80}
$$

These only represent two of the four equations of Maxwell in (1.51). To obtain the other two, we note that given the field strength tensor $F_{\mu\nu}$, we can define the dual field strength tensor as

$$\widetilde{F}_{\mu\nu} = \frac{1}{2}\,\epsilon_{\mu\nu\lambda\rho}F^{\lambda\rho} = \frac{1}{2}\,\epsilon_{\mu\nu}{}^{\lambda\rho}F_{\lambda\rho}. \tag{1.81}$$

Here $\epsilon^{\mu\nu\lambda\rho}$ (or $\epsilon_{\mu\nu\lambda\rho}$) defines the four dimensional Levi-Civita tensor with $\epsilon^{0123} = 1$. The dual field strength interchanges the roles of the electric and the magnetic fields and we have $\widetilde{F}_{0i} = -\frac{1}{2}\epsilon_{ijk}F_{jk} = B_i$ and $\widetilde{F}_{ij} = \epsilon_{ijk}E_k$ (where we have used the fact that ϵ^{0ijk} coincides with the three dimensional Levi-Civita tensor ϵ_{ijk}). The other two equations of Maxwell in (1.51) can now be written as

$$\partial^\mu \widetilde{F}_{\mu\nu} = 0. \tag{1.82}$$

Equation (1.82) can, in fact, be explicitly solved to determine

$$F_{\mu\nu} = \partial_\mu A_\nu - \partial_\nu A_\mu, \tag{1.83}$$

where A_μ is known as the (four) vector potential. (It contains the three dimensional scalar and vector potentials as $A^\mu = (\phi, \mathbf{A})$.) Thus, the set of four Maxwell's equations in (1.51) can be written either as

$$\partial^\mu F_{\mu\nu} = 0 = \partial^\mu \widetilde{F}_{\mu\nu}, \tag{1.84}$$

or equivalently as

$$\partial^\mu F_{\mu\nu} = 0, \qquad F_{\mu\nu} = \partial_\mu A_\nu - \partial_\nu A_\mu. \tag{1.85}$$

Before we examine the Lorentz covariance of Maxwell's equations, let us discuss how various tensors transform under a Lorentz transformation. We already know how a (contravariant) four vector transforms (see (1.61)), namely,

$$x^\mu \to \Lambda^\mu{}_\nu x^\nu. \tag{1.86}$$

This immediately tells us that under a Lorentz transformation, a second rank contravariant tensor $U^{\mu\nu}$ would transform as

$$U^{\mu\nu} \to \Lambda^{\mu}{}_{\lambda}\Lambda^{\nu}{}_{\rho}U^{\lambda\rho}. \tag{1.87}$$

Similarly any higher rank contravariant tensor would transform as

$$U^{\mu\nu\lambda\rho\cdots} \to \Lambda^{\mu}{}_{\mu'}\Lambda^{\nu}{}_{\nu'}\Lambda^{\lambda}{}_{\lambda'}\Lambda^{\rho}{}_{\rho'}\cdots U^{\mu'\nu'\lambda'\rho'\cdots}. \tag{1.88}$$

This is interesting because we can now ask what information can be derived from the transformation property of the invariant (contravariant) metric tensor, namely,

$$\eta^{\mu\nu} \to \Lambda^{\mu}{}_{\lambda}\Lambda^{\nu}{}_{\rho}\eta^{\lambda\rho}. \tag{1.89}$$

Since the metric is an invariant tensor, it does not change under a Lorentz transformation. (Transformations which leave the metric of a manifold invariant are also known as isometries of the manifold and Lorentz transformations define isometries of the Minkowski spacetime.) This implies

$$\eta^{\mu\nu} \to \Lambda^{\mu}{}_{\lambda}\Lambda^{\nu}{}_{\rho}\eta^{\lambda\rho} = \eta^{\mu\nu}, \tag{1.90}$$

which is what we had seen earlier in (1.62). Alternatively, we can write this as

$$\Lambda^{\mu}{}_{\lambda}\Lambda^{\nu\lambda} = \eta^{\mu\nu}, \quad \text{or, equivalently} \quad \Lambda_{\lambda}{}^{\mu}\Lambda^{\lambda\nu} = \eta^{\mu\nu}. \tag{1.91}$$

Similarly, the inverse (covariant) metric $\eta_{\mu\nu}$ is also an invariant tensor and one can show that this leads to (this also follows from (1.91) by lowering the indices with the metric tensor)

$$\Lambda_{\mu\lambda}\Lambda_{\nu}{}^{\lambda} = \eta_{\mu\nu}, \quad \text{or, equivalently} \quad \Lambda_{\lambda\mu}\Lambda^{\lambda}{}_{\nu} = \eta_{\mu\nu}. \tag{1.92}$$

Once we know the transformation properties of contravariant tensors as well as the fact that the metric tensor is invariant, then we can calculate how the lower index (covariant) objects transform under a Lorentz transformation. For example,

$$x_\mu = \eta_{\mu\nu} x^\nu \quad \rightarrow \quad \eta_{\mu\nu} \Lambda^\nu{}_\lambda x^\lambda$$
$$= \Lambda_{\mu\lambda} x^\lambda = \Lambda_\mu{}^\nu x_\nu. \tag{1.93}$$

Remembering the relationship (see (1.91))

$$\Lambda^\mu{}_\lambda \Lambda^{\nu\lambda} = \eta^{\mu\nu},$$

or, $\quad \Lambda^\mu{}_\lambda \Lambda_\nu{}^\lambda = \delta^\mu{}_\nu \quad$ or, equivalently $\quad \Lambda_\lambda{}^\mu \Lambda^\lambda{}_\nu = \delta^\mu{}_\nu, \quad$ (1.94)

we see that, under a Lorentz transformation, the lower index (covariant) vectors transform in an inverse way compared to the upper index (contravariant) vectors. Therefore, we distinguish between them by saying that

$$x^\mu \quad \rightarrow \quad \text{contravariant vector,}$$
$$x_\mu \quad \rightarrow \quad \text{covariant vector,}$$
$$U^{\mu\nu\lambda\rho...} \quad \rightarrow \quad \text{contravariant tensor,}$$
$$U_{\mu\nu\lambda\rho...} \quad \rightarrow \quad \text{covariant tensor.} \tag{1.95}$$

We are now in a position to calculate how a covariant tensor transforms under a Lorentz transformation

$$U_{\mu\nu} \rightarrow \Lambda_\mu{}^\lambda \Lambda_\nu{}^\rho U_{\lambda\rho},$$
$$U_{\mu\nu\lambda\rho...} \rightarrow \Lambda_\mu{}^{\mu'} \Lambda_\nu{}^{\nu'} \Lambda_\lambda{}^{\lambda'} \Lambda_\rho{}^{\rho'} \cdots U_{\mu'\nu'\lambda'\rho'....} \tag{1.96}$$

In addition, we see that we can have tensors of mixed nature. A tensor with m contravariant indices and n covariant ones would transform under a Lorentz transformation as

$$U^{\mu_1\cdots\mu_m}{}_{\nu_1\cdots\nu_n} \rightarrow \Lambda^{\mu_1}{}_{\mu'_1} \cdots \Lambda^{\mu_m}{}_{\mu'_m} \Lambda_{\nu_1}{}^{\nu'_1} \cdots \Lambda_{\nu_n}{}^{\nu'_n} U^{\mu'_1\cdots\mu'_m}{}_{\nu'_1\cdots\nu'_n}. \tag{1.97}$$

We also see now that the length of a vector transforms as

$$
\begin{aligned}
x^2 = x_\mu x^\mu \;\rightarrow\; & \Lambda_\mu{}^\nu x_\nu \Lambda^\mu{}_\lambda x^\lambda \\
= & \Lambda_\mu{}^\nu \Lambda^\mu{}_\lambda x_\nu x^\lambda \\
= & \delta^\nu{}_\lambda x_\nu x^\lambda = x_\nu x^\nu = x^2,
\end{aligned} \tag{1.98}
$$

where we have used (1.94). Namely, the length of a four vector is a Lorentz scalar as we have observed earlier.

Furthermore, let us note that since

$$
x^\mu \rightarrow x'^\mu = \Lambda^\mu{}_\nu x^\nu, \tag{1.99}
$$

we can write

$$
x^\mu = \Lambda_\nu{}^\mu x'^\nu, \tag{1.100}
$$

which follows from (1.94), namely,

$$
x'^\mu = \Lambda^\mu{}_\nu x^\nu,
$$

or, $\quad \Lambda_\mu{}^\lambda x'^\mu = \Lambda_\mu{}^\lambda \Lambda^\mu{}_\nu x^\nu = \delta^\lambda{}_\nu x^\nu = x^\lambda,$

or, $\quad x^\lambda = \Lambda_\mu{}^\lambda x'^\mu,$

or, $\quad x^\mu = \Lambda_\nu{}^\mu x'^\nu. \tag{1.101}$

Similarly, since

$$
x_\mu \rightarrow x'_\mu = \Lambda_\mu{}^\nu x_\nu, \tag{1.102}
$$

using (1.94), it follows that

$$
x_\mu = \Lambda^\nu{}_\mu x'_\nu. \tag{1.103}
$$

We are now in a position to calculate how the four derivatives transform under a Lorentz transformation. For example,

$$\partial^\mu = \frac{\partial}{\partial x_\mu} \to \frac{\partial}{\partial x'_\mu} = \frac{\partial x_\nu}{\partial x'_\mu} \frac{\partial}{\partial x_\nu}$$

$$= \Lambda^\mu{}_\nu \frac{\partial}{\partial x_\nu} = \Lambda^\mu{}_\nu \partial^\nu, \tag{1.104}$$

where we have used (1.103). Similarly, using (1.100) we obtain

$$\partial_\mu = \frac{\partial}{\partial x^\mu} \to \frac{\partial}{\partial x'^\mu} = \frac{\partial x^\nu}{\partial x'^\mu} \frac{\partial}{\partial x^\nu}$$

$$= \Lambda_\mu{}^\nu \frac{\partial}{\partial x^\nu} = \Lambda_\mu{}^\nu \partial_\nu. \tag{1.105}$$

Thus we see that contragradient and the cogradient vectors transform like contravariant and covariant vectors respectively. From these, it now follows that the D'Alembertian (1.75) is invariant under a Lorentz transformation.

We are now in a position to discuss the Lorentz covariance of Maxwell's equations. First of all we note that since the field strengths are tensors of second rank, under a Lorentz transformation, they transform as (see (1.96))

$$F_{\mu\nu} \to \Lambda_\mu{}^\lambda \Lambda_\nu{}^\rho F_{\lambda\rho},$$

$$\widetilde{F}_{\mu\nu} \to \Lambda_\mu{}^\lambda \Lambda_\nu{}^\rho \widetilde{F}_{\lambda\rho}. \tag{1.106}$$

Thus,

$$\partial^\mu F_{\mu\nu} \to \partial'^\mu F'_{\mu\nu} = \Lambda^\mu{}_{\mu'} \partial^{\mu'} \Lambda_\mu{}^\lambda \Lambda_\nu{}^\rho F_{\lambda\rho}$$

$$= \Lambda^\mu{}_{\mu'} \Lambda_\mu{}^\lambda \Lambda_\nu{}^\rho \, \partial^{\mu'} F_{\lambda\rho}$$

$$= \delta^\lambda_{\mu'} \Lambda_\nu{}^\rho \, \partial^{\mu'} F_{\lambda\rho} = \Lambda_\nu{}^\rho \, \partial^\lambda F_{\lambda\rho}. \tag{1.107}$$

This shows that if

$$\partial^\mu F_{\mu\nu} = 0, \tag{1.108}$$

in one inertial frame of reference, the form of the equation does not change in a Lorentz transformed frame of reference. Similarly, we can also show that the equation for the dual (1.82)

$$\partial^\mu \widetilde{F}_{\mu\nu} = 0, \tag{1.109}$$

also maintains its form in any inertial Lorentz frame so that Maxwell's equations are Lorentz covariant. We should note here that since

$$F_{\mu\nu} \to \Lambda_\mu{}^\lambda \Lambda_\nu{}^\rho F_{\lambda\rho}, \tag{1.110}$$

and the transformation matrices $\Lambda_\mu{}^\lambda$ are not necessarily diagonal, Lorentz transformations (boosts) mix up different field components. This is another way of saying that under a Lorentz transformation, the electric and the magnetic fields mix with each other which we are all familiar with.

▶ **Example (Null basis vectors).** As we have seen in (1.67), in Minkowski space vectors can be time-like, space-like or light-like (null). Let us ask if it is possible to find four linearly independent null (light-like) vectors in four dimensional Minkowski space which are also orthogonal.

It is quite easy to find four linearly independent null four vectors. For example, we can have

$$e_1^\mu = (1, 1, 0, 0), \qquad\qquad e_2^\mu = (1, 0, 1, 0),$$

$$e_3^\mu = (1, 0, 0, 1), \qquad\qquad e_4^\mu = (1, -1, 0, 0). \tag{1.111}$$

This set is obviously linearly independent and each vector is easily checked to be light-like (null), namely, $\eta_{\mu\nu} e_i^\mu e_i^\nu = 0, i$ (not summed) $= 1, 2, 3, 4$. Such a set defines a basis for the Minkowski space. In fact, this is the light-cone basis for a photon. Note, however, that these basis vectors are not orthogonal.

It is not possible to find a set which is in addition orthogonal and null for the following reason. We will attempt to build an orthogonal basis out of a set of null vectors via the Graham-Schmidt procedure. To that end, take any of the null vector and label it v_1^μ. v_1^μ necessarily has a time component and a space component (it is the only way it can be null but non-zero) so we can write

$$v_1^\mu = (1, \mathbf{v}_1), \quad \mathbf{v}_1 \cdot \mathbf{v}_1 = 1. \tag{1.112}$$

To get the second vector v_2^μ we know it, too, must have a time and space component so up to a multiplicative constant we can write

$$v_2^\mu = (1, \mathbf{v}_2), \quad \mathbf{v}_2 \cdot \mathbf{v}_2 = 1. \tag{1.113}$$

If we now impose orthogonality between the two vectors (1.112) and (1.113), we obtain

$$v_1 \cdot v_2 = 1 - \mathbf{v}_1 \cdot \mathbf{v}_2 = 0, \tag{1.114}$$

which implies

$$\mathbf{v}_1 \cdot \mathbf{v}_2 = 1. \tag{1.115}$$

But, $\mathbf{v}_1$, $\mathbf{v}_2$ are (three dimensional) unit vectors so that (1.115) implies that $\mathbf{v}_1 = \mathbf{v}_2$ which would amount to their not being linearly independent. This proves that it is impossible to find a set of four linearly independent light-like (null) vectors in Minkowski space which are also orthogonal.

◄

CHAPTER 2

Relativistic dynamics

2.1 Relativistic point particle

It is clear that Newton's equation which is covariant (form invariant) under Galilean transformations, is not covariant under Lorentz transformations. So if we require Lorentz invariance to be the invariance group of the physical world, we must modify Newton's laws. Let us note, however, that since Newton's laws have been tested again and again in the laboratory for slowly moving particles, it is only for highly relativistic particles ($|\mathbf{v}| \simeq c = 1$) that the equation may require any modification.

We have already seen that in Minkowski space we can write the line element as

$$\mathrm{d}s^2 = \mathrm{d}\tau^2 = \eta_{\mu\nu}\mathrm{d}x^\mu\mathrm{d}x^\nu = \mathrm{d}t^2 - \mathrm{d}\mathbf{x}^2, \tag{2.1}$$

where $\mathrm{d}\tau$ is known as the "proper time" of the particle and is related to the proper length through the speed of light ($\mathrm{d}s = c\mathrm{d}\tau$) which we have set to unity. The proper time, like the proper length, is an invariant quantity (since the speed of light is an invariant) and hence can be used to characterize the trajectory of a particle. To describe a Lorentz covariant generalization of Newton's equation, let us define a force four vector f^μ such that

$$m\,\frac{\mathrm{d}^2x^\mu}{\mathrm{d}\tau^2} = f^\mu, \tag{2.2}$$

where m represents the rest mass of the particle. Clearly since the proper time does not change under a Lorentz transformation, equation (2.2) takes the same form in all Lorentz frames. Furthermore, if

we know the form of f^μ, we can calculate the trajectory of the particle. To understand the nature of this force four vector, we note that
the form of the relativistic force four vector can be obtained from
the observation that since (2.2) holds in all Lorentz frames, it must
be valid in the rest frame as well. In the rest frame of the particle
$(\frac{d\mathbf{x}}{dt} = 0)$

$$d\tau = dt, \tag{2.3}$$

and hence (2.2) takes the form

$$m \frac{d^2 x^i}{dt^2} = f^i_{(\text{rest})},$$
$$f^0_{(\text{rest})} = 0. \tag{2.4}$$

Furthermore, if we define the relativistic four velocity of a particle as

$$u^\mu = \frac{dx^\mu}{d\tau} = (\frac{dt}{d\tau}, \frac{d\mathbf{x}}{d\tau}) = \frac{dt}{d\tau}(1, \frac{d\mathbf{x}}{dt}) = \gamma(1, \mathbf{v}), \tag{2.5}$$

then in the rest frame of the particle we have $u^\mu_{(\text{rest})} = (1, 0, 0, 0)$.
Therefore, in the rest frame we obtain

$$f^\mu_{(\text{rest})} u^\nu_{(\text{rest})} \eta_{\mu\nu} = f^\mu_{(\text{rest})} u^{(\text{rest})}_\mu = 0. \tag{2.6}$$

On the other hand, this is a (Lorentz invariant) scalar equation and
hence must be true in any inertial reference frame. Secondly, we see
explicitly that

$$u^\mu_{(\text{rest})} u^\nu_{(\text{rest})} \eta_{\mu\nu} = u^\mu_{(\text{rest})} u^{(\text{rest})}_\mu = 1. \tag{2.7}$$

This must also remain true in all inertial reference frames as can be
seen from (2.5) using the definition of the Lorentz factor in (1.72).
That these two equations are compatible can be seen in the following
way. If

$$u^\mu u^\nu \eta_{\mu\nu} = 1, \tag{2.8}$$

it follows that

$$\frac{\mathrm{d}}{\mathrm{d}\tau}\left(u^\mu u^\nu \eta_{\mu\nu}\right) = 0,$$

$$\text{or,} \quad \frac{\mathrm{d}u^\mu}{\mathrm{d}\tau}\, u^\nu \eta_{\mu\nu} = 0,$$

$$\text{or,} \quad \frac{1}{m}\, f^\mu u^\nu \eta_{\mu\nu} = 0,$$

$$\text{or,} \quad f^\mu u^\nu \eta_{\mu\nu} = f^\mu u_\mu = 0. \tag{2.9}$$

Here we have used the fact that

$$f^\mu = m\,\frac{\mathrm{d}^2 x^\mu}{\mathrm{d}\tau^2} = m\,\frac{\mathrm{d}u^\mu}{\mathrm{d}\tau}. \tag{2.10}$$

Equation (2.9) shows that the relativistic force (acceleration) has to be orthogonal to the four velocity and, as a result, not all components of the force four vector are independent.

In analogy with the non-relativistic case, we can define the four momentum of a particle as

$$p^\mu = mu^\mu, \tag{2.11}$$

so that the dynamical equation (2.2) (or (2.10)) can also be written in the familiar form (the rest mass m is a constant)

$$\frac{\mathrm{d}p^\mu}{\mathrm{d}\tau} = f^\mu. \tag{2.12}$$

We note that the space components of the four velocity are given by (see (2.5))

$$\mathbf{u} = \frac{\mathrm{d}\mathbf{x}}{\mathrm{d}\tau} = \gamma\mathbf{v}, \tag{2.13}$$

where $\gamma = \left(1 - v^2\right)^{-\frac{1}{2}}$ is the Lorentz factor introduced in (1.72) (remember $c = 1$). It follows now that for slowly moving particles $(v = |\mathbf{v}| \ll 1)$

$$\mathbf{u} = \mathbf{v} + 0\left(v^3\right). \tag{2.14}$$

Furthermore,

$$u^0 = \frac{dt}{d\tau} = \gamma = \left(1 - v^2\right)^{-\frac{1}{2}} = 1 + \frac{1}{2}\,v^2 + 0\left(v^4\right). \tag{2.15}$$

Thus, for slowly moving particles, we obtain from (2.11)

$$\begin{aligned}
\mathbf{p} &= m\mathbf{u} \simeq m\mathbf{v}, \\
p^0 &= mu^0 \simeq m + \frac{1}{2}\,mv^2.
\end{aligned} \tag{2.16}$$

It is clear that p^0 is nothing other than the (non-relativistic) energy of the particle (including the rest energy) and (2.11) reduces to the familiar energy and momentum relations of non-relativistic physics (in the proper limit). Thus we see that the energy and momentum are unified in this picture into a single four vector p^μ such that

$$p^\mu = (E, \mathbf{p}). \tag{2.17}$$

It follows from this that

$$\begin{aligned}
p^2 = p^\mu p^\nu \eta_{\mu\nu} &= \left(mu^\mu\right)\left(mu^\nu\right)\eta_{\mu\nu} \\
&= m^2 u^\mu u^\nu \eta_{\mu\nu} = m^2,
\end{aligned} \tag{2.18}$$

where we have used (2.8). Therefore, the energy and momentum of a relativistic particle satisfy the condition

$$p^2 = p^\mu p_\mu = E^2 - \mathbf{p}^2 = m^2, \tag{2.19}$$

which we recognize to be the Einstein relation.

Once we know the components of the force in the rest frame, we can calculate them in any other frame. Thus, for example, denoting the rest frame components of the force by (see (2.6))

$$
\begin{aligned}
f^0_{(rest)} &= F^0 = 0, \\
f^i_{(rest)} &= F^i,
\end{aligned}
\tag{2.20}
$$

in an arbitrary frame the components of the force take the form

$$
\begin{aligned}
f^0 &= \gamma \mathbf{v} \cdot \mathbf{F}, \\
\mathbf{f} &= \mathbf{F} + (\gamma - 1)\mathbf{v}\, \frac{(\mathbf{v} \cdot \mathbf{F})}{v^2},
\end{aligned}
\tag{2.21}
$$

where $\mathbf{v}$ represents the instantaneous velocity of the particle (frame) and $v = |\mathbf{v}|$. Here we have used the fact that an arbitrary four vector A^μ transforms under a general Lorentz boost as

$$
\begin{aligned}
A'^{\,0} &= \gamma\left(A^0 + \mathbf{v} \cdot \mathbf{A}\right), \\
\mathbf{A}' &= \mathbf{A} + (\gamma - 1)\frac{\mathbf{v}(\mathbf{v} \cdot \mathbf{A})}{v^2} + \gamma \mathbf{v} A^0,
\end{aligned}
\tag{2.22}
$$

Recalling that $u^\mu = (\gamma, \gamma \mathbf{v})$ in an arbitrary frame (see (2.5)), we note that

$$
\begin{aligned}
f^\mu u_\mu &= f^0 u^0 - \mathbf{f} \cdot \mathbf{u} \\
&= \gamma^2 \mathbf{v} \cdot \mathbf{F} - \{\gamma \mathbf{v} \cdot \mathbf{F} + \gamma(\gamma - 1)\mathbf{v} \cdot \mathbf{F}\} \\
&= 0,
\end{aligned}
\tag{2.23}
$$

as we would expect (see (2.9)).

▶ **Example (Transformation of force).** From the Lorentz transformation matrix (1.61), we would like to derive that under an arbitrary Lorentz transformation (boost) a contravariant vector transforms as (see (2.22))

$$
A'^{\,0} = \gamma\left(A^0 + \mathbf{v} \cdot \mathbf{A}\right),
$$

$$
\mathbf{A}' = \mathbf{A} + (\gamma - 1)\frac{\mathbf{v}(\mathbf{v} \cdot \mathbf{A})}{v^2} + \gamma \mathbf{v} A^0,
\tag{2.24}
$$

where $\mathbf{v}$ denotes the boost velocity, $v = |\mathbf{v}|$ and the Lorentz factor is given by (see (1.72) in units $c = 1$)

$$\gamma = \left(1 - v^2\right)^{-\frac{1}{2}}. \tag{2.25}$$

Furthermore, using this as well as assuming that in the rest frame the force four vector has the form

$$f^0_{(\text{rest})} = 0, \quad f^i_{(\text{rest})} = F^i, \tag{2.26}$$

we would like to show that in a frame where the instantaneous velocity of the particle is $\mathbf{v} = \frac{d\mathbf{x}}{dt}$,

$$f^0 = \gamma \mathbf{v} \cdot \mathbf{F} = \gamma \mathbf{v} \cdot \mathbf{f}_{(\text{rest})},$$

$$\mathbf{f} = \mathbf{F} + (\gamma - 1)\, \mathbf{v}\, \frac{\mathbf{v} \cdot \mathbf{F}}{v^2}. \tag{2.27}$$

We recall (see (2.5)) that the proper velocity of a particle (in any frame) is defined as

$$u^\mu = \frac{dx^\mu}{d\tau}, \tag{2.28}$$

leading to the components

$$u^0 = \frac{dx^0}{d\tau} = \frac{dt}{d\tau} = \gamma,$$

$$\mathbf{u} = \frac{d\mathbf{x}}{d\tau} = \frac{dt}{d\tau}\frac{d\mathbf{x}}{dt} = \gamma \mathbf{v}, \tag{2.29}$$

where $\mathbf{v} = \frac{d\mathbf{x}}{dt}$ represents the instantaneous velocity of the particle and $\gamma = (1 - \mathbf{v}^2)^{-1/2}$ denotes the Lorentz factor. It follows from (2.29) that in the rest frame of the particle where $\mathbf{v} = 0$, we have

$$u^\mu_{(\text{rest})} = (1, 0, 0, 0). \tag{2.30}$$

Since the proper velocity is a four vector, we can relate its components in any frame to the rest frame through the Lorentz transformation (1.61)

$$u^\mu = \Lambda^\mu{}_\nu u^\nu_{(\text{rest})}, \tag{2.31}$$

and comparing (2.29) and (2.31) (with the use of (2.30)) it follows that

$$\Lambda^0{}_0 = \gamma, \quad \Lambda^i{}_0 = \gamma v^i. \tag{2.32}$$

Furthermore, from the orthogonality relation satisfied by the Lorentz transformation (1.62) (or (1.92))

$$\Lambda^{\mu}{}_{\lambda}\Lambda^{\nu\lambda} = \eta^{\mu\nu}, \tag{2.33}$$

we conclude that

$$\Lambda^{0}{}_{\lambda}\Lambda^{0\lambda} = 1,$$

$$\text{or,} \quad \Lambda^{0}{}_{0}\Lambda^{00} + \Lambda^{0}{}_{i}\Lambda^{0i} = 1,$$

$$\text{or,} \quad (\Lambda^{0}{}_{i})^{2} = (\Lambda^{0}{}_{0})^{2} - 1 = \gamma^{2} - 1 = \frac{\mathbf{v}^{2}}{1 - \mathbf{v}^{2}} = \gamma^{2}\mathbf{v}^{2},$$

$$\text{or,} \quad \Lambda^{0}{}_{i} = \gamma v^{i}, \tag{2.34}$$

where we have chosen the positive square root for simplicity ($\mathbf{v}$ can take positive as well as negative values). The other components of the transformation matrix can also be determined similarly. However, this is all we need to transform the force four vector.

We know that

$$f^{0}_{(\text{rest})} = 0, \quad f^{i}_{(\text{rest})} = F^{i}. \tag{2.35}$$

From the form of the transformation matrix, we can now determine that in any frame

$$f^{0} = \Lambda^{0}{}_{\nu}f^{\nu}_{(\text{rest})} = \Lambda^{0}{}_{0}f^{0}_{(\text{rest})} + \Lambda^{0}{}_{i}f^{i}_{(\text{rest})}$$

$$= 0 + \gamma\mathbf{v} \cdot \mathbf{F} = \gamma\mathbf{v} \cdot \mathbf{F}. \tag{2.36}$$

The spatial component of the force can be easily determined from the orthogonality relation

$$\eta_{\mu\nu}u^{\mu}f^{\nu} = u^{0}f^{0} - \mathbf{u} \cdot \mathbf{f} = 0,$$

$$\text{or,} \quad \gamma^{2}\mathbf{v} \cdot \mathbf{F} - \gamma\mathbf{v} \cdot \mathbf{f} = 0,$$

$$\text{or,} \quad \mathbf{v} \cdot (\gamma\mathbf{F} - \mathbf{f}) = 0. \tag{2.37}$$

This determines

$$\mathbf{f} = \gamma\mathbf{F} + \mathbf{f}_{\perp}, \tag{2.38}$$

where $\mathbf{f}_{\perp}$ denotes a component perpendicular to the velocity $\mathbf{v}$. The unique orthogonal vector that one can construct from two vectors $\mathbf{F}$ and $\mathbf{v}$ is given by (up to an overall multiplicative factor)

$$\mathbf{F} - \frac{\mathbf{v}(\mathbf{v} \cdot \mathbf{F})}{\mathbf{v}^2}. \tag{2.39}$$

As a result, we determine the unique spatial component of the force in an arbitrary frame, which also has the correct rest frame limit, to be

$$\mathbf{f} = \gamma \mathbf{F} + \mathbf{f}_\perp = \gamma \mathbf{F} + (1 - \gamma) \left(\mathbf{F} - \frac{\mathbf{v}(\mathbf{v} \cdot \mathbf{F})}{\mathbf{v}^2} \right)$$

$$= \mathbf{F} + (\gamma - 1) \frac{\mathbf{v}(\mathbf{v} \cdot \mathbf{F})}{\mathbf{v}^2}. \tag{2.40}$$

We can write (2.36) and (2.40) together as

$$f^0 = \gamma \mathbf{v} \cdot \mathbf{F} = \gamma \mathbf{v} \cdot \mathbf{f}_{(\text{rest})},$$

$$\mathbf{f} = \mathbf{F} + (\gamma - 1) \, \mathbf{v} \, \frac{\mathbf{v} \cdot \mathbf{F}}{v^2}, \tag{2.41}$$

as discussed in (2.21).

◄

▶ **Example (Combination of velocities).** Observer O_A moves with velocity $\mathbf{v}_B$ with respect to O_B who moves with a velocity $\mathbf{v}_C$ with respect to O_C. The coordinate systems are related by the Lorentz transformations

$$x_A^\mu = \Lambda^\mu{}_\nu(\mathbf{v}_B) x_B^\nu,$$

$$x_B^\mu = \Lambda^\mu{}_\nu(\mathbf{v}_C) x_C^\nu, \tag{2.42}$$

where $\Lambda^\mu{}_\nu(\mathbf{v})$'s represent the respective Lorentz transformations (boosts). We would like to determine the velocity of O_A with respect to O_C, namely, the addition formula if all the velocities are in the same direction, say along the x direction.

We note from (2.42) that we can write

$$x_A^\mu = \Lambda^\mu{}_\nu(\mathbf{v}_B) \Lambda^\mu{}_\nu(\mathbf{v}_C) x_C^\nu, \tag{2.43}$$

and without loss of generality we can assume that the particle is at rest in frame C so that

$$x_C^\mu = (t_C, 0, 0, 0). \tag{2.44}$$

From (2.34) we know that the Lorentz transformation matrix between any two frames moving with velocity $\mathbf{v}$ has the form ($\Lambda^i{}_j(v)$ can also be determined in a similar manner)

$$\Lambda^0{}_0(v) = \gamma(v) = 1/\sqrt{1 - v^2},$$

$$\Lambda^0{}_i(v) = \Lambda^i{}_0 = \gamma(v)v^i,$$

$$\Lambda^i{}_j(v) = \delta^i{}_j - (\gamma(v) - 1)\frac{v^i v_j}{v^2}. \tag{2.45}$$

Denoting the instantaneous velocity in the frame A to be $\mathbf{v}_A$, it follows now from (2.42) as well as (2.45) that

$$\frac{dt_A}{dt_C} = \Lambda^0{}_\mu(v_B)\Lambda^\mu{}_\nu(v_C)\frac{dx^\nu_C}{dt_C} = \Lambda^0{}_\mu(v_B)\Lambda^\mu{}_0(v_C)$$

$$= \Lambda^0{}_0(v_B)\Lambda^0{}_0(v_C) + \Lambda^0{}_i(v_B)\Lambda^i{}_0(v_C)$$

$$= \gamma(v_B)\gamma(v_C)(1 + \mathbf{v}_B \cdot \mathbf{v}_C),$$

$$v^i_A = \frac{dx^i_A}{dt_A} = \Lambda^i{}_\mu(v_B)\Lambda^\mu{}_\nu(v_C)\frac{dx^\nu_C}{dt_A} = \Lambda^i{}_\mu(v_B)\Lambda^\mu{}_0(v_C)\frac{dt_C}{dt_A}$$

$$= (\Lambda^i{}_0(v_B)\Lambda^0{}_0(v_C) + \Lambda^i{}_j(v_B)\Lambda^j{}_0(v_C))\frac{dt_C}{dt_A}$$

$$= \left[\gamma(v_B)\gamma(v_C)v^i_B + \left(\delta^i_j - (\gamma(v_B) - 1)\frac{v^i_B v_{Bj}}{v^2_B}\right)\gamma(v_C)v^j_C\right]\frac{dt_C}{dt_A}$$

$$= \left[\gamma(v_B)v^i_B + v^i_C + (\gamma(v_B) - 1)\frac{\mathbf{v}_B \cdot \mathbf{v}_C}{v^2_B}v^i_B\right]\frac{\gamma(v_C)}{\gamma(v_B)\gamma(v_C)(1 + \mathbf{v}_B \cdot \mathbf{v}_C)} \tag{2.46}$$

In vector notation the velocity $\mathbf{v}_A$ can be written in terms of $\mathbf{v}_B$ and $\mathbf{v}_C$ as

$$\mathbf{v}_A = \left[\left(1 + \frac{\mathbf{v}_B \cdot \mathbf{v}_C}{v^2_B}\right)\mathbf{v}_B + \frac{1}{\gamma(v_B)}\left(\mathbf{v}_C - \frac{\mathbf{v}_B \cdot \mathbf{v}_C}{v^2_B}\mathbf{v}_B\right)\right]\frac{1}{1 + \mathbf{v}_B \cdot \mathbf{v}_C}. \tag{2.47}$$

If the velocities are all in the same direction (say along the x-axis) then the second term in (2.47), which represents $\mathbf{v}_{C\perp}$ (with respect to $\mathbf{v}_B$), will drop out, leaving the familiar expression for the addition of velocities

$$v_A = \frac{v_B + v_C}{1 + v_B v_C}, \tag{2.48}$$

where the vector notation is no longer relevant since the problem is now one dimensional. We can recognize the transformation in (2.48) as being identical to the relation for the sum of hyperbolic tangent of angles,

$$\tanh(a + b) = \frac{\tanh a + \tanh b}{1 + \tanh a \tanh b}. \tag{2.49}$$

In fact, if we parameterize the velocities in terms of angular parameters (known as rapidities) as

$$v_A = \tanh \theta_A, \quad v_B = \tanh \theta_B, \quad v_C = \tanh \theta_C, \tag{2.50}$$

then it follows from (2.48) and (2.49) that

$$\tanh \theta_A = \frac{\tanh \theta_B + \tanh \theta_C}{1 + \tanh \theta_B \tanh \theta_C} = \tanh(\theta_B + \theta_C), \tag{2.51}$$

which leads to

$$\theta_A = \theta_B + \theta_C. \tag{2.52}$$

In other words, while the addition of velocities (see (2.50)) have a nonstandard form, rapidities or angles representing velocities have a standard composition. This is analogous to the fact that while the slopes in Euclidean space have a nonstandard composition, the angles representing the slopes simply add.

◀

▶ **Example (Proper acceleration).** Let us consider the relativistic motion of a particle along the x-direction where the particle undergoes a uniform (constant) linear acceleration. We would like to show that the motion in this case leads to a hyperbolic trajectory (hyperbolic motion) and compare this with the corresponding motion in the non-relativistic case.

Since we have motion along just one spatial direction (x-axis), we can write the proper velocity as (see (2.5))

$$u^\mu = \frac{dx^\mu}{d\tau} = \gamma(1, v) = (\gamma, \gamma v), \quad \mu = 0, 1, \tag{2.53}$$

where v denotes the instantaneous velocity of the particle

$$v = \frac{dx}{dt}. \tag{2.54}$$

Since the Lorentz factor (1.72) has the form

$$\gamma = \left(1 - v^2\right)^{-\frac{1}{2}}, \tag{2.55}$$

it follows that

$$1 + \gamma^2 v^2 = \gamma^2. \tag{2.56}$$

Furthermore, since we have

$$\frac{d\gamma}{dt} = \gamma^3 v \frac{dv}{dt}, \tag{2.57}$$

we obtain the proper acceleration to be

$$a^\mu = \frac{du^\mu}{d\tau} = \gamma \frac{d}{dt}(\gamma, \gamma v) = \gamma \left(\frac{d\gamma}{dt}, \frac{d\gamma v}{dt}\right)$$

$$= \left(\gamma^4 v \frac{dv}{dt}, \gamma^2(1 + \gamma^2 v^2)\frac{dv}{dt}\right) = (v, 1)\gamma^4 \frac{dv}{dt}, \tag{2.58}$$

where we have used (2.56). It follows from (2.58) that

$$a^2 = \eta_{\mu\nu} a^\mu a^\nu = (v^2 - 1)\gamma^8 \left(\frac{dv}{dt}\right)^2 = -\gamma^6 \left(\frac{dv}{dt}\right)^2 < 0. \tag{2.59}$$

In the instantaneous rest frame of the particle where $\gamma = 1$, we can write this as

$$a^2 = -\left(\frac{dv}{dt}\right)^2 = -\alpha^2, \tag{2.60}$$

where the instantaneous acceleration is defined as

$$\frac{d(\gamma v)}{dt} = \alpha. \tag{2.61}$$

So far our discussion has been quite general. We will now consider the case of constant acceleration ($\alpha =$ constant). In this case, (2.61) leads to

$$\frac{d(\gamma v)}{dt} = \alpha = \text{constant},$$

or, $\gamma v = \alpha t,$

or, $\alpha^2 t^2 = \gamma^2 v^2,$

or, $1 + \alpha^2 t^2 = 1 + \gamma^2 v^2 = \gamma^2 = (1 - v^2)^{-1},$

or, $v^2 = 1 - \dfrac{1}{1 + \alpha^2 t^2} = \dfrac{\alpha^2 t^2}{1 + \alpha^2 t^2},$

or, $v = \dfrac{\alpha t}{(1 + \alpha^2 t^2)^{1/2}}. \tag{2.62}$

Here we have used (2.56), set the constant of integration to zero as well as taken the positive square root for simplicity. This choice of the integration constant corresponds to choosing the initial velocity to vanish, namely, $v = 0$ when $t = 0$.

On the other hand, the instantaneous velocity is defined as (see (2.54))

$$v = \frac{dx}{dt}, \tag{2.63}$$

which together with (2.62) leads to

$$\frac{dx}{dt} = \frac{\alpha t}{(1 + \alpha^2 t^2)^{1/2}},$$

$$\text{or,} \quad \alpha dx = \frac{1}{2} \frac{d(\alpha^2 t^2)}{(1 + \alpha^2 t^2)^{1/2}}. \tag{2.64}$$

Equation (2.64) can be integrated to give (we assume $x = 0$ when $t = 0$)

$$\alpha x = (1 + \alpha^2 t^2)^{1/2} - 1,$$

$$\text{or,} \quad (\alpha x + 1)^2 = \alpha^2 x^2 + 2\alpha x + 1 = 1 + \alpha^2 t^2,$$

$$\text{or,} \quad \alpha x^2 + 2x - \alpha t^2 = 0, \tag{2.65}$$

which describes a hyperbola with asymptotes

$$x + \frac{1}{\alpha} = \pm t. \tag{2.66}$$

Equation (2.65) should be contrasted with the nonrelativistic parabolic motion of the form (with the same initial condition $x = 0 = v$ when $t = 0$)

$$x = \frac{1}{2} \alpha t^2. \tag{2.67}$$

Finally, we note that we can carry out the complete analysis in terms of the proper time as well and write all the variables as functions of the proper time as

$$v = \tanh \alpha \tau, \qquad\qquad \gamma = (1 - v^2)^{-1/2} = \cosh \alpha \tau,$$

$$t = \frac{\sinh \alpha \tau}{\alpha}, \qquad\qquad x = \frac{(\cosh \alpha \tau - 1)}{\alpha}. \tag{2.68}$$

◀

2.2 Current and charge densities

Let us consider the motion of n charged particles with charge q_n at the coordinate x_n. In this case, the current density associated with the system is given by

$$\mathbf{J}(x) = \sum_n q_n \frac{d\mathbf{x}_n(t)}{dt} \delta^3 (\mathbf{x} - \mathbf{x}_n(t)). \tag{2.69}$$

Similarly, the charge density of the system is given by

$$\rho(x) = \sum_n q_n \delta^3 \left(\mathbf{x} - \mathbf{x}_n(t) \right). \tag{2.70}$$

If we identify $x_n^0(t) = t$, then we note that we can write the charge density also as

$$
\begin{aligned}
\rho(x) &= \sum_n q_n \delta^3 (\mathbf{x} - \mathbf{x}_n(t)) \\
&= \sum_n q_n \int dt' \, \delta^3 (\mathbf{x} - \mathbf{x}_n(t')) \, \delta(t - t') \\
&= \sum_n q_n \int dt' \, \delta^4 \left(x - x_n(t') \right) \\
&= \sum_n q_n \int d\tau \, \frac{dt'}{d\tau} \, \delta^4 (x - x_n(\tau)) \\
&= \int d\tau \sum_n q_n \, \frac{dx_n^0(\tau)}{d\tau} \, \delta^4 \left(x - x_n(\tau) \right),
\end{aligned}
\tag{2.71}
$$

where in the intermediate step we have changed the integration over t' to that over the proper time τ. Similarly, the current density can be rewritten as

$$
\begin{aligned}
\mathbf{J}(x) &= \sum_n q_n \, \frac{d\mathbf{x}_n(t)}{dt} \, \delta^3 \left(\mathbf{x} - \mathbf{x}_n(t) \right) \\
&= \int d\tau \sum_n q_n \, \frac{d\mathbf{x}_n(\tau)}{d\tau} \, \delta^4 \left(x - x_n(\tau) \right).
\end{aligned}
\tag{2.72}
$$

It is suggestive, therefore, to combine both the current and the charge densities into a four vector of the form

$$
\begin{aligned}
J^\mu(x) &= (\rho(x), \mathbf{J}(x)) \\
&= \int d\tau \sum_n q_n \, \frac{dx_n^\mu(\tau)}{d\tau} \, \delta^4 \left(x - x_n(\tau) \right),
\end{aligned}
\tag{2.73}
$$

where the time component of the current density four vector in (2.73) is the charge density of the system while the space components yield the (three dimensional) current density. The four vector nature of the current density is obvious from the fact that the proper time does not change under a Lorentz transformation and

$$\delta^4(x) \rightarrow \delta^4(\Lambda x) = \frac{1}{|\det \Lambda|} \, \delta^4(x) = \delta^4(x), \qquad (2.74)$$

where in the last step we have used the fact that we are considering only proper Lorentz transformations with $\det \Lambda = 1$ (see (1.68)). Therefore, we see that J^μ has the same Lorentz transformation properties as x^μ which is a four vector.

Furthermore, we note that

$$
\begin{aligned}
\partial_\mu J^\mu(x) &= \frac{\partial}{\partial x^\mu} \, J^\mu(x) \\
&= \frac{\partial}{\partial x^\mu} \int d\tau \sum_n q_n \frac{dx_n^\mu(\tau)}{d\tau} \, \delta^4(x - x_n(\tau)) \\
&= \int d\tau \sum_n q_n \frac{dx_n^\mu(\tau)}{d\tau} \, \frac{\partial}{\partial x^\mu} \, \delta^4(x - x_n(\tau)) \\
&= \int d\tau \sum_n q_n \frac{dx_n^\mu(\tau)}{d\tau} \left(-\frac{\partial}{\partial x_n^\mu(\tau)} \, \delta^4(x - x_n(\tau)) \right) \\
&= \int d\tau \sum_n q_n \left(-\frac{d}{d\tau} \, \delta^4(x - x_n(\tau)) \right) \\
&= \int d\tau \sum_n \left(\frac{dq_n}{d\tau} \right) \delta^4(x - x_n(\tau)) = 0. \qquad (2.75)
\end{aligned}
$$

Thus we see that the current is conserved (q_n is a constant) and, written out explicitly, equation (2.75) has the form of the familiar continuity equation, namely,

$$\partial_\mu J^\mu = 0,$$

or, $$\partial_0 J^0 + \partial_i J^i = 0,$$

or, $$\frac{\partial \rho}{\partial t} + \mathbf{\nabla} \cdot \mathbf{J} = 0. \tag{2.76}$$

This shows that the continuity equation (2.75) or (2.76) is Lorentz invariant (holds in any Lorentz frame).

Once we have a conservation law (conserved current density), we can define a conserved charge as

$$Q = \int d^3x \, J^0(\mathbf{x}, t) = Q(t). \tag{2.77}$$

That this is a constant in time (conserved) can be seen as follows,

$$\begin{aligned}
\frac{dQ}{dt} &= \int d^3x \, \frac{\partial J^0(\mathbf{x}, t)}{\partial t} \\
&= \int d^3x \, \frac{\partial \rho(\mathbf{x}, t)}{\partial t} = \int d^3x \, (-\mathbf{\nabla} \cdot \mathbf{J}(\mathbf{x}, t)) = 0, \tag{2.78}
\end{aligned}$$

where we have used the continuity equation (2.76) and the right hand side vanishes because of Gauss' theorem if the currents vanish at large separations. Thus, we obtain

$$\frac{dQ}{dt} = 0. \tag{2.79}$$

We can further show that the conserved charge Q is not only constant but is also a Lorentz scalar in the following way. Since

$$Q = \int d^3x \, J^0(\mathbf{x}, t), \tag{2.80}$$

and since J^0 behaves like the time component of a four vector under a Lorentz transformation, the charge must transform as

$$Q \sim \int d^4x = \text{Lorentz invariant.} \tag{2.81}$$

A more rigorous way of seeing this is as follows. Let us consider the integral

$$I = \int d^4x \, \partial_\mu \left(J^\mu(\mathbf{x}, t)\theta(n \cdot x) \right), \tag{2.82}$$

where n^μ is a time-like unit vector and

$$\theta(z) = \begin{cases} 1 & \text{if } z > 0, \\ 0 & \text{if } z < 0, \end{cases} \tag{2.83}$$

represents the step function. Therefore, writing out explicitly we have

$$I = \int d^4x \left[\partial_0 \left(J^0(\mathbf{x}, t)\theta(n \cdot x) \right) + \partial_i \left(J^i(\mathbf{x}, t)\theta(n \cdot x) \right) \right]. \tag{2.84}$$

The spatial divergence, of course, vanishes because of Gauss' theorem so that we can write

$$I = \int d^4x \, \partial_0 \left(J^0(\mathbf{x}, t)\theta(n \cdot x) \right). \tag{2.85}$$

If we now identify $n^\mu = (1, 0, 0, 0)$ and do the time integral then we have

$$\begin{aligned} I &= \int dt \, d^3x \, \partial_0 \left(J^0(\mathbf{x}, t)\theta(t) \right) \\ &= \int d^3x \, J^0(\mathbf{x}, t)\theta(t) \Big|_{t=-\infty}^{t=\infty} \\ &= \int d^3x \, J^0(\mathbf{x}, t = \infty) = Q(\infty) = Q, \end{aligned} \tag{2.86}$$

since Q is a constant. Thus we conclude that we can write

$$Q = \int d^4x \, \partial_\mu \left(J^\mu(\mathbf{x}, t) \theta \left(n \cdot x \right) \right)$$

$$= \int d^4x \, \left(\partial_\mu J^\mu \theta(n \cdot x) + J^\mu \partial_\mu \theta(n \cdot x) \right)$$

$$= \int d^4x \, J^\mu(x) \partial_\mu \theta(n \cdot x). \tag{2.87}$$

The Lorentz scalar nature of Q is now manifest since both d^4x and $\theta(n \cdot x)$ are Lorentz invariant quantities.

2.3 Maxwell's equations in the presence of sources

If there are charges and currents present, then we know that Maxwell's equations take the form $(c = 1)$

$$\boldsymbol{\nabla} \cdot \mathbf{E} = \rho,$$

$$\boldsymbol{\nabla} \cdot \mathbf{B} = 0,$$

$$\boldsymbol{\nabla} \times \mathbf{E} = -\frac{\partial \mathbf{B}}{\partial t},$$

$$\boldsymbol{\nabla} \times \mathbf{B} = \frac{\partial \mathbf{E}}{\partial t} + \mathbf{J}, \tag{2.88}$$

where ρ and $\mathbf{J}$ denote the charge density and the current density respectively (produced by the charged matter). We have already seen in (2.73) that ρ and $\mathbf{J}$ can be combined into a four vector

$$J^\mu = (\rho, \mathbf{J}). \tag{2.89}$$

Furthermore, we have also seen in (1.76) and (1.77) that the electric field $\mathbf{E}$ and the magnetic field $\mathbf{B}$ can be combined into the field strength tensor $F_{\mu\nu}$ such that

$$F_{0i} = E_i,$$

$$F_{ij} = -\epsilon_{ijk} B_k. \tag{2.90}$$

The dual field strength tensor in (1.81) leads to the components

$$
\begin{aligned}
\widetilde{F}_{0i} &= B_i, \\
\widetilde{F}_{ij} &= \epsilon_{ijk} E_k.
\end{aligned}
\tag{2.91}
$$

As a result, in the presence of sources we can write Maxwell's equations (2.88) as

$$
\partial^\mu F_{\mu\nu} = J_\nu, \qquad \partial^\mu \widetilde{F}_{\mu\nu} = 0.
\tag{2.92}
$$

That these give the correct Maxwell's equations can be seen as follows. Let $\nu = 0$ in the first of the two equations in (2.92) which leads to

$$
\partial^\mu F_{\mu 0} = J_0,
$$

or, $\quad \partial^i F_{i0} = \rho,$

or, $\quad (-\boldsymbol{\nabla}) \cdot (-\mathbf{E}) = \boldsymbol{\nabla} \cdot \mathbf{E} = \rho.$
$$\tag{2.93}$$

Similarly, choosing $\nu = j$, we obtain

$$
\partial^\mu F_{\mu j} = J_j,
$$

or, $\quad \partial^0 F_{0j} + \partial^i F_{ij} = J_j,$

or, $\quad \dfrac{\partial E_j}{\partial t} + \partial^i \left(-\epsilon_{ijk} B_k\right) = J_j,$

or, $\quad \dfrac{\partial \mathbf{E}}{\partial t} + (-\boldsymbol{\nabla}) \times \mathbf{B} = -\mathbf{J},$

or, $\quad \boldsymbol{\nabla} \times \mathbf{B} = \dfrac{\partial \mathbf{E}}{\partial t} + \mathbf{J}.$
$$\tag{2.94}$$

These give two of Maxwell's equations with sources in (2.88). The other two equations which are unaltered in the presence of charges and currents (compare with (1.51)) are still given by the equation involving the dual field strength tensor (as in (1.82))

$$\partial^\mu \widetilde{F}_{\mu\nu} = 0, \tag{2.95}$$

which can also be written as the Bianchi identity

$$\partial_\mu F_{\nu\lambda} + \partial_\nu F_{\lambda\mu} + \partial_\lambda F_{\mu\nu} = 0. \tag{2.96}$$

(The Bianchi identity can be explicitly checked to be true from the definition of the field strength tensor in (1.83).)

The Lorentz covariance of the equations can be seen from the fact that

$$
\begin{aligned}
\partial^\mu F_{\mu\nu} \ \to\ \partial'^\mu F'_{\mu\nu} &= \left(\Lambda^\mu{}_{\mu'}\partial^{\mu'}\right)\left(\Lambda_\mu{}^\lambda \Lambda_\nu{}^\rho\right) F_{\lambda\rho} \\
&= \Lambda^\mu{}_{\mu'}\Lambda_\mu{}^\lambda \Lambda_\nu{}^\rho\, \partial^{\mu'} F_{\lambda\rho} \\
&= \delta_{\mu'}^\lambda \Lambda_\nu{}^\rho\, \partial^{\mu'} F_{\lambda\rho} \\
&= \Lambda_\nu{}^\rho\, \partial^\lambda F_{\lambda\rho} = \Lambda_\nu{}^\rho\, \partial^\mu F_{\mu\rho}.
\end{aligned}
\tag{2.97}
$$

Similarly,

$$J_\nu \to J'_\nu = \Lambda_\nu{}^\rho J_\rho. \tag{2.98}$$

Therefore, it is clear that the equation

$$\partial^\mu F_{\mu\nu} = J_\nu, \tag{2.99}$$

maintains its form in all Lorentz frames. In other words, if the above equation is true in one frame, then in a different Lorentz frame the equation has the same form, namely,

$$\partial'^\mu F'_{\mu\nu} = J'_\nu. \tag{2.100}$$

2.4 Motion of a charged particle in EM field

If a charged particle with charge q is moving with a non-relativistic velocity $\mathbf{v}$ in an electromagnetic field, then we know that it experiences a force (Lorentz force) given by ($c = 1$)

$$\mathbf{F}_{(\text{NR})} = q\left(\mathbf{E} + \mathbf{v} \times \mathbf{B}\right). \tag{2.101}$$

If the particle is at rest then, of course, the force is purely electrostatic and is given by

$$\mathbf{F}_{(\text{rest})} = q\mathbf{E}. \tag{2.102}$$

We note that we can generalize this to the relativistic electromagnetic force experienced by a particle in the four vector notation as

$$f^{\mu} = qF^{\mu}{}_{\nu}\,\frac{\mathrm{d}x^{\nu}}{\mathrm{d}\tau} = qF^{\mu}{}_{\nu}u^{\nu}. \tag{2.103}$$

For example, this force is by construction orthogonal to the four velocity (as we would expect, see (2.9))

$$f^{\mu}u_{\mu} = qF^{\mu}{}_{\nu}u^{\nu}u_{\mu} = qF_{\mu\nu}u^{\mu}u^{\nu} = 0, \tag{2.104}$$

which follows from the anti-symmetry of the field strength tensor. Furthermore, in the rest frame of the particle ($\frac{\mathrm{d}\mathbf{x}}{\mathrm{d}t} = 0$), it is clear that

$$
\begin{aligned}
f^{0}_{(\text{rest})} &= qF^{0}{}_{\nu}\,\frac{\mathrm{d}x^{\nu}}{\mathrm{d}\tau} = qF^{0}{}_{i}\,\frac{\mathrm{d}x^{i}}{\mathrm{d}\tau} = 0, \\
f^{i}_{(\text{rest})} &= qF^{i}{}_{\nu}\,\frac{\mathrm{d}x^{\nu}}{\mathrm{d}\tau} \\
&= q\left(F^{i}{}_{0}\,\frac{\mathrm{d}x^{0}}{\mathrm{d}\tau} + F^{i}{}_{j}\,\frac{\mathrm{d}x^{j}}{\mathrm{d}\tau}\right) \\
&= qF^{i}{}_{0}\,\frac{\mathrm{d}x^{0}}{\mathrm{d}\tau} = qF^{i}{}_{0} = -qF_{i0} = qF_{0i} = qE_{i},
\end{aligned}
$$
$$\text{or,} \quad \mathbf{f}_{(\text{rest})} = q\mathbf{E}, \tag{2.105}$$

as we would expect. Here we have used the fact that in the rest frame of the particle $x^0 = t = \tau$. For a non-relativistic particle ($v \ll 1$), on the other hand, we note from (2.103) that

$$
\begin{aligned}
f^i_{(NR)} &= qF^i{}_\nu \frac{dx^\nu}{d\tau} \\
&= q\left(F^i{}_0 \frac{dx^0}{d\tau} + F^i{}_j \frac{dx^j}{d\tau}\right) \\
&= q\left(F_{0i} \frac{dx^0}{dt} \frac{dt}{d\tau} - F_{ij} \frac{dx^j}{dt} \frac{dt}{d\tau}\right) \\
&= q\frac{dt}{d\tau}\left(F_{0i} + \epsilon_{ijk}B_k v^j\right),
\end{aligned}
$$

$$
\text{or,}\quad \mathbf{f}_{(NR)} = q\frac{dt}{d\tau}(\mathbf{E} + \mathbf{v} \times \mathbf{B}) = q\gamma(\mathbf{E} + \mathbf{v} \times \mathbf{B})
$$

$$
= q(\mathbf{E} + \mathbf{v} \times \mathbf{B}) + 0\left(v^3\right). \tag{2.106}
$$

Furthermore, the time component of the force has the form

$$
\begin{aligned}
f^0_{(NR)} &= qF^0{}_\nu \frac{dx^\nu}{d\tau} \\
&= qF^0{}_i \frac{dx^i}{d\tau} \\
&= qF_{0i} \frac{dx^i}{dt} \frac{dt}{d\tau} \\
&= q\,\mathbf{v}\cdot\mathbf{E}\,\gamma \\
&= \mathbf{v}\cdot\mathbf{f}_{(NR)} = \mathbf{v}\cdot\mathbf{f}_{(rest)} + 0\left(v^3\right), \tag{2.107}
\end{aligned}
$$

which we can also identify with $\gamma\mathbf{v}\cdot\mathbf{f}_{(rest)}$. This, of course, also gives the time rate of change of the energy of the particle in an electromagnetic field. Thus, we see that the force four vector (2.103) reduces to the appropriate limits and, therefore, we conclude that the correct relativistic form of the equation of motion for a charged particle in an electromagnetic field is given by

$$
m\frac{d^2x^\mu}{d\tau^2} = m\frac{du^\mu}{d\tau} = \frac{dp^\mu}{d\tau} = f^\mu = qF^\mu{}_\nu \frac{dx^\nu}{d\tau} = qF^\mu{}_\nu u^\nu. \tag{2.108}
$$

The non-relativistic form of the equation can be obtained in the appropriate limit by using the components of the forces derived in (2.106) and (2.107).

2.5 Energy-momentum tensor

Just as for a distribution of charged particles we can define a charge density and a current density and combine them into a four vector, similarly for a system of particles with nonzero momentum we can define the following densities,

$$
\begin{aligned}
T^{\mu 0}_{(\text{matter})} &= \sum_n p_n^\mu \delta^3 \left(\mathbf{x} - \mathbf{x}_n(t) \right) \\
&= \int d\tau \sum_n p_n^\mu \frac{dx_n^0(\tau)}{d\tau} \delta^3 \left(\mathbf{x} - \mathbf{x}_n(\tau) \right) \delta \left(t - x_n^0(\tau) \right) \\
&= \int d\tau \sum_n p_n^\mu \frac{dx_n^0(\tau)}{d\tau} \delta^4 \left(x - x_n(\tau) \right), \qquad (2.109) \\
T^{\mu i}_{(\text{matter})} &= \sum_n p_n^\mu \frac{dx_n^i(t)}{dt} \delta^3 \left(\mathbf{x} - \mathbf{x}_n(t) \right) \\
&= \int d\tau \sum_n p_n^\mu \frac{dx_n^i(\tau)}{d\tau} \delta^3 \left(\mathbf{x} - \mathbf{x}_n(\tau) \right) \delta \left(t - x_n^0(\tau) \right) \\
&= \int d\tau \sum_n p_n^\mu \frac{dx_n^i(\tau)}{d\tau} \delta^4 \left(x - x_n(\tau) \right). \qquad (2.110)
\end{aligned}
$$

Therefore, we see that we can combine the two expressions in (2.109) and (2.110) to form a second rank tensor density

$$
T^{\mu\nu}_{(\text{matter})} = \int d\tau \sum_n p_n^\mu \frac{dx_n^\nu}{d\tau} \delta^4 \left(x - x_n(\tau) \right). \qquad (2.111)
$$

Furthermore, if we use the definition of the momentum for a relativistic particle (see (2.11))

$$
p_n^\mu = m \frac{dx_n^\mu}{d\tau}, \qquad (2.112)
$$

we see that the tensor $T^{\mu\nu}$ in (2.111) is a symmetric second rank tensor density and can be written as

$$T^{\mu\nu}_{(\text{matter})} = m \int d\tau \sum_n \frac{dx^{\mu}_n}{d\tau} \frac{dx^{\nu}_n}{d\tau} \delta^4 (x - x_n(\tau))$$

$$= T^{\nu\mu}_{(\text{matter})}. \tag{2.113}$$

This is known as the stress tensor density or the energy momentum tensor density of the system of particles.

Let us next note that

$$\partial_\mu T^{\mu\nu}_{(\text{matter})} = \frac{\partial}{\partial x^\mu} T^{\mu\nu}_{(\text{matter})}$$

$$= \frac{\partial}{\partial x^\mu} m \int d\tau \sum_n \frac{dx^{\mu}_n}{d\tau} \frac{dx^{\nu}_n}{d\tau} \delta^4 (x - x_n(\tau))$$

$$= m \int d\tau \sum_n \frac{dx^{\mu}_n}{d\tau} \frac{dx^{\nu}_n}{d\tau} \frac{\partial}{\partial x^\mu} \delta^4 (x - x_n(\tau))$$

$$= m \int d\tau \sum_n \frac{dx^{\mu}_n}{d\tau} \frac{dx^{\nu}_n}{d\tau} \left(-\frac{\partial}{\partial x^{\mu}_n(\tau)} \delta^4 (x - x_n(\tau)) \right)$$

$$= m \int d\tau \sum_n \frac{dx^{\nu}_n}{d\tau} \left(-\frac{d}{d\tau} \delta^4 (x - x_n(\tau)) \right)$$

$$= \sum_n \int d\tau\, p^{\nu}_n \left(-\frac{d}{d\tau} \delta^4 (x - x_n(\tau)) \right)$$

$$= \sum_n \int d\tau\, \frac{dp^{\nu}_n}{d\tau} \delta^4 (x - x_n(\tau))$$

$$= \int d\tau \sum_n f^{\nu}_n \delta^4 (x - x_n(\tau)). \tag{2.114}$$

We note that, unlike the current density in (2.73) where the coefficients q_n are time independent (see also (2.75)), here p^{μ}_n's are, in general, time dependent. Therefore, only if the total force acting on the individual particles of the system is zero, then we have

$$\partial_\mu T^{\mu\nu}_{(\text{matter})} = 0. \tag{2.115}$$

Namely, the energy momentum tensor for matter in (2.111) or (2.113) is conserved only in the absence of external forces and in this case we can define the charge associated with this tensor density as

$$P^\mu_{(\text{matter})}(t) = \int d^3x \, T^{0\mu}_{(\text{matter})} = \int d^3x \, T^{\mu 0}_{(\text{matter})}, \tag{2.116}$$

which is easily seen to be conserved (see (2.78)), namely,

$$\frac{dP^\mu_{(\text{matter})}}{dt} = 0. \tag{2.117}$$

We note from (2.111) or (2.113) that

$$
\begin{aligned}
P^\mu_{(\text{matter})} &= \int d^3x \, T^{0\mu}_{(\text{matter})} = \int d^3x \, T^{\mu 0}_{(\text{matter})} \\
&= \int d^3x \int d\tau \sum_n p^\mu_n \frac{dx^0_n}{d\tau} \, \delta^4\left(x - x_n(\tau)\right) \\
&= \sum_n \int d^4x \, p^\mu_n \, \delta^4\left(x - x_n(\tau)\right) \\
&= \sum_n p^\mu_n.
\end{aligned}
\tag{2.118}
$$

In other words, the conserved charge in this case is the sum of the momenta of individual particles (total momentum of the system) and it is clear that under a Lorentz transformation it transforms like a contravariant four vector.

On the other hand, if there are external forces acting on the system, we have seen in (2.114) that the divergence of the energy momentum tensor density (stress tensor density) of the system of particles is not zero and is given by

$$\partial_\mu T^{\mu\nu}_{(\text{matter})} = \int d\tau \sum_n f^\nu_n \delta^4\left(x - x_n(\tau)\right). \tag{2.119}$$

Let us assume, for example, that the particles carry individual charges q_n and that the force they experience is due to the presence of an external electromagnetic field. In this case, we can write (see (2.103))

$$
\begin{aligned}
\partial_\mu T^{\mu\nu}_{(\text{matter})} &= \int d\tau \sum_n f_n^{\;\nu} \delta^4 \left(x - x_n(\tau)\right) \\
&= \int d\tau \sum_n q_n F^\nu_{\;\lambda} \frac{dx_n^{\;\lambda}}{d\tau} \delta^4 \left(x - x_n(\tau)\right) \\
&= F^\nu_{\;\lambda} \int d\tau \sum_n q_n \frac{dx_n^{\;\lambda}}{d\tau} \delta^4 \left(x - x_n(\tau)\right), \qquad (2.120)
\end{aligned}
$$

where we have used the fact that the external field depends only on x and, therefore, can be taken outside the sum and the integral. The rest of the expression together with the integral sign is simply the current density which we have defined earlier in (2.73) and hence we obtain

$$
\partial_\mu T^{\mu\nu}_{(\text{matter})} = F^\nu_{\;\lambda} J^\lambda. \qquad (2.121)
$$

To understand better what this means, let us define a second rank symmetric tensor from the electromagnetic field strength tensor (so that it is gauge invariant) in the following way

$$
T^{\mu\nu}_{(\text{em})} = T^{\nu\mu}_{(\text{em})} = -F^\mu_{\;\lambda} F^{\nu\lambda} + \frac{1}{4} \eta^{\mu\nu} F_{\lambda\delta} F^{\lambda\delta}. \qquad (2.122)
$$

We can work out the explicit forms of the components of this tensor density associated with the electromagnetic field from its definition in (2.122)

$$
\begin{aligned}
T^{00}_{(\text{em})} &= -F^{0\mu} F^0_{\;\mu} + \frac{1}{4} \eta^{00} F_{\mu\nu} F^{\mu\nu} \\
&= -F^{0i} F^0_{\;i} + \frac{1}{4} \left(F_{0i} F^{0i} + F_{i0} F^{i0} + F_{ij} F^{ij}\right) \\
&= F_{0i} F_{0i} + \frac{1}{4} \left(-2 F_{0i} F_{0i} + (-\epsilon_{ijk} B_k)(-\epsilon_{ij\ell} B_\ell)\right)
\end{aligned}
$$

$$\begin{aligned}
&= E_i E_i - \frac{1}{2} E_i E_i + \frac{1}{2} B_i B_i \\
&= \frac{1}{2} \left(\mathbf{E}^2 + \mathbf{B}^2 \right), \\
T^{0i}_{(\text{em})} &= -F^{0\mu} F^i{}_\mu + \frac{1}{4} \eta^{0i} F_{\mu\nu} F^{\mu\nu} \\
&= -F^{0j} F^i{}_j \\
&= -\left(-F_{0j}\right)\left(-F_{ij}\right) = -F_{0j} F_{ij} \\
&= -E_j \left(-\epsilon_{ijk} B_k\right) = \epsilon_{ijk} E_j B_k = (\mathbf{E} \times \mathbf{B})_i.
\end{aligned} \tag{2.123}$$

This shows that the expressions for $T^{00}_{(\text{em})}$ and $T^{0i}_{(\text{em})}$ lead respectively to the correct energy and momentum (recall the Poynting vector) densities for the electromagnetic field as we know from electrodynamics. Therefore, we can identify $T^{\mu\nu}_{(\text{em})}$ in (2.122) with the stress tensor density of the electromagnetic field.

It follows now from its definition in (2.122) that

$$\begin{aligned}
\partial_\mu T^{\mu\nu}_{(\text{em})} &= -(\partial_\mu F^\mu{}_\lambda) F^{\nu\lambda} - F^\mu{}_\lambda \partial_\mu F^{\nu\lambda} + \frac{1}{2} \eta^{\mu\nu} (\partial_\mu F^{\lambda\rho}) F_{\lambda\rho} \\
&= -(\partial_\mu F^\mu{}_\lambda) F^{\nu\lambda} - F_{\mu\lambda} \partial^\mu F^{\nu\lambda} + \frac{1}{2} (\partial^\nu F^{\lambda\rho}) F_{\lambda\rho} \\
&= -(\partial_\mu F^\mu{}_\lambda) F^{\nu\lambda} + \frac{1}{2} F_{\lambda\rho} \left(\partial^\rho F^{\nu\lambda} + \partial^\lambda F^{\rho\nu} \right) + \frac{1}{2} (\partial^\nu F^{\lambda\rho}) F_{\lambda\rho} \\
&= -(\partial_\mu F^\mu{}_\lambda) F^{\nu\lambda} + \frac{1}{2} F_{\lambda\rho} \left(\partial^\nu F^{\lambda\rho} + \partial^\lambda F^{\rho\nu} + \partial^\rho F^{\nu\lambda} \right) \\
&= -(\partial_\mu F^\mu{}_\lambda) F^{\nu\lambda} = -J_\lambda F^{\nu\lambda} = -F^\nu{}_\lambda J^\lambda.
\end{aligned} \tag{2.124}$$

Here, in the intermediate steps we have used the Bianchi identity (see the discussion following (2.95))

$$\partial_\mu F_{\nu\lambda} + \partial_\nu F_{\lambda\mu} + \partial_\lambda F_{\mu\nu} = 0, \tag{2.125}$$

as well as Maxwell's equations (2.92) in the presence of sources

$$\partial_\mu F^{\mu\nu} = J^\nu. \tag{2.126}$$

Thus, we see from (2.121) and (2.124) that

$$\partial_\mu T^{\mu\nu}_{(\text{matter})} = F^\nu_{\ \lambda} J^\lambda = -\partial_\mu T^{\mu\nu}_{(\text{em})}, \tag{2.127}$$

which shows that if we define the total energy momentum tensor density of the system of particles as well as the electromagnetic radiation field as

$$
\begin{aligned}
T^{\mu\nu}_{(\text{total})} &= T^{\mu\nu}_{(\text{matter})} + T^{\mu\nu}_{(\text{em})} \\
&= \int d\tau \sum_n p_n^{\ \mu} \frac{dx_n^{\ \nu}}{d\tau} \delta^4 \left(x - x_n(\tau) \right) \\
&\quad - F^{\mu\lambda} F^\nu_{\ \lambda} + \frac{1}{4} \eta^{\mu\nu} F^{\lambda\rho} F_{\lambda\rho},
\end{aligned}
\tag{2.128}
$$

then we have

$$\partial_\mu T^{\mu\nu}_{(\text{total})} = 0. \tag{2.129}$$

Namely, the total stress tensor density associated with the system including matter and radiation is conserved. (The energy momentum tensor density of an isolated system is divergence free because of Poincaré invariance.)

2.6 Angular momentum

Once we know that the energy momentum tensor density of an isolated system is conserved, we can construct other tensor densities from this, which would also be conserved. For example, given

$$\partial_\mu T^{\mu\nu} = 0 = \partial_\nu T^{\mu\nu}, \quad T^{\mu\nu} = T^{\nu\mu}, \tag{2.130}$$

let us define

$$M^{\lambda\mu\nu} \equiv x^\mu T^{\nu\lambda} - x^\nu T^{\mu\lambda}. \tag{2.131}$$

We note that by construction this tensor density is anti-symmetric in the last two indices, namely,

$$M^{\lambda\mu\nu} = -M^{\lambda\nu\mu}. \tag{2.132}$$

We also recognize that it is divergence free in the first index, namely,

$$
\begin{aligned}
\partial_\lambda M^{\lambda\mu\nu} &= \frac{\partial}{\partial x^\lambda} \left(x^\mu T^{\nu\lambda} - x^\nu T^{\mu\lambda} \right) \\
&= \left(\delta^\mu_\lambda T^{\nu\lambda} + x^\mu \partial_\lambda T^{\nu\lambda} - \delta^\nu_\lambda T^{\mu\lambda} - x^\nu \partial_\lambda T^{\mu\lambda} \right) \\
&= T^{\nu\mu} - T^{\mu\nu} = 0,
\end{aligned} \tag{2.133}
$$

where we have used the fact that the stress tensor is symmetric and conserved (see (2.130)). Furthermore, since $M^{\lambda\mu\nu}$ is conserved in its first index, we can construct the charge associated with this tensor density as

$$J^{\mu\nu} = \int \mathrm{d}^3x \; M^{0\mu\nu} = -J^{\nu\mu}. \tag{2.134}$$

We can show (in the standard manner, see (2.78)) that this tensor is constant in time (conserved) and transforms like an anti-symmetric second rank tensor under a Lorentz transformation. To see what such a conserved charge corresponds to, let us look at a system of free particles for which

$$
\begin{aligned}
M^{0ij} &= x^i T^{j0} - x^j T^{i0} \\
&= x^i \sum_n p_n^j \delta^3(\mathbf{x} - \mathbf{x}_n(t)) - x^j \sum_n p_n^i \delta^3(\mathbf{x} - \mathbf{x}_n(t)) \\
&= \sum_n \left(x_n^i p_n^j - x_n^j p_n^i \right) \delta^3(\mathbf{x} - \mathbf{x}_n(t)) \\
&= \sum_n \epsilon_{ijk} L_n^k \delta^3(\mathbf{x} - \mathbf{x}_n(t)).
\end{aligned} \tag{2.135}
$$

In other words, the space components of the tensor density $M^{0\mu\nu}$ correspond to the angular momentum density of the system of particles. Hence we can identify $J^{\mu\nu} = -J^{\nu\mu}$ as the generalized relativistic angular momentum of the system.

The angular momentum of a particle, of course, consists of two parts, the orbital angular momentum and the internal angular momentum. The internal angular momentum is isolated by defining the Pauli-Lubanski spin four vector as

$$s_\mu = \frac{1}{2} \, \epsilon_{\mu\nu\lambda\rho} J^{\nu\lambda} u^\rho, \tag{2.136}$$

where $J^{\nu\lambda}, u^\rho$ denote the generalized angular momentum (2.134) and the four velocity (2.5) respectively. s_μ is a constant (time independent) four vector for a free particle which can be seen from

$$\frac{\mathrm{d} s_\mu}{\mathrm{d} t} = \frac{1}{2} \, \epsilon_{\mu\nu\lambda\rho} J^{\nu\lambda} \, \frac{\mathrm{d} u^\rho}{\mathrm{d} t} = 0, \tag{2.137}$$

where we have used the fact that $J^{\mu\nu}$ is time independent and that in the absence of forces, the four velocity does not depend on time either. We note that in the rest frame of the particle

$$u^\mu = (1, 0, 0, 0), \tag{2.138}$$

the components of s_μ take the forms (in the rest frame, there is no orbital angular momentum)

$$s_i^{(\text{rest})} = \frac{1}{2} \, \epsilon_{ijk} J^{jk} = \text{spin angular momentum}, \tag{2.139}$$

$$s_0^{(\text{rest})} = 0. \tag{2.140}$$

Furthermore, although s_μ may appear to have four independent components, it actually has three since it has to satisfy the constraint

$$u^\mu s_\mu = 0, \tag{2.141}$$

in any frame of reference which follows from its definition in (2.136). (This is like the constraint on the relativistic force four vector derived in (2.9).)

CHAPTER 3

Principle of general covariance

3.1 Principle of equivalence

Let us consider a particle moving in a static, uniform gravitational field as well as under the influence of other (translation invariant) inertial forces. The equation of motion according to Newton, in this case, would be given by

$$m\,\frac{d^2\mathbf{x}}{dt^2} = m\mathbf{g} + \sum_n \mathbf{F}_n(\mathbf{x} - \mathbf{x}_n), \qquad (3.1)$$

where the acceleration due to gravity $\mathbf{g}$ is assumed to be a space and time independent constant and the $\mathbf{F}_n(\mathbf{x} - \mathbf{x}_n)$'s represent inertial forces. As a consequence of the fact that the inertial and the gravitational masses are the same (see (1.43)), we conclude from (3.1) that we can transform away the gravitational force, in the present case, by a simple coordinate redefinition, namely, if we let

$$\mathbf{x} \to \mathbf{x} - \frac{1}{2}\,\mathbf{g}t^2, \qquad (3.2)$$

we can write the equation of motion (3.1) as

$$m\,\frac{d^2\mathbf{x}}{dt^2} = \sum_n \mathbf{F}_n(\mathbf{x} - \mathbf{x}_n). \qquad (3.3)$$

In other words, in this case we can always find a coordinate system where the effects of a static uniform gravitational field would disappear. (In this case, the new frame of reference would correspond to a uniformly accelerated frame.) In this sense, we can think of a static uniform gravitational force as being an apparent force or a pseudoforce.

Let us see physically what this example implies. Consider a person (observer) in an elevator (laboratory) in space (say, in a space ship) who performs the following experiment. He releases various masses and finds that they all fall down with the same acceleration. He wishes to find out the cause of the acceleration except that he is not allowed to look outside the elevator. Clearly from all the experiments that he can perform inside the elevator, he can conclude that there exist only two possibilities for the source of acceleration:

1. There is a large gravitational mass attached to the bottom of the elevator which pulls all particles downwards.

2. The other possibility is, of course, that the elevator is being accelerated upwards uniformly so that all particles appear to be falling downwards.

However, he cannot differentiate between the two possibilities. This shows that the acceleration due to a static uniform gravitational force is equivalent to the uniform acceleration due to an inertial force. Therefore, by a suitable change of the coordinate system we can transform away the effects of gravitation and the equations of motion would take the same form as in an inertial reference frame without gravitation.

There are two interesting issues that arise at this point. If the above analogy is correct we can do the following experiment. Let one side of the elevator be fitted with a source that emits a pulse of light. Let the opposite side of the elevator be fixed with a detector which receives the pulse. If the elevator is being accelerated upwards we can calculate and show that the light ray would define a parabolic trajectory as shown in Fig. 3.1. That is, its trajectory would bend. However, if this is equivalent to the presence of a gravitational mass, then we would conclude that a light ray would bend in the presence of a large gravitational mass. This was, of course, a phenomenon unheard of in classical physics. However, Einstein was so strongly convinced of the equivalence principle that he made the daring prediction that light rays would bend in the presence of large gravitational masses. This was later verified experimentally by Eddington. (We would do this calculation later in the course of the lectures.)

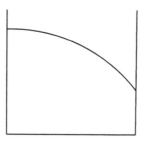

Figure 3.1: Bending of light in a space ship.

The second question we need to ask is what happens to this equivalence in the presence of a non-static and nonuniform gravitational field. Pulsars and other astrophysical objects provide physical examples of non-static gravitational sources. Our own earth provides an example of a nonuniform gravitational field since all particles are attracted radially towards the center of the earth. It is not possible to balance such nonuniform accelerations by any linear inertial acceleration as shown in Fig. 3.2. Therefore, in such a case the two situations would be distinguishable. In the background of a non-static and nonuniform gravitational force particles would feel tidal forces which would lead their trajectories to deviate from the vertical straight line motion. These geodesic deviations, of course, would allow one to detect the presence of gravitation. However, if we restrict ourselves to very small regions of space (and time) then the gravitational field would be uniform and in such regions it would again be impossible to detect the presence of gravitation.

These observations led to the formulation of the principle of equivalence. It says that at every space-time point in an arbitrary gravitational field, it is always possible to choose a locally inertial frame such that within a sufficiently small region around the point in question, the laws of nature take the same form as in inertial Cartesian coordinate systems in the absence of gravitation.

Some comments are in order here. First of all, note that this is what is usually referred to as the strong principle of equivalence. If

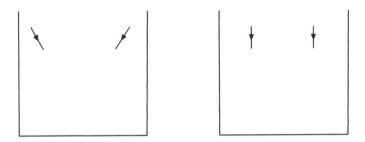

Figure 3.2: The effect of gravitation is represented on the left diagram while the right denotes linear acceleration.

the phrase "laws of nature" is replaced by "laws of motion of freely falling (moving) particles", then it is referred to as the weak principle of equivalence. Furthermore, we are yet to specify what we mean by the form of the laws of nature. If we choose the laws of nature as those invariant under Lorentz transformations then we are led to what is known as the Einstein's theory of gravity or the general theory of relativity. However, if we choose the laws of nature in the Cartesian frame to satisfy Galilean invariance, we obtain the Newton-Cartan theory of gravity. We note that these theories have essentially very different characters. In the case of Lorentz invariance we can define a four dimensional invariant length ($c = 1$)

$$ds^2 = d\tau^2 = dt^2 - d\mathbf{x}^2, \tag{3.4}$$

so that such theories admit a nontrivial metric and lead to metric (Riemannian) theories of gravitation. However, in the case of Galilean invariance we cannot define a four dimensional invariant length. Theories with Galilean invariance, therefore, do not admit a metric and hence lead to non-metric (non-Riemannian) theories of gravitation. We will concern ourselves only with laws of nature that remain invariant under a Lorentz transformation of the inertial frames.

The principle of equivalence further emphasizes that gravitation is a kinematic effect. One of the interpretations of this is that a gravitational field would not produce any new phenomenon that we cannot observe in a Lorentz frame. However, the physical interpretations of the equivalence principle are still open to discussions. Furthermore, let us also note the similarity between the equivalence principle and the observations of Gauss. As we have already discussed, Gauss had observed that even in a curved manifold we can establish a locally Euclidean coordinate system (see the discussion around (1.8)). Principle of equivalence, on the other hand, tells us that even in the presence of a gravitational field we can find a locally inertial Cartesian coordinate system. Correspondingly, we can think of a gravitational field as producing curvature in the space-time manifold and the effect of gravitation on a particle can then be simply thought of as being equivalent to a particle moving in a curved geometry. That this is plausible can be seen with the following examples.

Let us consider a sheet of rubber attached to some boundary so that it is flat. If we now roll a steel ball on to the sheet, it is clear that it would create some dip thereby producing a curvature in the manifold of the rubber sheet. If we roll another steel ball onto the rubber sheet it would again make a dip in the rubber sheet and it would move towards the other steel ball giving the impression that the two balls are attracted towards each other as indicated in Fig. 3.3. This shows that it is plausible to think of the effect of gravitation entirely as producing curvature in the space-time manifold. Gravitational attraction between masses can then be thought of as the motion of a particle in the curved manifold which is produced by the distribution of all other masses.

Figure 3.3: Masses giving rise to curvature in a rubber sheet.

A second example is the movement of bugs on the surface of a spherical apple. Let us allow two bugs at the equator to crawl towards the north pole along straight paths as shown in Figure. 3.4. Over a very small region (locally), they would, of course, appear to be moving parallel to each other. However, over a larger region the paths would appear converging. This is what is known as the geodesic deviation and we have noted this in connection with the experiment in the elevator in Fig. 3.2. Furthermore, note that if the apple has a stem, then the shape of the apple is not completely spherical. Therefore, when the bugs crawl near the dimple caused by the stem, there would be a large deflection in their paths. This is similar to the deflection caused in the path of a comet by the sun. It would appear as if the bugs are attracted by the stem and that is the cause of the deflection in their paths.

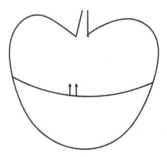

Figure 3.4: Motion on the surface of an apple.

These examples illustrate that in Einstein's view space-time is curved due to a distribution of matter (masses) in the universe and on this curved manifold freely moving particles follow the straightest possible paths also known as geodesics. Locally the paths look like the paths in a flat manifold. However, over a finite region it is the geodesic deviation which exhibits the curved nature of the manifold. The way in which the dimple around the stem gives the impression of attraction corresponds to the fact that massive bodies modify the space-time curvature in their vicinity and this modification affects the geodesics in such a way as to give the impression that the free particles are being acted on by a force whereas in reality they are simply following the straightest paths in the curved space-time.

3.2 Principle of general covariance

In the case of Lorentz transformations we have observed that the length interval $(c = 1)$

$$ds^2 = d\tau^2 = \eta_{\mu\nu}dx^\mu dx^\nu, \tag{3.5}$$

remains invariant under a Lorentz transformation. We also know that a Lorentz frame is a special Cartesian coordinate system. In general for a curved manifold with a metric, the length interval is defined as (recall (1.9))

$$ds^2 = d\tau^2 = g_{\mu\nu}(x)dx^\mu dx^\nu. \tag{3.6}$$

Furthermore, we can use different coordinate systems to describe events but the essential point is that the length interval between any two infinitesimally close events remains invariant under a general coordinate transformation (change of coordinates). The principle of general covariance further says that all physical equations are be covariant under a general coordinate transformation. That is, they retain their form under a general change of coordinates.

Let us examine the consequences of a general coordinate transformation on tensors in some detail. Let us define an arbitrary coordinate transformation of the form

$$x^\mu \rightarrow x'^{\,\mu} = x'^{\,\mu}(x), \tag{3.7}$$

so that

$$dx^\mu \rightarrow dx'^{\,\mu} = \frac{\partial x'^{\,\mu}}{\partial x^\nu}\,dx^\nu. \tag{3.8}$$

Any quantity $V^\mu(x)$ which transforms like the coordinate differential dx^μ is called a contravariant vector. We note that in a curved manifold, coordinates x^μ are merely labels (and are not vectors unlike the flat Minkowski space) and, therefore, their transformation simply defines four functions. It is worth emphasizing here that, unlike flat space-time, here there is no concept of a coordinate x_μ (since x^μ is not a vector, we cannot lower its index). It is the coordinate differentials that behave like vectors. That is, under an arbitrary coordinate transformation

$$x^\mu \rightarrow x'^{\,\mu} = x'^{\,\mu}(x), \tag{3.9}$$

a contravariant vector $V^\mu(x)$ transforms as

$$V^\mu(x) \to V'^\mu(x') = \frac{\partial x'^\mu}{\partial x^\nu} V^\nu(x). \tag{3.10}$$

Furthermore, under such a coordinate transformation

$$\partial_\mu = \frac{\partial}{\partial x^\mu} \to \partial'_\mu = \frac{\partial}{\partial x'^\mu} = \frac{\partial x^\nu}{\partial x'^\mu} \partial_\nu. \tag{3.11}$$

The transformation matrix in (3.11) is, in fact, the inverse of the transformation matrix in (3.10) and a vector $A_\mu(x)$ which transforms like the gradient in (3.11) is called a covariant vector. Namely, under a general coordinate transformation, a covariant vector transforms as

$$A_\mu(x) \to A'_\mu(x') = \frac{\partial x^\nu}{\partial x'^\mu} A_\nu(x). \tag{3.12}$$

The transformation

$$x^\mu \to x'^\mu = x'^\mu(x), \tag{3.13}$$

implies an inverse relation of the form (the coordinate transformation is assumed to be invertible)

$$x^\mu = x^\mu(x'), \tag{3.14}$$

from which we can calculate the transformation matrix $\frac{\partial x^\nu}{\partial x'^\mu}$ in (3.12). Note that unlike Lorentz transformations, these coordinate transformations do not have to be linear.

Once we know how covariant and contravariant vectors transform under a coordinate change we can look for many other interesting consequences. First of all, we note that under a general coordinate change the scalar product of a covariant and a contravariant vector would transform as

$$A_\mu(x)B^\mu(x) \rightarrow A'_\mu(x')B'^\mu(x')$$

$$= \frac{\partial x^\nu}{\partial x'^\mu} A_\nu(x) \frac{\partial x'^\mu}{\partial x^\lambda} B^\lambda(x)$$

$$= \frac{\partial x^\nu}{\partial x'^\mu} \frac{\partial x'^\mu}{\partial x^\lambda} A_\nu(x)B^\lambda(x)$$

$$= \delta^\nu_\lambda A_\nu(x)B^\lambda(x) = A_\lambda(x)B^\lambda(x). \tag{3.15}$$

In other words, the scalar product of two vectors and, therefore, the length of a vector behaves like a scalar (invariant) under an arbitrary change of coordinates. Furthermore, since the scalar product of two vectors can also be written in terms of the metric tensor, we conclude that

$$A_\mu(x)B^\mu(x) = g_{\mu\nu}(x)A^\mu(x)B^\nu(x)$$

$$\rightarrow g'_{\mu\nu}(x')A'^\mu(x')B'^\nu(x')$$

$$= g'_{\mu\nu}(x') \frac{\partial x'^\mu}{\partial x^\lambda} A^\lambda(x) \frac{\partial x'^\nu}{\partial x^\rho} B^\rho(x)$$

$$= g'_{\mu\nu}(x') \frac{\partial x'^\mu}{\partial x^\lambda} \frac{\partial x'^\nu}{\partial x^\rho} A^\lambda(x)B^\rho(x)$$

$$= g_{\lambda\rho}(x)A^\lambda(x)B^\rho(x). \tag{3.16}$$

Therefore, this determines

$$g'_{\mu\nu}(x') \frac{\partial x'^\mu}{\partial x^\lambda} \frac{\partial x'^\nu}{\partial x^\rho} = g_{\lambda\rho}(x),$$

$$\text{or,} \quad g'_{\mu\nu}(x') = \frac{\partial x^\lambda}{\partial x'^\mu} \frac{\partial x^\rho}{\partial x'^\nu} g_{\lambda\rho}(x). \tag{3.17}$$

Note that we can write the scalar product of two vectors also as

$$
\begin{aligned}
A_\mu(x)B^\mu(x) &= g^{\mu\nu}(x)A_\mu(x)B_\nu(x) \\
&\to g'^{\mu\nu}(x')A'_\mu(x')B'_\nu(x') \\
&= g'^{\mu\nu}(x') \frac{\partial x^\lambda}{\partial x'^\mu} A_\lambda(x) \frac{\partial x^\rho}{\partial x'^\nu} B_\rho(x) \\
&= g'^{\mu\nu}(x') \frac{\partial x^\lambda}{\partial x'^\mu} \frac{\partial x^\rho}{\partial x'^\nu} A_\lambda(x)B_\rho(x) \\
&= g^{\lambda\rho}(x)A_\lambda(x)B_\rho(x).
\end{aligned}
\tag{3.18}
$$

Therefore, we obtain

$$
g'^{\mu\nu}(x') \frac{\partial x^\lambda}{\partial x'^\mu} \frac{\partial x^\rho}{\partial x'^\nu} = g^{\lambda\rho}(x),
$$

or,
$$
g'^{\mu\nu}(x') = \frac{\partial x'^\mu}{\partial x^\lambda} \frac{\partial x'^\nu}{\partial x^\rho} g^{\lambda\rho}(x),
\tag{3.19}
$$

so that the covariant and the contravariant metric tensors transform as second rank covariant and contravariant tensors respectively. They maintain their symmetry properties under a coordinate transformation. However, note that unlike the case of Lorentz transformations, here the metric tensors change under a coordinate transformation (they are not invariant tensors).

Note also that the Kronecker delta defined as the product of the contravariant and the covariant metric tensor transforms as

$$
\begin{aligned}
\delta^\mu_\nu &= g^{\mu\lambda}(x)g_{\lambda\nu}(x) \\
&\to g'^{\mu\lambda}(x')g'_{\lambda\nu}(x') \\
&= \frac{\partial x'^\mu}{\partial x^\rho} \frac{\partial x'^\lambda}{\partial x^\sigma} g^{\rho\sigma}(x) \frac{\partial x^\tau}{\partial x'^\lambda} \frac{\partial x^\delta}{\partial x'^\nu} g_{\tau\delta}(x) \\
&= \frac{\partial x'^\mu}{\partial x^\rho} \frac{\partial x'^\lambda}{\partial x^\sigma} \frac{\partial x^\tau}{\partial x'^\lambda} \frac{\partial x^\delta}{\partial x'^\nu} g^{\rho\sigma}(x)g_{\tau\delta}(x) \\
&= \frac{\partial x'^\mu}{\partial x^\rho} \delta^\tau_\sigma \frac{\partial x^\delta}{\partial x'^\nu} g^{\rho\sigma}(x)g_{\tau\delta}
\end{aligned}
$$

$$= \frac{\partial x'^{\mu}}{\partial x^{\rho}} \frac{\partial x^{\delta}}{\partial x'^{\nu}} g^{\rho\sigma}(x) g_{\sigma\delta}(x)$$

$$= \frac{\partial x'^{\mu}}{\partial x^{\rho}} \frac{\partial x^{\delta}}{\partial x'^{\nu}} \delta_{\delta}^{\rho} = \frac{\partial x'^{\mu}}{\partial x^{\rho}} \frac{\partial x^{\rho}}{\partial x'^{\nu}} = \delta_{\nu}^{\mu}. \tag{3.20}$$

This shows that the Kronecker delta which is a mixed second rank tensor is really an invariant tensor under an arbitrary change of the coordinate system.

It is clear now that, under a general coordinate transformation, a contravariant tensor of rank n would transform as

$$T^{\mu_1\mu_2\cdots\mu_n}(x) \quad \to \quad T'^{\mu_1\mu_2\cdots\mu_n}(x')$$

$$= \frac{\partial x'^{\mu_1}}{\partial x^{\nu_1}} \frac{\partial x'^{\mu_2}}{\partial x^{\nu_2}} \cdots \frac{\partial x'^{\mu_n}}{\partial x^{\nu_n}} T^{\nu_1\nu_2\cdots\nu_n}(x). \tag{3.21}$$

Similarly a covariant tensor of rank n would transform as

$$T_{\mu_1\mu_2\cdots\mu_n}(x) \quad \to \quad T'_{\mu_1\mu_2\cdots\mu_n}(x')$$

$$= \frac{\partial x^{\nu_1}}{\partial x'^{\mu_1}} \frac{\partial x^{\nu_2}}{\partial x'^{\mu_2}} \cdots \frac{\partial x^{\nu_n}}{\partial x'^{\mu_n}} T_{\nu_1\nu_2\cdots\nu_n}(x), \tag{3.22}$$

while for a mixed tensor, we have

$$T^{\mu_1\cdots\mu_m}{}_{\nu_1\cdots\nu_n}(x) \to T'^{\mu_1\cdots\mu_m}{}_{\nu_1\cdots\nu_n}(x')$$

$$= \frac{\partial x'^{\mu_1}}{\partial x^{\lambda_1}} \frac{\partial x'^{\mu_2}}{\partial x^{\lambda_2}} \cdots \frac{\partial x'^{\mu_m}}{\partial x^{\lambda_m}} \frac{\partial x^{\rho_1}}{\partial x'^{\nu_1}} \cdots \frac{\partial x^{\rho_n}}{\partial x'^{\nu_n}} T^{\lambda_1\cdots\lambda_m}{}_{\rho_1\cdots\rho_n}(x). \tag{3.23}$$

We note that a tensor retains its symmetry properties under a general coordinate transformation. Furthermore, it is worth recognizing here that the transformation properties in (3.21)-(3.23) have the same forms as we have seen in Minkowski space, (1.88), (1.96), (1.97), if we identify

$$\Lambda^{\mu}{}_{\nu}(x) = \frac{\partial x'^{\mu}}{\partial x^{\nu}}, \quad \Lambda_{\mu}{}^{\nu}(x) = \frac{\partial x^{\nu}}{\partial x'^{\mu}}, \tag{3.24}$$

and allow the transformation matrices to be coordinate dependent. Furthermore, it is important to recognize that the transformation matrices in (3.24) should not be thought of as tensors in a curved manifold.

3.3 Tensor densities

In addition to tensors, in a curved manifold we also come across tensor densities. The most familiar example of a density is, of course, the volume element (of the manifold). We note that in flat space, the volume element remains invariant under a proper Lorentz transformation, namely,

$$\mathrm{d}^4 x \to \mathrm{d}^4 x' = \left| \frac{\partial x'^{\mu}}{\partial x^{\nu}} \right| \mathrm{d}^4 x = (\det \Lambda^{\mu}{}_{\nu}) \, \mathrm{d}^4 x = \mathrm{d}^4 x, \qquad (3.25)$$

where $\left| \frac{\partial x'}{\partial x} \right|$ is the Jacobian of the transformation and we have used the fact that, for a proper Lorentz transformation, (see (1.68))

$$\det \Lambda^{\mu}{}_{\nu} = 1. \qquad (3.26)$$

(In fact, we have already used the invariance of the volume element in (2.87) to show that the conserved charge is a Lorentz scalar.) However, under a general coordinate transformation in a curved manifold

$$\mathrm{d}^4 x \to \mathrm{d}^4 x' = \left| \frac{\partial x'}{\partial x} \right| \mathrm{d}^4 x = \left(\det \frac{\partial x'^{\mu}}{\partial x^{\nu}} \right) \mathrm{d}^4 x, \qquad (3.27)$$

where the Jacobian is no longer unity and we see that the volume element in a curved manifold behaves like a scalar except for this determinant factor. We call such a quantity a scalar density of weight $+1$ where the weight is determined from the power of the determinant of the transformation matrix $\frac{\partial x'^{\mu}}{\partial x^{\nu}}$ that arises under a coordinate transformation. It is, of course, important to know that the four volume is a scalar density because to form a scalar action (which is invariant under a general coordinate transformation) for a dynamical system

$$S = \int d^4x \, \mathcal{L}, \tag{3.28}$$

we then need to choose a Lagrangian density $\mathcal{L}$ for the dynamical system such that it transforms like a scalar density of weight -1 under a general coordinate transformation.

Another scalar density that frequently arises in the study of gravitation (metric spaces) is the determinant of the covariant metric tensor

$$g = \det \, g_{\mu\nu}. \tag{3.29}$$

Since under a coordinate transformation (see (3.17))

$$g_{\mu\nu}(x) \to g'_{\mu\nu}(x') = \frac{\partial x^\lambda}{\partial x'^\mu} \frac{\partial x^\rho}{\partial x'^\nu} \, g_{\lambda\rho}(x), \tag{3.30}$$

it follows that

$$
\begin{aligned}
g \to g' \; &= \; \det \, g'_{\mu\nu}(x') \\
&= \; \det \frac{\partial x^\lambda}{\partial x'^\mu} \, \det \frac{\partial x^\rho}{\partial x'^\nu} \, \det g_{\lambda\rho} \\
&= \; \left| \frac{\partial x}{\partial x'} \right|^2 g \\
&= \; \left| \frac{\partial x'}{\partial x} \right|^{-2} g.
\end{aligned}
\tag{3.31}
$$

Therefore, the determinant of the covariant metric $g_{\mu\nu}$ transforms as a scalar density of weight -2. The last step in (3.31) follows from the observation (through the use of chain rule) that

$$\frac{\partial x'^\lambda}{\partial x^\mu} \frac{\partial x^\mu}{\partial x'^\nu} = \frac{\partial x'^\lambda}{\partial x'^\nu} = \delta^\lambda_\nu, \tag{3.32}$$

and, therefore,

$$\left|\frac{\partial x'}{\partial x}\right| \left|\frac{\partial x}{\partial x'}\right| = 1,$$

or, $$\left|\frac{\partial x}{\partial x'}\right| = \left|\frac{\partial x'}{\partial x}\right|^{-1}. \tag{3.33}$$

Since $g^{\mu\nu}$ and $g_{\mu\nu}$ transform inversely under a general coordinate transformation, we have (see (3.19))

$$\det g^{\mu\nu}(x) \;\rightarrow\; \det g'^{\,\mu\nu}(x')$$

$$= \;\det \frac{\partial x'^{\,\mu}}{\partial x^{\lambda}}\, \det \frac{\partial x'^{\,\nu}}{\partial x^{\rho}}\, \det g^{\lambda\rho}(x)$$

$$= \;\left|\frac{\partial x'}{\partial x}\right|^{2} \det g^{\lambda\rho}(x), \tag{3.34}$$

so that $\det g^{\mu\nu}(x)$ transforms like a scalar density of weight $+2$. This tells us that we can obtain a scalar action from a scalar Lagrangian density (with weight zero) as

$$S = \int \mathrm{d}^4 x \,\sqrt{-g}\, \mathcal{L}. \tag{3.35}$$

In other words $\sqrt{-g}\,\mathrm{d}^4 x$ defines the invariant volume element in a curved manifold ($(-g)$ is positive in any even dimension with our choice of signatures for the metric (1.55)).

These are, of course, all scalar densities. But there is also a tensor density that is quite useful in curved manifolds. It is the Levi-Civita tensor density $\epsilon^{\mu\nu\lambda\rho}$ (in four dimensions), which is completely anti-symmetric and has constant components in any reference frame (this is what we normally know as the Levi-Civita tensor in flat space-time)

$$\epsilon^{\mu\nu\lambda\rho} = \begin{cases} +1 & \text{if } \mu\nu\lambda\rho \text{ are even permutations of 0,1,2,3,} \\ -1 & \text{if } \mu\nu\lambda\rho \text{ are odd permutations,} \\ 0 & \text{if any two indices are the same.} \end{cases}$$

$$\tag{3.36}$$

We note that if it were a true tensor, under a change of coordinates, it would transform as

$$\epsilon^{\mu\nu\lambda\rho} \to \epsilon'^{\mu\nu\lambda\rho} = \frac{\partial x'^\mu}{\partial x^\kappa} \frac{\partial x'^\nu}{\partial x^\sigma} \frac{\partial x'^\lambda}{\partial x^\eta} \frac{\partial x'^\rho}{\partial x^\xi} \epsilon^{\kappa\sigma\eta\xi}. \tag{3.37}$$

But by definition of a determinant the right hand side is nothing other than $\epsilon^{\mu\nu\lambda\rho}$ multiplied by the Jacobian of the transformation. Namely,

$$\begin{aligned} \epsilon^{\mu\nu\lambda\rho} \ &\to \ \frac{\partial x'^\mu}{\partial x^\kappa} \frac{\partial x'^\nu}{\partial x^\sigma} \frac{\partial x'^\lambda}{\partial x^\eta} \frac{\partial x'^\rho}{\partial x^\xi} \epsilon^{\kappa\sigma\eta\xi} \\ &= \ \left(\det \frac{\partial x'^\kappa}{\partial x^\sigma} \right) \epsilon^{\mu\nu\lambda\rho} = \left| \frac{\partial x'}{\partial x} \right| \epsilon^{\mu\nu\lambda\rho}. \end{aligned} \tag{3.38}$$

Consequently, we see that if $\epsilon^{\mu\nu\lambda\rho}$ were a true tensor then it would transform as in (3.38) and, in such a case, it cannot be constant in any arbitrary frame (it will have different coordinate dependent values for its components in different frames). In other words, we can no longer say that in the new frame

$$\epsilon^{\mu\nu\lambda\rho} = \begin{cases} +1 & \text{for even permutations,} \\ -1 & \text{for odd permutations,} \\ 0 & \text{if any two indices are the same.} \end{cases}$$

For this to be true, $\epsilon^{\mu\nu\lambda\rho}$ must transform under a coordinate change as

$$\begin{aligned} \epsilon^{\mu\nu\lambda\rho} \ &\to \ \left| \frac{\partial x}{\partial x'} \right| \frac{\partial x'^\mu}{\partial x^\kappa} \frac{\partial x'^\nu}{\partial x^\sigma} \frac{\partial x'^\lambda}{\partial x^\eta} \frac{\partial x'^\rho}{\partial x^\xi} \epsilon^{\mu\nu\lambda\rho} \\ &= \ \left| \frac{\partial x}{\partial x'} \right| \left| \frac{\partial x'}{\partial x} \right| \epsilon^{\mu\nu\lambda\rho} = \epsilon^{\mu\nu\lambda\rho}. \end{aligned} \tag{3.39}$$

In other words the Levi-Civita tensor density must transform as a tensor density of weight -1. Similarly the covariant Levi-Civita tensor density must transform as a tensor density of weight $+1$, namely,

$$\epsilon_{\mu\nu\lambda\rho} \;\; \rightarrow \;\; \left| \frac{\partial x'}{\partial x} \right| \frac{\partial x^\kappa}{\partial x'^\mu} \frac{\partial x^\sigma}{\partial x'^\nu} \frac{\partial x^\eta}{\partial x'^\lambda} \frac{\partial x^\xi}{\partial x'^\rho} \; \epsilon_{\kappa\sigma\eta\xi}$$

$$= \;\; \left| \frac{\partial x'}{\partial x} \right| \left| \frac{\partial x}{\partial x'} \right| \; \epsilon_{\mu\nu\lambda\rho} = \epsilon_{\mu\nu\lambda\rho}. \tag{3.40}$$

It is clear now that given any tensor density one can form a true tensor in the following way. Let $t^{\mu_1\cdots\mu_n}$ be a tensor density of rank n and weight w so that under a coordinate transformation

$$t^{\mu_1\cdots\mu_n} \;\; \rightarrow \;\; \left| \frac{\partial x'}{\partial x} \right|^w \frac{\partial x'^{\mu_1}}{\partial x^{\nu_1}} \cdots \frac{\partial x'^{\mu_n}}{\partial x^{\nu_n}} \; t^{\nu_1\cdots\nu_n}. \tag{3.41}$$

Let us now define the quantity

$$T^{\mu_1\cdots\mu_n} = \left(\sqrt{-g} \right)^w t^{\mu_1\cdots\mu_n}. \tag{3.42}$$

Then clearly under a coordinate change

$$T^{\mu_1\cdots\mu_n} \;\; \rightarrow \;\; \left(\left| \frac{\partial x'}{\partial x} \right|^{-1} \sqrt{-g} \right)^w \left| \frac{\partial x'}{\partial x} \right|^w \frac{\partial x'^{\mu_1}}{\partial x^{\nu_1}} \cdots \frac{\partial x'^{\mu_n}}{\partial x^{\nu_n}} \; t^{\nu_1\cdots\nu_n}$$

$$= \;\; \frac{\partial x'^{\mu_1}}{\partial x^{\nu_1}} \cdots \frac{\partial x'^{\mu_n}}{\partial x^{\nu_n}} \left(\sqrt{-g} \right)^w t^{\nu_1\cdots\nu_n}$$

$$= \;\; \frac{\partial x'^{\mu_1}}{\partial x^{\nu_1}} \cdots \frac{\partial x'^{\mu_n}}{\partial x^{\nu_n}} \; T^{\nu_1\cdots\nu_n}. \tag{3.43}$$

In other words, $T^{\mu_1\cdots\mu_n}$ defined in (3.42) would behave like a true tensor of rank n under a change of coordinates.

We now see that we can define the following tensors from the Levi-Civita tensor densities

$$\bar{\epsilon}^{\mu\nu\lambda\rho} \;\; = \;\; \left(\sqrt{-g} \right)^{-1} \epsilon^{\mu\nu\lambda\rho},$$

$$\bar{\epsilon}_{\mu\nu\lambda\rho} \;\; = \;\; \sqrt{-g} \; \epsilon_{\mu\nu\lambda\rho}, \tag{3.44}$$

which will transform as true fourth rank anti-symmetric tensors under a general coordinate transformation, namely,

$$\bar{\epsilon}^{\mu\nu\lambda\rho} \;\rightarrow\; \frac{\partial x'^{\mu}}{\partial x^{\kappa}} \frac{\partial x'^{\nu}}{\partial x^{\sigma}} \frac{\partial x'^{\lambda}}{\partial x^{\eta}} \frac{\partial x'^{\rho}}{\partial x^{\xi}} \, \bar{\epsilon}^{\kappa\sigma\eta\xi},$$

$$\bar{\epsilon}_{\mu\nu\lambda\rho} \;\rightarrow\; \frac{\partial x^{\kappa}}{\partial x'^{\mu}} \frac{\partial x^{\sigma}}{\partial x'^{\nu}} \frac{\partial x^{\eta}}{\partial x'^{\lambda}} \frac{\partial x^{\xi}}{\partial x'^{\rho}} \, \bar{\epsilon}_{\kappa\sigma\eta\xi}. \tag{3.45}$$

However, it is clear that, unlike the Levi-Civita tensor densities, the tensors in (3.44) would not have the form (3.36) in arbitrary coordinate frames. Let us note from the transformation properties of the Levi-Civita tensor densities, (3.39) and (3.40), that they can also be used to form a scalar action out of a Lagrangian density involving the anti-symmetric contravariant Levi-Civita tensor density (such as in the case of spin $\frac{3}{2}$ fields). Namely, this provides an alternative to forming a scalar Lagrangian density with $\sqrt{-g}$ as in (3.35).

CHAPTER 4

Affine connection and covariant derivative

4.1 Parallel transport of a vector

In a curved manifold various geometric concepts are inherently more involved than in a flat manifold. For example, let us discuss the idea of parallel transport of vectors in a curved manifold. We need to parallel transport a vector in order to define the derivative (of the vector) which transforms covariantly under a general coordinate transformation. We note that the derivative of a vector is conventionally defined as

$$\partial_\nu A^\mu(x) = \lim_{\epsilon \to 0} \frac{A^\mu(x + \epsilon n) - A^\mu(x)}{\epsilon}, \qquad (4.1)$$

where ϵ represents an infinitesimal (scalar) parameter while n^μ denotes a unit vector along the fixed direction x^ν (along which the derivative is taken). We see from this definition that the derivative in (4.1) involves comparing the vector at two different (neighboring) coordinates and, consequently, the difference is not a vector in general. To compare the vector at two coordinates and thereby define a derivative with covariant properties, we have to parallel transport $A^\mu(x)$ to the point $x + \epsilon n$ such that the difference of the two will be a vector.

Let us define the problem of parallel transport with a simple example. Let us consider the surface of a sphere and take a vector at the north pole pointing south along a given longitude. Let us move it parallel to itself along the longitude. The naive idea of parallel transport is, of course, to take the vector and put its tip on the arrow so that its direction is always along the south and length the same. When we reach the equator we move along the equator by an angle θ keeping the vector pointing south all along and then move back up to the north pole along a different longitude. In this entire operation we keep the vector pointing south throughout. However, we note that when we reach the starting point, the transported vector makes an angle θ with the starting vector. That is, we note that if we follow our naive idea of parallel transport, then when we move around a closed curve the initial and the final vectors are no longer parallel. This is because the surface of a sphere is a curved manifold and the parallel transport of a vector leads to new features.

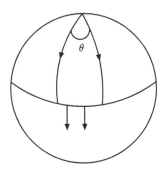

Figure 4.1: Parallel transport of a vector on the surface of a sphere.

Of course, the naive notion of parallel transport has to be modified even in the absence of curvature if we are using curvilinear coordinates. We illustrate this with a simple example using polar coordinates in the flat two dimensional space. First, let us look at two vectors $\mathbf{A}$ and $\mathbf{B}$ at equidistant points from the origin in two dimensions as shown in Fig. 4.2. If we use Cartesian coordinates, then we can write

$$\mathbf{A} = (A_1, A_2), \qquad \mathbf{B} = (B_1, B_2). \tag{4.2}$$

We say that $\mathbf{B}$ is the parallel transport of the vector $\mathbf{A}$ if the components are equal, namely, if

$$A_1 = B_1, \qquad A_2 = B_2, \tag{4.3}$$

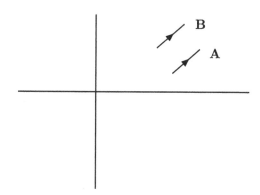

Figure 4.2: Parallel transport of a vector on the plane to an equidistant point from the origin in Cartesian coordinates.

B represents the parallel transport of the vector **A** and it follows from (4.3) that $|\mathbf{A}| = |\mathbf{B}|$ (under parallel transport the length remains invariant).

Let us now look at the same system in the polar coordinates shown in Fig. 4.3. Then, clearly the components of the vector and its parallel transport in polar coordinates are given by

$$
\mathbf{A} = (A^r, A^\theta) = (A\cos\xi, \frac{A}{r}\sin\xi),
$$

$$
\mathbf{B} = (B^r, B^\theta) = (B\cos(\xi - d\theta), \frac{B}{r}\sin(\xi - d\theta))
$$

$$
\simeq (B(\cos\xi + d\theta\sin\xi), \frac{B}{r}(\sin\xi - d\theta\cos\xi)), \tag{4.4}
$$

where r denotes the radial coordinate of the two vectors (they are equidistant from the origin). Here we have used the fact that in polar coordinates, the angular component is divided by the radial coordinate (recall that the angular component of the metric in polar coordinates is r^2) and have identified $A = |\mathbf{A}|, B = |\mathbf{B}|$. We see that the components of the two vectors **A** and **B** are no longer identical in polar coordinates (even if we assume $A = B$) even though **B** is the parallel transport of **A**. That is, a vector being the parallel transport of another does not necessarily imply that the components of the two

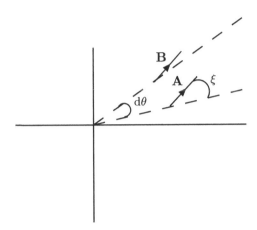

Figure 4.3: Parallel transport of a vector on the plane to an equidistant point in polar coordinates.

vectors have to be equal in a general curvilinear coordinate system. In fact, even the coordinate axes change as we move from point to point on the manifold.

That having equal components does not lead to parallel transport in a curved manifold (or in curvilinear coordinates) can also be seen in the following way. Let $A^\mu(x)$ be a vector at the point x^μ and let us assume that $A^\mu(x')$ defines the vector at x'^μ such that

$$A^\mu(x) = A^\mu(x'). \tag{4.5}$$

Namely, let us assume that the components of the vectors are equal at the two points so that we may naively conclude that $A^\mu(x')$ is the vector $A^\mu(x)$ parallel transported to the point x'^μ. However, if under parallel transport the lengths of the two vectors are the same, namely,

$$g_{\mu\nu}(x)A^\mu(x)A^\nu(x) = g_{\mu\nu}(x')A^\mu(x')A^\nu(x'), \tag{4.6}$$

we conclude that this relation can be true only if

$$g_{\mu\nu}(x) = g_{\mu\nu}(x'). \tag{4.7}$$

Namely, only if the manifold is flat or only if the metric does not change with the coordinates of the manifold will the equality of components of two vectors imply the lengths of the two vectors to be the same. In general, on a curved manifold (or in curvilinear coordinates)

$$g_{\mu\nu}(x) \neq g_{\mu\nu}(x'), \tag{4.8}$$

and, therefore, parallel transport of vectors would imply

$$A^{\mu}(x) \neq A^{\mu}(x'). \tag{4.9}$$

A constant vector is also against the spirit of general covariance. For example, let us assume that in a coordinate system $\bar{x}^{\mu}$ the vector $\bar{A}^{\mu}(\bar{x})$ has constant components. Then in a new coordinate system x^{μ} we have

$$A^{\mu}(x) = \frac{\partial x^{\mu}}{\partial \bar{x}^{\lambda}} \, \bar{A}^{\lambda}(\bar{x}), \tag{4.10}$$

and clearly this cannot have constant components in all coordinate systems since the transformation matrix is in general coordinate dependent. (This is like the example of the vector in polar coordinates that we have discussed in (4.4).) Let us, therefore, find out how a constant vector in one coordinate system $\bar{x}^{\mu}$ changes in a different coordinate system x^{μ} as we move from one point to a neighboring point on a curve parameterized by an invariant parameter τ (proper time). In this case, we can represent the vector along the trajectory $x^{\mu}(\tau)$ as

$$A^{\mu}(x) = A^{\mu}(x(\tau)), \tag{4.11}$$

from which it follows that

$$\frac{\mathrm{d}A^\mu(x(\tau))}{\mathrm{d}\tau} = \frac{\mathrm{d}}{\mathrm{d}\tau}\left(\frac{\partial x^\mu}{\partial \overline{x}^\lambda} \overline{A}^\lambda(\overline{x})\right)$$

$$= \frac{\partial^2 x^\mu}{\partial \overline{x}^\lambda \partial \overline{x}^\rho} \frac{\mathrm{d}\overline{x}^\rho}{\mathrm{d}\tau} \overline{A}^\lambda(\overline{x})$$

$$= \frac{\partial^2 x^\mu}{\partial \overline{x}^\lambda \partial \overline{x}^\rho} \frac{\partial \overline{x}^\rho}{\partial x^\sigma} \frac{\mathrm{d}x^\sigma}{\mathrm{d}\tau} \frac{\partial \overline{x}^\lambda}{\partial x^\nu} A^\nu(x(\tau))$$

$$= \left(\frac{\partial^2 x^\mu}{\partial \overline{x}^\lambda \partial \overline{x}^\rho} \frac{\partial \overline{x}^\rho}{\partial x^\sigma} \frac{\partial \overline{x}^\lambda}{\partial x^\nu}\right) \frac{\mathrm{d}x^\sigma}{\mathrm{d}\tau} A^\nu(x(\tau))$$

$$= \Gamma^\mu_{\sigma\nu} \frac{\mathrm{d}x^\sigma}{\mathrm{d}\tau} A^\nu(x(\tau)), \tag{4.12}$$

where we have used (4.10) and have defined

$$\Gamma^\mu_{\sigma\nu} = \left(\frac{\partial^2 x^\mu}{\partial \overline{x}^\lambda \partial \overline{x}^\rho} \frac{\partial \overline{x}^\rho}{\partial x^\sigma} \frac{\partial \overline{x}^\lambda}{\partial x^\nu}\right). \tag{4.13}$$

Although $\Gamma^\mu_{\sigma\nu}$ is by definition symmetric in σ and ν, let us ignore its actual form derived above and write the differential increment of the vector $A^\mu(x(\tau))$ as we move along a curve to a neighboring point as

$$\mathrm{d}A^\mu = \Gamma^\mu_{\sigma\nu}\mathrm{d}x^\sigma A^\nu, \tag{4.14}$$

where we assume that $\Gamma^\mu_{\sigma\nu}$ is a matrix of arbitrary form to be determined. This is known as the affine connection of the manifold and is clearly needed to compare vectors at neighboring points along a curve if we are in a curved manifold or if we are using curvilinear coordinates.

What the above relation tells us is that if we have a vector $A^\mu(x)$ at the point x, then its value when parallel transported to a neighboring point $x + \mathrm{d}x$ is given by

$$A^{*\,\mu}(x + \mathrm{d}x) = A^\mu(x) + \mathrm{d}A^\mu(x)$$

$$= A^\mu(x) + \Gamma^\mu_{\sigma\nu}(x)\mathrm{d}x^\sigma A^\nu(x). \tag{4.15}$$

Furthermore, if we want this to be a tensor relation valid in all co-ordinate frames then we expect

$$A^{*\prime\mu}(x' + \mathrm{d}x') = A'^{\mu}(x') + \Gamma'^{\mu}_{\sigma\nu}(x')\mathrm{d}x'^{\sigma}A'^{\nu}(x'). \tag{4.16}$$

If the parallel displacement (transport) does not change the vector nature of A^{μ}, then we have

$$A'^{\mu}(x') = \frac{\partial x'^{\mu}}{\partial x^{\lambda}} A^{\lambda}(x),$$

$$A^{*\prime\mu}(x' + \mathrm{d}x') = \frac{\partial x'^{\mu}}{\partial x^{\lambda}}\bigg|_{x+\mathrm{d}x} A^{*\lambda}(x + \mathrm{d}x). \tag{4.17}$$

Using these, eq, (4.16) leads to

$$\frac{\partial x'^{\mu}}{\partial x^{\lambda}}\bigg|_{x+\mathrm{d}x} A^{*\lambda}(x + \mathrm{d}x) = \frac{\partial x'^{\mu}}{\partial x^{\lambda}}A^{\lambda}(x) + \Gamma'^{\mu}_{\sigma\nu}(x')\mathrm{d}x'^{\sigma}A'^{\nu}(x'), \tag{4.18}$$

which can be written out explicitly as (see (4.15))

$$\frac{\partial x'^{\mu}}{\partial x^{\lambda}}\bigg|_{x+\mathrm{d}x} \left(A^{\lambda}(x) + \Gamma^{\lambda}_{\kappa\eta}(x)\mathrm{d}x^{\kappa}A^{\eta}(x)\right)$$

$$= \frac{\partial x'^{\mu}}{\partial x^{\lambda}}A^{\lambda}(x) + \Gamma'^{\mu}_{\sigma\nu}(x')\mathrm{d}x'^{\sigma}A'^{\nu}(x'). \tag{4.19}$$

Furthermore, we can expand

$$\frac{\partial x'^{\mu}}{\partial x^{\lambda}}\bigg|_{x+\mathrm{d}x} = \frac{\partial x'^{\mu}}{\partial x^{\lambda}} + \frac{\partial^2 x'^{\mu}}{\partial x^{\eta}\partial x^{\lambda}} \mathrm{d}x^{\eta} + \cdots, \tag{4.20}$$

and keeping up to terms only linear in the increment $\mathrm{d}x$ we obtain

$$\frac{\partial x'^{\mu}}{\partial x^{\lambda}} \left(A^{\lambda}(x) + \Gamma^{\lambda}_{\kappa\eta}(x) dx^{\kappa} A^{\eta}(x) \right) + \frac{\partial^2 x'^{\mu}}{\partial x^{\eta} \partial x^{\lambda}} \, dx^{\eta} A^{\lambda}(x)$$

$$= \frac{\partial x'^{\mu}}{\partial x^{\lambda}} A^{\lambda}(x) + \Gamma'^{\mu}_{\sigma\nu}(x') dx'^{\sigma} A'^{\nu}(x'),$$

or, $\Gamma'^{\mu}_{\sigma\nu}(x') dx'^{\sigma} A'^{\nu}(x') = \dfrac{\partial x'^{\mu}}{\partial x^{\lambda}} \Gamma^{\lambda}_{\kappa\eta} dx^{\kappa} A^{\eta}(x) + \dfrac{\partial^2 x'^{\mu}}{\partial x^{\eta} \partial x^{\lambda}} \, dx^{\eta} A^{\lambda}(x)$

$$= \frac{\partial x'^{\mu}}{\partial x^{\lambda}} \Gamma^{\lambda}_{\kappa\eta}(x) \frac{\partial x^{\kappa}}{\partial x'^{\sigma}} \, dx'^{\sigma} \frac{\partial x^{\eta}}{\partial x'^{\nu}} A'^{\nu}(x')$$

$$+ \frac{\partial^2 x'^{\mu}}{\partial x^{\eta} \partial x^{\lambda}} \frac{\partial x^{\eta}}{\partial x'^{\sigma}} \, dx'^{\sigma} \frac{\partial x^{\lambda}}{\partial x'^{\nu}} A'^{\nu}(x'),$$

or, $\Gamma'^{\mu}_{\sigma\nu}(x') dx'^{\sigma} A'^{\nu}(x')$

$$= \left(\frac{\partial x'^{\mu}}{\partial x^{\lambda}} \frac{\partial x^{\kappa}}{\partial x'^{\sigma}} \frac{\partial x^{\eta}}{\partial x'^{\nu}} \Gamma^{\lambda}_{\kappa\eta}(x) + \frac{\partial^2 x'^{\mu}}{\partial x^{\eta} \partial x^{\lambda}} \frac{\partial x^{\eta}}{\partial x'^{\sigma}} \frac{\partial x^{\lambda}}{\partial x'^{\nu}} \right) dx'^{\sigma} A'^{\nu}(x').$$

$$(4.21)$$

This determines

$$\Gamma'^{\mu}_{\sigma\nu}(x') = \frac{\partial x'^{\mu}}{\partial x^{\lambda}} \frac{\partial x^{\kappa}}{\partial x'^{\sigma}} \frac{\partial x^{\eta}}{\partial x'^{\nu}} \Gamma^{\lambda}_{\kappa\eta}(x) + \frac{\partial^2 x'^{\mu}}{\partial x^{\eta} \partial x^{\lambda}} \frac{\partial x^{\eta}}{\partial x'^{\sigma}} \frac{\partial x^{\lambda}}{\partial x'^{\nu}}. \quad (4.22)$$

Thus, we see that even if we may not know the exact form of the affine connection, the requirement of general covariance determines the transformation properties of the connection under a coordinate transformation. Furthermore, we note that the affine connection does not behave like a true tensor. It is the second (inhomogeneous) term in (4.22) which spoils the tensor nature of the connection.

With these conditions, we see that parallel transport of a vector is defined through the relation (namely, how the components of a vector change under an infinitesimal parallel transport)

$$dA^{\mu} = \Gamma^{\mu}_{\sigma\nu} dx^{\sigma} A^{\nu}, \quad (4.23)$$

where the affine connection has the transformation property under a coordinate transformation as derived in (4.22). The transformation (4.23) (namely, the change under a parallel transport) is also known as an affine transformation. We note here that affine transformations are quite common in physics although we may not always recognize them. For example, we note that in the case of the two dimensional example involving polar coordinates in (4.4), we can write ($|\mathbf{B}| = |\mathbf{A}| = A$)

$$A^r = A\cos\xi, \quad A^\theta = \frac{A}{r}\sin\xi,$$

$$\mathrm{d}A^r = A\mathrm{d}\theta\sin\xi = \mathrm{d}\theta A\sin\xi = r\mathrm{d}\theta A^\theta,$$

$$\mathrm{d}A^\theta = -\frac{A}{r}\mathrm{d}\theta\cos\xi = -\frac{1}{r}\mathrm{d}\theta A\cos\xi = -\frac{1}{r}\mathrm{d}\theta A^r. \tag{4.24}$$

Consequently, we conclude that this corresponds to an affine transformation with the identification (the components of the connection can actually be calculated explicitly once we know its definition in terms of the metric which we will discuss in the next section)

$$\Gamma^r_{rr} = \Gamma^r_{\theta r} = \Gamma^r_{r\theta} = 0,$$

$$\Gamma^r_{\theta\theta} = r,$$

$$\Gamma^\theta_{rr} = \Gamma^\theta_{\theta\theta},$$

$$\Gamma^\theta_{\theta r} = \Gamma^\theta_{r\theta} = -\frac{1}{r}, \tag{4.25}$$

so that we can express (4.24) as

$$\mathrm{d}A^i = \Gamma^i_{jk}\mathrm{d}x^j A^k, \tag{4.26}$$

where $i, j, k = r, \theta$ and $\mathrm{d}x^r = \mathrm{d}r = 0$ (the vectors are equidistant), while $\mathrm{d}x^\theta = \mathrm{d}\theta$.

Some comments are in order here. First of all we note that manifolds where parallel transport of vectors is defined are called affine spaces. We simply need the notion of a connection to define parallel transport. A metric is not necessary for this in general. A space where a metric is also defined is called a metric space. Furthermore, although the connection does not transform like a tensor, we note that under a coordinate redefinition, the difference between two affine connections transforms like a genuine tensor (since the inhomogeneous term in (4.22) cancels). Namely,

$$\Gamma^\mu_{\sigma\nu}(x) - \overline{\Gamma}^\mu_{\sigma\nu}(x) \rightarrow \Gamma'^\mu_{\sigma\nu}(x') - \overline{\Gamma}'^\mu_{\sigma\nu}(x')$$

$$= \frac{\partial x'^\mu}{\partial x^\lambda} \frac{\partial x^\kappa}{\partial x'^\sigma} \frac{\partial x^\eta}{\partial x'^\nu} \Gamma^\lambda_{\kappa\eta} + \frac{\partial^2 x'^\mu}{\partial x^\kappa \partial x^\eta} \frac{\partial x^\kappa}{\partial x'^\sigma} \frac{\partial x^\eta}{\partial x'^\nu}$$

$$- \frac{\partial x'^\mu}{\partial x^\lambda} \frac{\partial x^\kappa}{\partial x'^\sigma} \frac{\partial x^\eta}{\partial x'^\nu} \overline{\Gamma}^\lambda_{\kappa\eta} - \frac{\partial^2 x'^\mu}{\partial x^\kappa \partial x^\eta} \frac{\partial x^\kappa}{\partial x'^\sigma} \frac{\partial x^\eta}{\partial x'^\nu}$$

$$= \frac{\partial x'^\mu}{\partial x^\lambda} \frac{\partial x^\kappa}{\partial x'^\sigma} \frac{\partial x^\eta}{\partial x'^\nu} \left(\Gamma^\lambda_{\kappa\eta}(x) - \overline{\Gamma}^\lambda_{\kappa\eta}(x) \right). \tag{4.27}$$

Although initially within the context in which the connection was introduced, it was symmetric in its lower indices (i.e., in the study of gauge symmetries etc.) both Weyl and Einstein pointed out that for the purposes of defining parallel transport of a vector, no symmetry property of the affine connection is necessary. In other words, one should allow for the fact that the affine connection can be asymmetric, namely,

$$\Gamma^\mu_{\sigma\nu} = \frac{1}{2} \left(\Gamma^\mu_{\sigma\nu} + \Gamma^\mu_{\nu\sigma} \right) + \frac{1}{2} \left(\Gamma^\mu_{\sigma\nu} - \Gamma^\mu_{\nu\sigma} \right), \tag{4.28}$$

where the anti-symmetric part is not necessarily zero. In this case the anti-symmetric part of the connection leads to what is known as the torsion tensor. However, for the purpose of discussions in classical gravity, we will assume that the connection is symmetric in its lower indices. Let us, however, point out the following two useful general theorems.

If there exists a coordinate system in which the connection is asymmetric, i.e., $\Gamma^\mu_{\sigma\nu} \neq \Gamma^\mu_{\nu\sigma}$,

1. *then the connection remains asymmetric in any other coordinate frame.*

2. *then it is impossible to find a coordinate system in which all components of the connection vanish at a given point.*

The proof of the first theorem can be given in the following way. Let us assume that there exists a coordinate system in which the connection is symmetric so that

$$\Gamma^\mu_{\sigma\nu}(x) = \Gamma^\mu_{\nu\sigma}(x). \tag{4.29}$$

We know that under a coordinate change the connection transforms as (see (4.22))

$$\Gamma'^\mu_{\sigma\nu}(x') = \frac{\partial x'^\mu}{\partial x^\lambda} \frac{\partial x^\kappa}{\partial x'^\sigma} \frac{\partial x^\eta}{\partial x'^\nu} \Gamma^\lambda_{\kappa\eta}(x) + \frac{\partial^2 x'^\mu}{\partial x^\kappa \partial x^\eta} \frac{\partial x^\kappa}{\partial x'^\sigma} \frac{\partial x^\eta}{\partial x'^\nu}, \tag{4.30}$$

so that the transformed connection would again be symmetric and this symmetry property would be true in all coordinate frames. In other words, the connection cannot develop an anti-symmetric part through a coordinate change. Conversely, it follows that if the connection is asymmetric (has an anti-symmetric part) in one coordinate frame it must be asymmetric in all frames. (We cannot get rid off the anti-symmetric part of the connection through a coordinate transformation.)

The proof of the second theorem follows from the fact that if all the components of the connection are zero at a gven point in one coordinate frame, then under a coordinate transformation it would have the form (see (4.22))

$$\Gamma'^\mu_{\sigma\nu}(x') = \frac{\partial^2 x'^\mu}{\partial x^\kappa \partial x^\eta} \frac{\partial x^\kappa}{\partial x'^\sigma} \frac{\partial x^\eta}{\partial x'^\nu} = \Gamma'^\mu_{\nu\sigma}(x'). \tag{4.31}$$

Namely, the connection would have a symmetric form in all coordinates. (It cannot have an anti-symmetric part in any coordinate frame.) Conversely, if the connection is asymmetric (has an anti-symmetric part), all of its components cannot vanish at a given point in any coordinate frame since the anti-symmetric part cannot be set to zero through a coordinate transformation. Let us recall that the principle of equivalence says that locally we can always find a flat Cartesian coordinate system. This requires that all the coefficients of the connection must vanish in that frame at that point which, therefore, restricts the form of the connection to be symmetric in an arbitrary frame in light of these theorems. This fact is expressed as the following theorem.

The necessary and sufficient condition for the existence of a local coordinate system in which the components of a vector are not altered by an infinitesimal affine transformation is that the components of the affine connection be symmetric in the lower indices.

To prove the necessary condition, let us assume that there exists a coordinate system where the vector components are unaltered by parallel transport, namely,

$$\mathrm{d}A^{\mu} = \Gamma^{\mu}_{\sigma\nu}\mathrm{d}x^{\sigma}A^{\nu} = 0. \tag{4.32}$$

Since both $\mathrm{d}x^{\sigma}$ and A^{ν} are arbitrary, this implies that in this coordinate system

$$\Gamma^{\mu}_{\sigma\nu}(x) = 0. \tag{4.33}$$

In any other coordinate system, therefore, the connection would be

$$\Gamma'^{\mu}_{\sigma\nu}(x') = \frac{\partial^2 x'^{\mu}}{\partial x^{\kappa}\partial x^{\eta}}\,\frac{\partial x^{\kappa}}{\partial x'^{\sigma}}\,\frac{\partial x^{\eta}}{\partial x'^{\nu}} = \Gamma'^{\mu}_{\nu\sigma}(x'). \tag{4.34}$$

which is symmetric in the lower indices.

To prove the sufficient condition we assume that the components of the connection are symmetric in the lower indices, namely,

$$\Gamma^{\mu}_{\sigma\nu}(x) = \Gamma^{\mu}_{\nu\sigma}(x). \tag{4.35}$$

Let us consider a point at the origin of the coordinate system x^{μ}. Furthermore, let us consider the following coordinate transformation

$$x'^{\mu} = x^{\mu} + \frac{1}{2}\,A^{\mu}_{\sigma\nu}x^{\sigma}x^{\nu}, \tag{4.36}$$

where $A^{\mu}_{\sigma\nu}$ is an arbitrary constant matrix (only the symmetric part in the lower indices participates in this coordinate definition by symmetry). This gives

$$\left.\frac{\partial x'^{\mu}}{\partial x^{\lambda}}\right|_{x=0} = \delta^{\mu}_{\lambda},$$

$$\left.\frac{\partial^2 x'^{\mu}}{\partial x^{\sigma} \partial x^{\nu}}\right|_{x=0} = \frac{1}{2}\left(A^{\mu}_{\sigma\nu} + A^{\mu}_{\nu\sigma}\right). \tag{4.37}$$

Therefore, under this coordinate transformation (see (4.22))

$$\Gamma'^{\mu}_{\sigma\nu}(x')\big|_{x'=0} = \left(\frac{\partial x'^{\mu}}{\partial x^{\lambda}} \frac{\partial x^{\kappa}}{\partial x'^{\sigma}} \frac{\partial x^{\eta}}{\partial x'^{\nu}} \Gamma^{\lambda}_{\kappa\eta} + \frac{\partial^2 x'^{\mu}}{\partial x^{\kappa} \partial x^{\eta}} \frac{\partial x^{\kappa}}{\partial x'^{\sigma}} \frac{\partial x^{\eta}}{\partial x'^{\nu}}\right)_{x=0}$$

$$= \delta^{\mu}_{\lambda}\delta^{\kappa}_{\sigma}\delta^{\eta}_{\nu} \Gamma^{\lambda}_{\kappa\eta}(0) + \frac{1}{2}\left(A^{\mu}_{\kappa\eta} + A^{\mu}_{\eta\kappa}\right)\delta^{\kappa}_{\sigma}\delta^{\eta}_{\nu}$$

$$= \Gamma^{\mu}_{\sigma\nu}(0) + \frac{1}{2}\left(A^{\mu}_{\sigma\nu} + A^{\mu}_{\nu\sigma}\right). \tag{4.38}$$

In other words, under this coordinate change the connection retains its symmetric form as it should. Furthermore, if we choose the transformation matrix such that

$$\frac{1}{2}\left(A^{\mu}_{\sigma\nu} + A^{\mu}_{\nu\sigma}\right) = -\Gamma^{\mu}_{\sigma\nu}(0), \tag{4.39}$$

then, in the new coordinate frame we will have

$$\Gamma'^{\mu}_{\sigma\nu}(0) = 0. \tag{4.40}$$

Hence in this coordinate system we have

$$\mathrm{d}A'^{\mu} = \Gamma'^{\mu}_{\sigma\nu}\mathrm{d}x'^{\sigma}A'^{\nu} = 0. \tag{4.41}$$

We note that if $\Gamma^{\mu}_{\sigma\nu}(0)$ has an anti-symmetric part, (4.39) cannot be satisfied and it would not be possible to find a (locally flat) coordinate system with $\Gamma'^{\mu}_{\sigma\nu}(0) = 0$. Therefore, in the study of classical gravitation, where we frequently make contact with a (locally) flat Minkowski frame, we assume that the affine connection is symmetric in its lower indices.

4.2 Christoffel symbol

We have seen so far that a vector can be transported parallel to itself in an affine space if we know the components of the connection. We do not need the metric at all for this purpose. However, if in addition we also require the length of a vector to remain constant in the process, then the notion of the metric also has to be brought in (in defining length). As we will see this determines the connection uniquely in terms of the metric tensor.

Let us take two vectors A^μ and V^μ defined on a curve $x^\mu(\tau)$ parameterized by the proper time τ. The scalar product of the vectors along the curve is defined to be

$$g_{\mu\nu}(x(\tau))A^\mu(x(\tau))V^\nu(x(\tau)). \tag{4.42}$$

If we require that the scalar product of the two vectors remains invariant as we parallel transport them to neighboring points (if $A^\mu = V^\mu$ this defines the length of the vector), then we conclude

$$\frac{\mathrm{d}}{\mathrm{d}\tau}\left(g_{\mu\nu}A^\mu V^\nu\right) = 0,$$

$$\text{or,} \quad \frac{\mathrm{d}g_{\mu\nu}}{\mathrm{d}\tau}A^\mu V^\nu + g_{\mu\nu}\frac{\mathrm{d}A^\mu}{\mathrm{d}\tau}V^\nu + g_{\mu\nu}A^\mu\frac{\mathrm{d}V^\nu}{\mathrm{d}\tau} = 0. \tag{4.43}$$

If we now use (4.13) for the parallel transport of the two vectors (as well as the chain rule of differentiation) we obtain

$$\partial_\lambda g_{\mu\nu}\frac{\mathrm{d}x^\lambda}{\mathrm{d}\tau}A^\mu V^\nu + g_{\mu\nu}\Gamma^\mu_{\lambda\rho}\frac{\mathrm{d}x^\lambda}{\mathrm{d}\tau}A^\rho V^\nu + g_{\mu\nu}A^\mu\Gamma^\nu_{\lambda\rho}\frac{\mathrm{d}x^\lambda}{\mathrm{d}\tau}V^\rho = 0,$$

$$\text{or,} \quad \partial_\lambda g_{\mu\nu}\frac{\mathrm{d}x^\lambda}{\mathrm{d}\tau}A^\mu V^\nu + \left(g_{\sigma\nu}\Gamma^\sigma_{\lambda\mu} + g_{\mu\sigma}\Gamma^\sigma_{\lambda\nu}\right)\frac{\mathrm{d}x^\lambda}{\mathrm{d}\tau}A^\mu V^\nu = 0,$$

$$\text{or,} \quad \left(\partial_\lambda g_{\mu\nu} + g_{\sigma\nu}\Gamma^\sigma_{\lambda\mu} + g_{\sigma\mu}\Gamma^\sigma_{\nu\lambda}\right)\frac{\mathrm{d}x^\lambda}{\mathrm{d}\tau}A^\mu V^\nu = 0,$$

$$\text{or,} \quad \partial_\lambda g_{\mu\nu} + g_{\sigma\nu}\Gamma^\sigma_{\lambda\mu} + g_{\sigma\mu}\Gamma^\sigma_{\nu\lambda} = 0, \tag{4.44}$$

where we have used the symmetry of the metric tensor as well as the affine connection in the intermediate steps, namely,

$$g_{\mu\nu} = g_{\nu\mu}, \qquad \Gamma^{\mu}_{\sigma\nu} = \Gamma^{\mu}_{\nu\sigma}. \tag{4.45}$$

Furthermore, if we cyclically permute the indices λ, μ, ν, we obtain

$$\partial_{\mu}g_{\nu\lambda} + g_{\sigma\lambda}\Gamma^{\sigma}_{\mu\nu} + g_{\sigma\nu}\Gamma^{\sigma}_{\lambda\mu} = 0, \tag{4.46}$$

$$\partial_{\nu}g_{\lambda\mu} + g_{\sigma\mu}\Gamma^{\sigma}_{\nu\lambda} + g_{\sigma\lambda}\Gamma^{\sigma}_{\mu\nu} = 0. \tag{4.47}$$

Let us add (4.46) and (4.47) which leads to

$$\partial_{\mu}g_{\nu\lambda} + \partial_{\nu}g_{\lambda\mu} + 2g_{\sigma\lambda}\Gamma^{\sigma}_{\mu\nu} + g_{\sigma\nu}\Gamma^{\sigma}_{\lambda\mu} + g_{\sigma\mu}\Gamma^{\sigma}_{\nu\lambda} = 0. \tag{4.48}$$

If we subtract (4.44) from the sum in (4.48), we obtain

$$\partial_{\mu}g_{\nu\lambda} + \partial_{\nu}g_{\lambda\mu} - \partial_{\lambda}g_{\mu\nu} + 2g_{\sigma\lambda}\Gamma^{\sigma}_{\mu\nu} + g_{\sigma\nu}\Gamma^{\sigma}_{\lambda\mu} + g_{\sigma\mu}\Gamma^{\sigma}_{\nu\lambda}$$

$$-g_{\sigma\nu}\Gamma^{\sigma}_{\lambda\mu} - g_{\sigma\mu}\Gamma^{\sigma}_{\nu\lambda} = 0,$$

or, $\quad \partial_{\mu}g_{\nu\lambda} + \partial_{\nu}g_{\lambda\mu} - \partial_{\lambda}g_{\mu\nu} + 2g_{\sigma\lambda}\Gamma^{\sigma}_{\mu\nu} = 0,$

or, $\quad \Gamma^{\sigma}_{\mu\nu} = -\dfrac{1}{2}\, g^{\sigma\lambda}\left(\partial_{\mu}g_{\nu\lambda} + \partial_{\nu}g_{\lambda\mu} - \partial_{\lambda}g_{\mu\nu}\right). \tag{4.49}$

This is known as the Christoffel symbol and defines the connection as the unique function of the metric in (4.49). First of all we note from this definition that the connection or the Christoffel symbol is explicitly symmetric in the lower indices

$$\Gamma^{\sigma}_{\mu\nu} = \Gamma^{\sigma}_{\nu\mu}. \tag{4.50}$$

Secondly we had noted earlier that the connection can be thought of as describing the effect of curvature of the manifold which is due to gravitational forces. Hence the connection can be thought of as the potential for gravitation. But we now see that the connection is uniquely expressible in terms of the metric and hence we can think of the metric tensor as the true potential of gravitation. Furthermore, if we can find a (locally) flat Cartesian coordinate system, then the connection would vanish at that point since the metric tensor at that point has constant value.

▶ **Example (Parallel transport in polar coordinates).** Let us consider the line element on the plane in polar coordinates given by

$$d\tau^2 = dr^2 + r^2 d\theta^2. \tag{4.51}$$

The covariant and the contravariant metric components for the space follow from (4.51) to be

$$g_{rr} = 1, \qquad g_{\theta\theta} = r^2, \qquad g^{rr} = 1, \qquad g^{\theta\theta} = \frac{1}{r^2}. \tag{4.52}$$

We note that the metric components in (4.52) depend at the most on the radial coordinate r which leads to

$$\partial_\mu g_{\nu\lambda} = 2r\delta_{\mu 0}\delta_{\nu\lambda}\delta_{\lambda 1}, \quad \mu, \nu, \lambda = r, \theta. \tag{4.53}$$

Using this, we can now determine the nontrivial components of the connection from the definition (4.49)

$$\Gamma^\sigma_{\mu\nu} = -\frac{1}{2}g^{\sigma\lambda}(\partial_\mu g_{\nu\lambda} + \partial_\nu g_{\mu\lambda} - \partial_\lambda g_{\mu\nu}), \tag{4.54}$$

which have the forms

$$\Gamma^\theta_{r\theta} = \Gamma^\theta_{\theta r} = -\frac{1}{r}, \qquad \Gamma^r_{\theta\theta} = r. \tag{4.55}$$

These can, in fact, be compared with (4.25).

The general formula for parallel transport (4.14)

$$dA^\mu = \Gamma^\mu_{\nu\sigma}dx^\nu A^\sigma, \tag{4.56}$$

in this case, leads to

$$dA^r = \Gamma^r_{\theta\theta}d\theta A^\theta = rA^\theta d\theta,$$

$$dA^\theta = \Gamma^\theta_{r\theta}dr A^\theta + \Gamma^\theta_{\theta r}d\theta A^r = -\frac{1}{r}(A^\theta dr + A^r d\theta). \tag{4.57}$$

Setting $dr = 0$ we recover the result in (4.24).

◀

▶ **Example (Parallel transport in spherical coordinates).** Let us consider the three dimensional space parameterized in spherical coordinates so that the line element is given by

$$d\tau^2 = dr^2 + r^2(d\theta^2 + \sin^2\theta d\phi^2). \tag{4.58}$$

It follows from (4.58) that the nontrivial components of the covariant metric tensor have the forms

$$g_{rr} = 1, \qquad g_{\theta\theta} = r^2, \qquad g_{\phi\phi} = r^2 \sin^2 \theta. \tag{4.59}$$

Since this is a diagonal matrix the contravariant metric tensor is trivially determined from the inverse to be

$$g^{rr} = 1, \qquad g^{\theta\theta} = \frac{1}{r^2}, \qquad g^{\phi\phi} = \frac{1}{r^2 \sin^2 \theta}. \tag{4.60}$$

Let us note from (4.59) that only r and θ derivatives of the metric tensor can be nonzero and we can write ($\mu, \nu, \lambda = r, \theta, \phi$)

$$\partial_\mu g_{\nu\lambda} = 2r\, \delta_{\mu r} \delta_{\nu\theta} \delta_{\lambda\theta} + 2(r \sin^2 \theta + r^2 \sin \theta \cos \theta) \delta_{\mu\theta} \delta_{\nu\phi} \delta_{\lambda\phi}. \tag{4.61}$$

Using this as well as (4.60) we can now determine the components of the connection from the definition (4.49) which take the forms

$$\Gamma^r_{\mu\nu} = r(\delta_{\mu\theta} \delta_{\nu\theta} + \sin^2 \theta\, \delta_{\mu\phi} \delta_{\nu\phi}),$$

$$\Gamma^\theta_{\mu\nu} = -\frac{1}{r} \left(\delta_{\mu r} \delta_{\nu\theta} + \delta_{\mu\theta} \delta_{\nu r} \right) + \sin \theta \cos \theta \delta_{\mu\phi} \delta_{\nu\phi},$$

$$\Gamma^\phi_{\mu\nu} = -\frac{1}{r} \left(\delta_{\mu r} \delta_{\nu\phi} + \delta_{\mu\phi} \delta_{\nu r} \right) - \cot \theta (\delta_{\mu\theta} \delta_{\nu\phi} + \delta_{\mu\phi} \delta_{\nu\theta}). \tag{4.62}$$

The parallel transport of any vector A^μ can now be obtained from (4.14) to be

$$dA^r = r(d\theta\, A^\theta + \sin^2 \theta\, d\phi\, A^\phi),$$

$$dA^\theta = -\frac{1}{r} \left(dr\, A^\theta + d\theta\, A^r \right) + \sin \theta \cos \theta\, d\phi\, A^\phi,$$

$$dA^\phi = -\frac{1}{r} \left(dr\, A^\phi + d\phi\, A^r \right) - \cot \theta \left(d\theta\, A^\phi + d\phi\, A^\theta \right). \tag{4.63}$$

◀

▶ **Example (Parallel transport in cylindrical coordinates).** Let us consider the three dimensional space parameterized in cylindrical coordinates so that the line element is given by

$$d\tau^2 = dr^2 + r^2 d\theta^2 + dz^2. \tag{4.64}$$

It follows now that the nontrivial covariant and contravariant metric components of the space are given respectively by

$$g_{rr} = 1, \qquad\qquad g_{\theta\theta} = r^2, \qquad\qquad g_{zz} = 1,$$

$$g^{rr} = 1, \qquad\qquad g^{\theta\theta} = \frac{1}{r^2}, \qquad\qquad g^{zz} = 1. \qquad (4.65)$$

Comparing with (4.51) and (4.52), we recognize that this problem is quite similar to the analysis carried out in the polar coordinates. Although the problem involves an extra dimension, since the metric $g_{zz} = 1$ and the other metric components are independent of the coordinate z, the nontrivial components of the connection remain the same as in (4.55), namely,

$$\Gamma^{\theta}_{r\theta} = \Gamma^{\theta}_{\theta r} = -\frac{1}{r}, \quad \Gamma^{r}_{\theta\theta} = r, \qquad\qquad (4.66)$$

The parallel transport of a vector is now determined to be

$$dA^r = \Gamma^r_{\theta\theta} d\theta A^\theta = r A^\theta d\theta,$$

$$dA^\theta = \Gamma^\theta_{r\theta} dr A^\theta + \Gamma^\theta_{\theta r} d\theta A^r = -\frac{1}{r}(A^\theta dr + A^r d\theta),$$

$$dA^z = 0. \qquad\qquad (4.67)$$

We note that the z-component of the vector does not change under a parallel transport because the connection vanishes along the direction z, namely,

$$\Gamma^z_{\mu\nu} = 0, \quad \mu, \nu = r, \theta, z. \qquad\qquad (4.68)$$

◀

Let us next show that the Christoffel symbol transforms correctly (see (4.22)) under a coordinate change. We note from (4.49) that under a coordinate change

$$\Gamma^\sigma_{\mu\nu} \to \Gamma'^\sigma_{\mu\nu} = -\frac{1}{2} g'^{\sigma\lambda} \left(\partial'_\mu g'_{\nu\lambda} + \partial'_\nu g'_{\lambda\mu} - \partial'_\lambda g'_{\mu\nu} \right). \qquad (4.69)$$

Let us simplify the expression inside the parenthesis in (4.69)

$$\partial'_\mu g'_{\nu\lambda} + \partial'_\nu g'_{\lambda\mu} - \partial'_\lambda g'_{\mu\nu}$$

$$= \partial'_\mu \left(\frac{\partial x^{\nu_1}}{\partial x'^\nu} \frac{\partial x^{\lambda_1}}{\partial x'^\lambda} \, g_{\nu_1\lambda_1} \right) + \partial'_\nu \left(\frac{\partial x^{\lambda_1}}{\partial x'^\lambda} \frac{\partial x^{\mu_1}}{\partial x'^\mu} \, g_{\lambda_1\mu_1} \right)$$

$$\qquad -\partial'_\lambda \left(\frac{\partial x^{\mu_1}}{\partial x'^\mu} \frac{\partial x^{\nu_1}}{\partial x'^\nu} \, g_{\mu_1\nu_1} \right)$$

$$= \frac{\partial x^{\nu_1}}{\partial x'^\nu} \frac{\partial x^{\lambda_1}}{\partial x'^\lambda} \, \partial'_\mu g_{\nu_1\lambda_1} + \frac{\partial x^{\lambda_1}}{\partial x'^\lambda} \frac{\partial x^{\mu_1}}{\partial x'^\mu} \, \partial'_\nu g_{\lambda_1\mu_1}$$

$$\qquad - \frac{\partial x^{\mu_1}}{\partial x'^\mu} \frac{\partial x^{\nu_1}}{\partial x'^\nu} \, \partial'_\lambda g_{\mu_1\nu_1}$$

$$\qquad + \left(\frac{\partial^2 x^{\nu_1}}{\partial x'^\mu \partial x'^\nu} \frac{\partial x^{\lambda_1}}{\partial x'^\lambda} + \frac{\partial x^{\nu_1}}{\partial x'^\nu} \frac{\partial^2 x^{\lambda_1}}{\partial x'^\mu \partial x'^\lambda} \right) g_{\nu_1\lambda_1}$$

$$\qquad + \left(\frac{\partial^2 x^{\lambda_1}}{\partial x'^\nu \partial x'^\lambda} \frac{\partial x^{\mu_1}}{\partial x'^\mu} + \frac{\partial x^{\lambda_1}}{\partial x'^\lambda} \frac{\partial^2 x^{\mu_1}}{\partial x'^\nu \partial x'^\mu} \right) g_{\lambda_1\mu_1}$$

$$\qquad - \left(\frac{\partial^2 x^{\mu_1}}{\partial x'^\lambda \partial x'^\mu} \frac{\partial x^{\nu_1}}{\partial x'^\nu} + \frac{\partial x^{\mu_1}}{\partial x'^\mu} \frac{\partial^2 x^{\nu_1}}{\partial x'^\lambda \partial x'^\nu} \right) g_{\mu_1\nu_1}$$

$$= \frac{\partial x^{\nu_1}}{\partial x'^\nu} \frac{\partial x^{\lambda_1}}{\partial x'^\lambda} \frac{\partial x^{\mu_1}}{\partial x'^\mu} \, \partial_{\mu_1} g_{\nu_1\lambda_1} + \frac{\partial x^{\lambda_1}}{\partial x'^\lambda} \frac{\partial x^{\mu_1}}{\partial x'^\mu} \frac{\partial x^{\nu_1}}{\partial x'^\nu} \, \partial_{\nu_1} g_{\lambda_1\mu_1}$$

$$\qquad - \frac{\partial x^{\mu_1}}{\partial x'^\mu} \frac{\partial x^{\nu_1}}{\partial x'^\nu} \frac{\partial x^{\lambda_1}}{\partial x'^\lambda} \, \partial_{\lambda_1} g_{\mu_1\nu_1}$$

$$\qquad + \frac{\partial x^{\lambda_1}}{\partial x'^\lambda} \left(g_{\nu_1\lambda_1} \frac{\partial^2 x^{\nu_1}}{\partial x'^\mu \partial x'^\nu} + g_{\lambda_1\mu_1} \frac{\partial^2 x^{\mu_1}}{\partial x'^\nu \partial x'^\mu} \right)$$

$$\qquad + \frac{\partial x^{\nu_1}}{\partial x'^\nu} \left(g_{\nu_1\lambda_1} \frac{\partial^2 x^{\lambda_1}}{\partial x'^\mu \partial x'^\lambda} - g_{\mu_1\nu_1} \frac{\partial^2 x^{\mu_1}}{\partial x'^\lambda \partial x'^\mu} \right)$$

$$\qquad + \frac{\partial x^{\mu_1}}{\partial x'^\mu} \left(g_{\lambda_1\mu_1} \frac{\partial^2 x^{\lambda_1}}{\partial x'^\nu \partial x'^\lambda} - g_{\mu_1\nu_1} \frac{\partial^2 x^{\nu_1}}{\partial x'^\lambda \partial x'^\nu} \right)$$

$$= \frac{\partial x^{\mu_1}}{\partial x'^\mu} \frac{\partial x^{\nu_1}}{\partial x'^\nu} \frac{\partial x^{\lambda_1}}{\partial x'^\lambda} \left(\partial_{\mu_1} g_{\nu_1\lambda_1} + \partial_{\nu_1} g_{\lambda_1\mu_1} - \partial_{\lambda_1} g_{\mu_1\nu_1} \right)$$

$$+2g_{\nu_1\lambda_1} \frac{\partial x^{\lambda_1}}{\partial x'^{\lambda}} \frac{\partial^2 x^{\nu_1}}{\partial x'^{\mu}\partial x'^{\nu}}$$

$$= \frac{\partial x^{\lambda_1}}{\partial x'^{\lambda}} \left[\frac{\partial x^{\mu_1}}{\partial x'^{\mu}} \frac{\partial x^{\nu_1}}{\partial x'^{\nu}} \left(\partial_{\mu_1} g_{\nu_1\lambda_1} + \partial_{\nu_1} g_{\lambda_1\mu_1} - \partial_{\lambda_1} g_{\mu_1\nu_1}\right) \right.$$

$$\left. +2g_{\nu_1\lambda_1} \frac{\partial^2 x^{\nu_1}}{\partial x'^{\mu}\partial x'^{\nu}} \right], \tag{4.70}$$

where, in the intermediate steps we have simplified the parenthesis using relations such as

$$g_{\nu_1\lambda_1} \frac{\partial^2 x^{\lambda_1}}{\partial x'^{\mu}\partial x'^{\lambda}} - g_{\mu_1\nu_1} \frac{\partial^2 x^{\mu_1}}{\partial x'^{\lambda}\partial x'^{\mu}}$$

$$= g_{\nu_1\lambda_1} \frac{\partial^2 x^{\lambda_1}}{\partial x'^{\mu}\partial x'^{\lambda}} - g_{\lambda_1\nu_1} \frac{\partial^2 x^{\lambda_1}}{\partial x'^{\mu}\partial x'^{\lambda}} = 0. \tag{4.71}$$

Using (4.70) in (4.69) we obtain

$$\Gamma'^{\sigma}_{\mu\nu} = -\frac{1}{2} g'^{\sigma\lambda} \left(\partial'_{\mu} g'_{\nu\lambda} + \partial'_{\nu} g'_{\lambda\mu} - \partial'_{\lambda} g'_{\mu\nu}\right)$$

$$= -\frac{1}{2} \frac{\partial x'^{\sigma}}{\partial x^{\sigma_1}} \frac{\partial x'^{\lambda}}{\partial x^{\lambda_2}} g^{\sigma_1\lambda_2} \frac{\partial x^{\lambda_1}}{\partial x'^{\lambda}}$$

$$\times \left[\frac{\partial x^{\mu_1}}{\partial x'^{\mu}} \frac{\partial x^{\nu_1}}{\partial x'^{\nu}} \left(\partial_{\mu_1} g_{\nu_1\lambda_1} + \partial_{\nu_1} g_{\lambda_1\mu_1} - \partial_{\lambda_1} g_{\mu_1\nu_1}\right) \right.$$

$$\left. +2g_{\nu_1\lambda_1} \frac{\partial^2 x^{\nu_1}}{\partial x'^{\mu}\partial x'^{\nu}} \right]$$

$$= -\frac{1}{2} \delta^{\lambda_1}_{\lambda_2} \frac{\partial x'^{\sigma}}{\partial x^{\sigma_1}} g^{\sigma_1\lambda_2}$$

$$\times \left[\frac{\partial x^{\mu_1}}{\partial x'^{\mu}} \frac{\partial x^{\nu_1}}{\partial x'^{\nu}} \left(\partial_{\mu_1} g_{\nu_1\lambda_1} + \partial_{\nu_1} g_{\lambda_1\mu_1} - \partial_{\lambda_1} g_{\mu_1\nu_1}\right) \right.$$

$$\left. +2g_{\nu_1\lambda_1} \frac{\partial^2 x^{\nu_1}}{\partial x'^{\mu}\partial x'^{\nu}} \right]$$

$$= -\frac{1}{2} g^{\sigma_1\lambda_1} \left[\frac{\partial x'^{\sigma}}{\partial x^{\sigma_1}} \frac{\partial x^{\mu_1}}{\partial x'^{\mu}} \frac{\partial x^{\nu_1}}{\partial x'^{\nu}} \left(\partial_{\mu_1} g_{\nu_1\lambda_1} + \partial_{\nu_1} g_{\lambda_1\mu_1} - \partial_{\lambda_1} g_{\mu_1\nu_1}\right) \right.$$

$$+ 2g_{\nu_1\lambda_1} \frac{\partial x'^\sigma}{\partial x^{\sigma_1}} \frac{\partial^2 x^{\nu_1}}{\partial x'^\mu \partial x'^\nu} \Bigg]$$

$$= \frac{\partial x'^\sigma}{\partial x^{\sigma_1}} \frac{\partial x^{\mu_1}}{\partial x'^\mu} \frac{\partial x^{\nu_1}}{\partial x'^\nu} \Gamma^{\sigma_1}_{\mu_1\nu_1} - \frac{\partial x'^\sigma}{\partial x^\lambda} \frac{\partial^2 x^\lambda}{\partial x'^\mu \partial x'^\nu}. \tag{4.72}$$

This shows that the Christoffel symbol does not transform like a true tensor. However, the inhomogeneous term in (4.72) looks different from the one we have derived earlier in (4.22). The fact that they are the same can be seen as follows,

$$-\frac{\partial x'^\sigma}{\partial x^\lambda} \frac{\partial^2 x^\lambda}{\partial x'^\mu \partial x'^\nu} = -\frac{\partial}{\partial x'^\mu}\left(\frac{\partial x'^\sigma}{\partial x^\lambda} \frac{\partial x^\lambda}{\partial x'^\nu}\right) + \frac{\partial}{\partial x'^\mu}\left(\frac{\partial x'^\sigma}{\partial x^\lambda}\right)\frac{\partial x^\lambda}{\partial x'^\nu}$$

$$= -\frac{\partial}{\partial x'^\mu}\left(\delta^\sigma_\nu\right) + \frac{\partial x^\rho}{\partial x'^\mu}\frac{\partial}{\partial x^\rho}\left(\frac{\partial x'^\sigma}{\partial x^\lambda}\right)\frac{\partial x^\lambda}{\partial x'^\nu}$$

$$= \frac{\partial^2 x'^\sigma}{\partial x^\rho \partial x^\lambda}\frac{\partial x^\rho}{\partial x'^\mu}\frac{\partial x^\lambda}{\partial x'^\nu} = \frac{\partial^2 x'^\sigma}{\partial x^\lambda \partial x^\rho}\frac{\partial x^\lambda}{\partial x'^\mu}\frac{\partial x^\rho}{\partial x'^\nu}. \tag{4.73}$$

Therefore, the Christoffel symbol (4.49) transforms like the affine connection in (4.22) and we conclude, through the principle of equivalence, that once we know the metric tensor, the parallel transport of a vector is simply given by the relation

$$\mathrm{d}A^\sigma = \Gamma^\sigma_{\mu\nu}\mathrm{d}x^\mu A^\nu, \tag{4.74}$$

where we can identify

$$\Gamma^\sigma_{\mu\nu} = -\frac{1}{2}\, g^{\sigma\lambda}\left(\partial_\mu g_{\nu\lambda} + \partial_\nu g_{\lambda\mu} - \partial_\lambda g_{\mu\nu}\right). \tag{4.75}$$

Let us end this discussion by discussing an example from mechanics to show that the Christoffel symbol occurs very frequently in the study of equations of motion, although we may not recognize them. For example, if $x^i(t)$ and $\dot{x}^i(t)$ denote the generalized coordinates and velocities of a classical, non-relativistic system, then we can write down the Lagrangian of the system as (a dot denotes a derivative with respect to t)

$$L = \frac{1}{2} \, g_{ij}(x)\dot{x}^i \dot{x}^j - V\left(x_i\right), \tag{4.76}$$

where we have set the mass of the particle to unity for simplicity and $V(x_i)$ represents the potential in which the particle is moving. This is a very general form of the Lagrangian. For example, for a particle moving in a given potential in two dimensional Euclidean space, the Lagrangian has the form

$$L = \frac{1}{2} \, \left(\dot{x}^2 + \dot{y}^2\right) - V(x, y), \tag{4.77}$$

which in polar coordinates becomes

$$
\begin{aligned}
L &= \frac{1}{2} \, \left(\dot{r}^2 + r^2 \dot{\theta}^2\right) - V(r, \theta) \\
&= \frac{1}{2} \, g_{ij}\dot{x}^i \dot{x}^j - V\left(x_i\right),
\end{aligned} \tag{4.78}
$$

with the identifications

$$
\begin{aligned}
x^1 &= r, \qquad x^2 = \theta, \\
g_{11} &= 1, \qquad g_{22} = r^2, \qquad g_{12} = g_{21} = 0.
\end{aligned} \tag{4.79}
$$

The equation of motion (Euler-Lagrange equation) for this general system is given by

$$\frac{\mathrm{d}}{\mathrm{d}t} \frac{\partial L}{\partial \dot{x}^i} - \frac{\partial L}{\partial x^i} = 0,$$

or, $\quad \dfrac{\mathrm{d}}{\mathrm{d}t} \left(g_{ij}\dot{x}^j\right) - \dfrac{1}{2} \dfrac{\partial g_{jk}}{\partial x^i} \, \dot{x}^j \dot{x}^k + \dfrac{\partial V}{\partial x^i} = 0,$

or, $\quad g_{ij}\ddot{x}^j + \dfrac{\partial g_{ij}}{\partial x^k} \, \dot{x}^k \dot{x}^j - \dfrac{1}{2} \dfrac{\partial g_{jk}}{\partial x^i} \, \dot{x}^j \dot{x}^k = -\dfrac{\partial V}{\partial x^i} = F_i,$

or, $\quad g_{ij}\ddot{x}^j + \left(\partial_k g_{ji} - \dfrac{1}{2} \, \partial_i g_{jk}\right) \dot{x}^j \dot{x}^k = F_i,$

or, $g_{ij}\ddot{x}^j + \dfrac{1}{2} \left(\partial_k g_{ji} + \partial_j g_{ik} - \partial_i g_{jk} \right) \dot{x}^j \dot{x}^k = F_i,$

or, $\ddot{x}^i + \dfrac{1}{2} g^{i\ell} \left(\partial_k g_{j\ell} + \partial_j g_{\ell k} - \partial_\ell g_{jk} \right) \dot{x}^j \dot{x}^k = F^i,$

or, $\ddot{x}^i - \Gamma^i_{jk} \dot{x}^j \dot{x}^k = F^i.$ \hfill (4.80)

We note that if we write the generalized (non-relativistic) velocity as

$$u^i = \frac{\mathrm{d}x^i}{\mathrm{d}t} = \dot{x}^i, \tag{4.81}$$

then in the force free case, i.e., when $F^i = 0$, the equation of motion (4.80) takes the form

$$\frac{\mathrm{d}u^i}{\mathrm{d}t} - \Gamma^i_{jk} \frac{\mathrm{d}x^j}{\mathrm{d}t} u^k = 0, \tag{4.82}$$

which is nothing other than the equation for the parallel transport of the velocity vector u^i (see (4.13)). Namely, the equation of motion in this (force free) case simply implies that the velocity vector is transported parallel to itself along the trajectory $x^i(t)$.

4.3 Covariant derivative of contravariant tensors

We have seen how scalars, vectors and tensors transform under a coordinate change. We have also seen that the gradients themselves transform like covariant vectors under a change of coordinates (in fact, their transformation defines covariant vectors, see discussion following (3.11)). Let us now see how the gradients acting on various tensors transform under a general coordinate transformation. For example, for a scalar function (tensor of rank 0) we have

$$\phi(x) \;\rightarrow\; \phi'(x') = \phi(x),$$

$$\partial_\mu \phi(x) \;\rightarrow\; \partial'_\mu \phi'(x') = \frac{\partial x^{\mu_1}}{\partial x'^\mu} \, \partial_{\mu_1} \phi(x),$$

$$\partial^\mu \phi(x) \;\rightarrow\; \partial'^\mu \phi'(x') = \frac{\partial x'^\mu}{\partial x^{\mu_1}} \, \partial^{\mu_1} \phi(x). \tag{4.83}$$

Therefore, we see that the cogradient and the contragradient of scalar functions transform like covariant and contravariant vectors respectively.

Let us next look at the transformation of the derivative acting on a contravariant vector (contravariant tensor of rank 1),

$$A^\mu(x) \quad \to \quad A'^\mu(x') = \frac{\partial x'^\mu}{\partial x^{\mu_1}} A^{\mu_1}(x),$$

$$\partial_\sigma A^\mu(x) \quad \to \quad \partial'_\sigma A'^\mu(x') = \frac{\partial x^{\sigma_1}}{\partial x'^\sigma} \partial_{\sigma_1} \left(\frac{\partial x'^\mu}{\partial x^{\mu_1}} A^{\mu_1}(x) \right)$$

$$= \frac{\partial x^{\sigma_1}}{\partial x'^\sigma} \frac{\partial^2 x'^\mu}{\partial x^{\sigma_1} \partial x^{\mu_1}} A^{\mu_1} + \frac{\partial x^{\sigma_1}}{\partial x'^\sigma} \frac{\partial x'^\mu}{\partial x^{\mu_1}} \partial_{\sigma_1} A^{\mu_1}(x). \quad (4.84)$$

Thus we see that the cogradient of a contravariant vector does not transform like a pure tensor. (This is what we had pointed out earlier in motivating the concept of parallel transport of a vector.) The first term which spoils the tensor nature is, however, reminiscent of the transformation property of the affine connection or the Christoffel symbol. Namely, we know that under a general coordinate transformation (see (4.22))

$$\Gamma^\mu_{\sigma\lambda}(x) \to \Gamma'^\mu_{\sigma\lambda}(x')$$

$$= \frac{\partial x'^\mu}{\partial x^{\mu_1}} \frac{\partial x^{\sigma_1}}{\partial x'^\sigma} \frac{\partial x^{\lambda_1}}{\partial x'^\lambda} \Gamma^{\mu_1}_{\sigma_1\lambda_1} + \frac{\partial^2 x'^\mu}{\partial x^{\sigma_1} \partial x^{\lambda_1}} \frac{\partial x^{\sigma_1}}{\partial x'^\sigma} \frac{\partial x^{\lambda_1}}{\partial x'^\lambda}. \quad (4.85)$$

As a result, we see that under a general coordinate transformation,

$$\Gamma^\mu_{\sigma\lambda} A^\lambda(x) \to \Gamma'^\mu_{\sigma\lambda} A'^\lambda(x')$$

$$= \left(\frac{\partial x'^\mu}{\partial x^{\mu_1}} \frac{\partial x^{\sigma_1}}{\partial x'^\sigma} \frac{\partial x^{\lambda_1}}{\partial x'^\lambda} \Gamma^{\mu_1}_{\sigma_1\lambda_1} + \frac{\partial^2 x'^\mu}{\partial x^{\sigma_1} \partial x^{\lambda_1}} \frac{\partial x^{\sigma_1}}{\partial x'^\sigma} \frac{\partial x^{\lambda_1}}{\partial x'^\lambda} \right)$$

$$\times \frac{\partial x'^\lambda}{\partial x^{\lambda_2}} A^{\lambda_2}(x)$$

$$= \delta^{\lambda_1}_{\lambda_2} \left(\frac{\partial x'^\mu}{\partial x^{\mu_1}} \frac{\partial x^{\sigma_1}}{\partial x'^\sigma} \Gamma^{\mu_1}_{\sigma_1\lambda_1} + \frac{\partial^2 x'^\mu}{\partial x^{\sigma_1} \partial x^{\lambda_1}} \frac{\partial x^{\sigma_1}}{\partial x'^\sigma} \right) A^{\lambda_2}$$

$$= \frac{\partial x'^\mu}{\partial x^{\mu_1}} \frac{\partial x^{\sigma_1}}{\partial x'^\sigma} \Gamma^{\mu_1}_{\sigma_1 \lambda_1} A^{\lambda_1} + \frac{\partial^2 x'^\mu}{\partial x^{\sigma_1} \partial x^{\lambda_1}} \frac{\partial x^{\sigma_1}}{\partial x'^\sigma} A^{\lambda_1}. \tag{4.86}$$

Consequently, we note from (4.84) and (4.86) that if we define a generalized derivative of a contravariant vector as

$$D_\sigma A^\mu(x) = \partial_\sigma A^\mu(x) - \Gamma^\mu_{\sigma\lambda}(x) A^\lambda(x), \tag{4.87}$$

it would transform under a general coordinate transformation as

$$D_\sigma A^\mu(x) \quad \rightarrow \quad D'_\sigma A'^\mu(x') = \partial'_\sigma A'^\mu(x') - \Gamma'^\mu_{\sigma\lambda}(x') A'^\lambda(x')$$

$$= \frac{\partial x^{\sigma_1}}{\partial x'^\sigma} \frac{\partial^2 x'^\mu}{\partial x^{\sigma_1} \partial x^{\lambda_1}} A^{\lambda_1}(x) + \frac{\partial x^{\sigma_1}}{\partial x'^\sigma} \frac{\partial x'^\mu}{\partial x^{\mu_1}} \partial_{\sigma_1} A^{\mu_1}(x)$$

$$- \frac{\partial x'^\mu}{\partial x^{\mu_1}} \frac{\partial x^{\sigma_1}}{\partial x'^\sigma} \Gamma^{\mu_1}_{\sigma_1 \lambda} A^\lambda(x) - \frac{\partial^2 x'^\mu}{\partial x^{\sigma_1} \partial x^{\lambda_1}} \frac{\partial x^{\sigma_1}}{\partial x'^\sigma} A^{\lambda_1}(x)$$

$$= \frac{\partial x^{\sigma_1}}{\partial x'^\sigma} \frac{\partial x'^\mu}{\partial x^{\mu_1}} \left(\partial_{\sigma_1} A^{\mu_1}(x) - \Gamma^{\mu_1}_{\sigma_1 \lambda}(x) A^\lambda(x) \right)$$

$$= \frac{\partial x^{\sigma_1}}{\partial x'^\sigma} \frac{\partial x'^\mu}{\partial x^{\mu_1}} D_{\sigma_1} A^{\mu_1}(x). \tag{4.88}$$

In other words, the generalized derivative of a contravariant vector defined in (4.87) transforms like a pure tensor under a change of coordinates. Such a derivative which has covariance properties under a transformation of the coordinates is called the covariant derivative of a contravariant vector.

To understand the physical meaning of a covariant derivative, let us recall what it means to take the derivative of a vector in the Euclidean space. In Cartesian coordinates we have

$$A^i(x + dx) - A^i(x) = dx^j \frac{dA^i}{dx^j} + O(dx^2). \tag{4.89}$$

Namely, we parallel transport the vector $A^i(x)$ at the point x to the point $x + dx$ and the difference between $A^i(x + dx)$ and the parallel transported vector in the limit of vanishing separation defines the derivative. (As we have already seen, in the Euclidean space, parallel

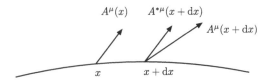

Figure 4.4: Parallel transport of a vector and the covariant derivative.

transport of a vector in Cartesian coordinates is trivial in the sense
that $A^{*i}(x + \mathrm{d}x) = A^i(x)$. The reason why parallel transport is
necessary is because $A^\mu(x + \mathrm{d}x) - A^\mu(x)$ is not a vector in general in
a curved space since the coefficient matrices for the transformation
of the two vectors at the two points will be different. On the other
hand, $A^\mu(x + \mathrm{d}x) - A^{*\mu}(x + \mathrm{d}x)$ where $A^{*\mu}(x + \mathrm{d}x)$ represents the
parallel transport of $A^\mu(x)$ to the point $x + \mathrm{d}x$ is a genuine vector
and hence can be used to define a derivative.)

If we now extend the above definition to a curved manifold as
shown in Fig. 4.4, we see that the definition of the parallel transport
leads to the definition of the derivative as (see (4.15))

$$
\begin{aligned}
& A^\mu(x + \mathrm{d}x) - A^{*\mu}(x + \mathrm{d}x) \\
=~ & A^\mu(x) + \mathrm{d}x^\nu \frac{\partial A^\mu}{\partial x^\nu} - A^\mu(x) - \Gamma^\mu_{\nu\lambda}\mathrm{d}x^\nu A^\lambda + O(\mathrm{d}x^2) \\
=~ & \mathrm{d}x^\nu \left(\frac{\partial A^\mu}{\partial x^\nu} - \Gamma^\mu_{\nu\lambda}A^\lambda \right) + O(\mathrm{d}x^2) \\
=~ & \mathrm{d}x^\nu \left(\partial_\nu A^\mu - \Gamma^\mu_{\nu\lambda}A^\lambda \right) + O(\mathrm{d}x^2) \\
=~ & \mathrm{d}x^\nu D_\nu A^\mu + O(\mathrm{d}x^2).
\end{aligned}
\tag{4.90}
$$

In other words, the covariant derivative is the natural extension of the
derivative to a curved manifold. Furthermore, note that if gravitation
is absent, the covariant derivative reduces to the normal derivative
in Cartesian coordinates. This also gives a method for introducing
gravity into a Lorentz invariant theory. Namely, we take a Lorentz
invariant theory and replace the Minkowski metric by the metric of
the curved manifold and the ordinary derivatives by the covariant
derivatives.

Let us next look at the contragradient derivative acting on a contravariant vector which would transform as

$$\partial^\sigma A^\mu \rightarrow \partial'^\sigma A'^\mu(x')$$

$$= \frac{\partial x'^\sigma}{\partial x^{\sigma_1}} \partial^{\sigma_1} \left(\frac{\partial x'^\mu}{\partial x^{\mu_1}} A^{\mu_1}(x) \right)$$

$$= \frac{\partial x'^\sigma}{\partial x^{\sigma_1}} g^{\sigma_1 \sigma_2} \frac{\partial^2 x'^\mu}{\partial x^{\sigma_2} \partial x^{\mu_1}} A^{\mu_1}(x) + \frac{\partial x'^\sigma}{\partial x^{\sigma_1}} \frac{\partial x'^\mu}{\partial x^{\mu_1}} \partial^{\sigma_1} A^{\mu_1}. \quad (4.91)$$

Again we see that this does not transform like a pure tensor. But we have also noted in (4.86) that under a general coordinate transformation

$$\Gamma^\mu_{\rho\lambda} A^\lambda \rightarrow \Gamma'^\mu_{\rho\lambda} A'^\lambda$$

$$= \frac{\partial x'^\mu}{\partial x^{\mu_1}} \frac{\partial x^{\rho_1}}{\partial x'^\rho} \Gamma^{\mu_1}_{\rho_1 \lambda} A^\lambda + \frac{\partial^2 x'^\mu}{\partial x^{\rho_1} \partial x^\lambda} \frac{\partial x^{\rho_1}}{\partial x'^\rho} A^\lambda, \quad (4.92)$$

which leads to

$$g'^{\sigma\rho} \Gamma^\mu_{\rho\lambda} A^\lambda \rightarrow g'^{\sigma\rho} \Gamma'^\mu_{\rho\lambda} A'^\lambda$$

$$= \frac{\partial x'^\sigma}{\partial x^{\sigma_1}} \frac{\partial x'^\rho}{\partial x^{\rho_2}} g^{\sigma_1 \rho_2} \left(\frac{\partial x'^\mu}{\partial x^{\mu_1}} \frac{\partial x^{\rho_1}}{\partial x'^\rho} \Gamma^{\mu_1}_{\rho_1 \lambda} A^\lambda + \frac{\partial^2 x'^\mu}{\partial x^{\rho_1} \partial x^\lambda} \frac{\partial x^{\rho_1}}{\partial x'^\rho} A^\lambda \right)$$

$$= \frac{\partial x'^\sigma}{\partial x^{\sigma_1}} \delta^{\rho_1}_{\rho_2} g^{\sigma_1 \rho_2} \left(\frac{\partial x'^\mu}{\partial x^{\mu_1}} \Gamma^{\mu_1}_{\rho_1 \lambda} A^\lambda + \frac{\partial^2 x'^\mu}{\partial x^{\rho_1} \partial x^\lambda} A^\lambda \right)$$

$$= \frac{\partial x'^\mu}{\partial x^{\mu_1}} \frac{\partial x'^\sigma}{\partial x^{\sigma_1}} \left(g^{\sigma_1 \rho_1} \Gamma^{\mu_1}_{\rho_1 \lambda} A^\lambda \right) + g^{\sigma_1 \rho_1} \frac{\partial x'^\sigma}{\partial x^{\sigma_1}} \frac{\partial^2 x'^\mu}{\partial x^{\rho_1} \partial x^\lambda} A^\lambda$$

$$= \frac{\partial x'^\mu}{\partial x^{\mu_1}} \frac{\partial x'^\sigma}{\partial x^{\sigma_1}} \left(g^{\sigma_1 \rho_1} \Gamma^{\mu_1}_{\rho_1 \lambda} A^\lambda \right) + \frac{\partial x'^\sigma}{\partial x^{\sigma_1}} g^{\sigma_1 \sigma_2} \frac{\partial^2 x'^\mu}{\partial x^{\sigma_2} \partial x^\lambda} A^\lambda. \quad (4.93)$$

It is clear now that the generalized derivative

$$D^\sigma A^\mu = \left(\partial^\sigma A^\mu - g^{\sigma\rho} \Gamma^\mu_{\rho\lambda} A^\lambda \right), \quad (4.94)$$

would transform under a general coordinate transformation as

$$
\begin{aligned}
D^\sigma A^\mu(x) &\to D'^\sigma A'^\mu(x') \\
&= \frac{\partial x'^\sigma}{\partial x^{\sigma_1}}\, g^{\sigma_1\sigma_2}\, \frac{\partial^2 x'^\mu}{\partial x^{\sigma_2}\partial x^\lambda}\, A^\lambda + \frac{\partial x'^\sigma}{\partial x^{\sigma_1}}\, \frac{\partial x'^\mu}{\partial x^{\mu_1}}\, \partial^{\sigma_1} A^{\mu_1} \\
&\quad - \frac{\partial x'^\sigma}{\partial x^{\sigma_1}}\, \frac{\partial x'^\mu}{\partial x^{\mu_1}}\, \left(g^{\sigma_1\rho_1}\Gamma^{\mu_1}_{\rho_1\lambda} A^\lambda \right) - \frac{\partial x'^\sigma}{\partial x^{\sigma_1}}\, g^{\sigma_1\sigma_2}\, \frac{\partial^2 x'^\mu}{\partial x^{\sigma_2}\partial x^\lambda}\, A^\lambda \\
&= \frac{\partial x'^\sigma}{\partial x^{\sigma_1}}\, \frac{\partial x'^\mu}{\partial x^{\mu_1}}\, \left(\partial^{\sigma_1} A^{\mu_1} - g^{\sigma_1\rho_1}\Gamma^{\mu_1}_{\rho_1} A^\lambda \right) \\
&= \frac{\partial x'^\sigma}{\partial x^{\sigma_1}}\, \frac{\partial x'^\mu}{\partial x^{\mu_1}}\, D^{\sigma_1} A^{\mu_1}(x).
\end{aligned}
\tag{4.95}
$$

This generalized derivative, therefore, transforms covariantly under a general transformation of coordinates and is known as the contravariant derivative of a contravariant vector. Furthermore, we note that

$$
\begin{aligned}
D^\sigma A^\mu &= \partial^\sigma A^\mu - g^{\sigma\rho}\Gamma^\mu_{\rho\lambda} A^\lambda \\
&= g^{\sigma\rho}\partial_\rho A^\mu - g^{\sigma\rho}\Gamma^\mu_{\rho\lambda} A^\lambda = g^{\sigma\rho}\left(\partial_\rho A^\mu - \Gamma^\mu_{\rho\lambda} A^\lambda \right) \\
&= g^{\sigma\rho} D_\rho A^\mu,
\end{aligned}
\tag{4.96}
$$

as we would expect. (In fact, the contravariant derivative can be simply introduced through the metric tensor as in (4.96) since each factor, $g^{\sigma\rho}$ and $(D_\rho A^\mu)$, transforms individually like a tensor.)

This shows that in the case of curved manifolds one has to replace the ordinary derivatives by covariant and contravariant derivatives. However, the exact form of these derivatives depends on the objects they act on. For example, for a scalar function (see (4.83))

$$
\begin{aligned}
D_\mu \phi(x) &= \partial_\mu \phi(x), \\
D^\mu \phi(x) &= \partial^\mu \phi(x),
\end{aligned}
\tag{4.97}
$$

since ordinary derivatives acting on scalar functions have covariant properties. However, for contravariant tensors, the covariant and contravariant derivatives are defined as,

$$D_\sigma T^{\mu_1\cdots\mu_n} = \partial_\sigma T^{\mu_1\cdots\mu_n} - \Gamma^{\mu_1}_{\sigma\lambda} T^{\lambda\mu_2\cdots\mu_n}$$

$$-\Gamma^{\mu_2}_{\sigma\lambda} T^{\mu_1\lambda\cdots\mu_n} \cdots - \Gamma^{\mu_n}_{\sigma\lambda} T^{\mu_1\cdots\mu_{n-1}\lambda}, \qquad (4.98)$$

and

$$D^\sigma T^{\mu_1\cdots\mu_n} = g^{\sigma\rho} D_\rho T^{\mu_1\cdots\mu_n}. \qquad (4.99)$$

That these have covariant transformation properties under a general coordinate transformation can be easily checked.

▶ **Example (Leibniz rule).** Let us recall that if we have a product of vectors $(A^\mu B^\nu)$, this behaves like a second rank tensor under a general coordinate transformation. Therefore, the covariant derivative acts on such a product as in (4.98)

$$D_\sigma(A^\mu B^\nu) = \partial_\sigma(A^\mu B^\nu) - \Gamma^\mu_{\sigma\rho}(A^\rho B^\nu) - \Gamma^\nu_{\sigma\rho}(A^\mu B^\rho)$$

$$= (\partial_\sigma A^\mu)B^\nu + A^\mu(\partial_\sigma B^\nu) - (\Gamma^\mu_{\sigma\rho}A^\rho)B^\nu - A^\mu(\Gamma^\nu_{\sigma\rho}B^\rho)$$

$$= \left(\partial_\sigma A^\mu - \Gamma^\mu_{\sigma\rho}A^\rho\right)B^\nu + A^\mu\left(\partial_\sigma B^\nu - \Gamma^\nu_{\sigma\rho}B^\rho\right)$$

$$= (D_\sigma A^\mu)B^\nu + A^\mu(D_\sigma B^\nu). \qquad (4.100)$$

Namely, the covariant derivative satisfies the Leibniz rule for a product of vectors (tensors).

◀

4.4 Metric compatibility

As a particular application of the results of the last section, let us analyze how the covariant derivatives would act on the metric tensors. From (4.98) we see that

$$D_\lambda g^{\mu\rho} = \partial_\lambda g^{\mu\rho} - \Gamma^\mu_{\lambda\sigma} g^{\sigma\rho} - \Gamma^\rho_{\lambda\sigma} g^{\mu\sigma}. \qquad (4.101)$$

Furthermore, if we use the expression for the Christoffel symbol defined in (4.49), namely,

$$\Gamma^\mu_{\lambda\sigma} = -\frac{1}{2} g^{\mu\nu}\left(\partial_\lambda g_{\sigma\nu} + \partial_\sigma g_{\nu\lambda} - \partial_\nu g_{\lambda\sigma}\right). \qquad (4.102)$$

we see from (4.101) that

$$
\begin{aligned}
D_\lambda g^{\mu\rho} &= \partial_\lambda g^{\mu\rho} + \frac{1}{2}\, g^{\mu\nu} \left(\partial_\lambda g_{\sigma\nu} + \partial_\sigma g_{\nu\lambda} - \partial_\nu g_{\lambda\sigma} \right) g^{\sigma\rho} \\
&\quad + \frac{1}{2}\, g^{\rho\nu} \left(\partial_\lambda g_{\sigma\nu} + \partial_\sigma g_{\nu\lambda} - \partial_\nu g_{\lambda\sigma} \right) g^{\mu\sigma} \\
&= \partial_\lambda g^{\mu\rho} + \frac{1}{2}\, g^{\mu\nu} \left(\partial_\lambda g_{\sigma\nu} + \partial_\sigma g_{\nu\lambda} - \partial_\nu g_{\lambda\sigma} \right) g^{\sigma\rho} \\
&\quad + \frac{1}{2}\, g^{\mu\nu} \left(\partial_\lambda g_{\nu\sigma} + \partial_\nu g_{\sigma\lambda} - \partial_\sigma g_{\lambda\nu} \right) g^{\sigma\rho} \\
&= \partial_\lambda g^{\mu\rho} + g^{\mu\nu} \left(\partial_\lambda g_{\sigma\nu} \right) g^{\sigma\rho} \\
&= \partial_\lambda g^{\mu\rho} + g^{\mu\nu} \left(\partial_\lambda \left(g_{\sigma\nu} g^{\sigma\rho} \right) - g_{\sigma\nu} \partial_\lambda g^{\sigma\rho} \right) \\
&= \partial_\lambda g^{\mu\rho} - \delta^\mu_\sigma \partial_\lambda g^{\sigma\rho} = \partial_\lambda g^{\mu\rho} - \partial_\lambda g^{\mu\rho} = 0, \qquad (4.103)
\end{aligned}
$$

where we have used the symmetry properties of the metric tensor. This shows that the covariant derivative of the contravariant metric tensor vanishes. Furthermore, since

$$
g^{\mu\rho} g_{\rho\nu} = \delta^\mu_\nu, \tag{4.104}
$$

it follows that (the covariant derivative, like the ordinary derivative, satisfies the Leibniz rule, see previous example)

$$
D_\sigma \left(g^{\mu\rho} g_{\rho\nu} \right) = 0,
$$

or, $\left(D_\sigma g^{\mu\rho} \right) g_{\rho\nu} + g^{\mu\rho} D_\sigma g_{\rho\nu} = 0,$

or, $g^{\mu\rho} D_\sigma g_{\rho\nu} = 0. \tag{4.105}$

Since this is true for an arbitrary metric we conclude that

$$
D_\sigma g_{\rho\nu} = 0. \tag{4.106}
$$

In a similar manner we can show

$$D^\sigma g^{\mu\nu} = 0, \qquad D^\sigma g_{\mu\nu} = 0. \tag{4.107}$$

In other words, the covariant derivatives of the metric tensor identically vanish. (Metric tensor is covariantly flat.) It is a constant tensor in the absolute sense in a curved manifold. Of course, this can be seen intuitively in the following way. In a locally flat Cartesian coordinate system the metric tensor reduces to the constant Minkowski metric. In such a frame the Christoffel symbol vanishes since the metric tensor is constant. (The other way of seeing this is that the effect of gravitation is zero in this frame.) In this frame, therefore,

$$D_\sigma g^{\mu\nu} = \partial_\sigma \eta^{\mu\nu} = 0. \tag{4.108}$$

Since this is a tensor equation, it must be true in all coordinate frames which leads to (4.103).

Of course, we could have reversed this argument and could have demanded

$$D_\sigma g^{\mu\nu} = 0, \tag{4.109}$$

and this would have led to the unique choice for the Christoffel symbol determined in (4.49). It is for this reason that this condition (covariant constancy of the metric) is also known as the metric compatibility condition. This simply means that the parallel transported metric tensor has the same value at $x + \mathrm{d}x$ as the metric at that point, namely,

$$g^*_{\mu\nu}(x + \mathrm{d}x) = g_{\mu\nu}(x + \mathrm{d}x). \tag{4.110}$$

Of all possible spaces, metric compatibility selects out a special class of spaces that is relevant in the study of gravitation. Furthermore, let us note that the metric compatibility condition rests on the fact that we can find a locally flat Cartesian coordinate system free of gravitation. This in turn requires the connections to be symmetric. In spaces with torsion, i.e., in spaces where the connection is not symmetric, metric compatibility raises special questions which we will not go into here.

▶ **Example (Christoffel symbol from metric compatibilty).** Let us consider the metric compatibility condition (4.105)

$$D_\sigma g_{\mu\nu} = 0, \tag{4.111}$$

which when written out explicitly has the form

$$\partial_\sigma g_{\mu\nu} + \Gamma^\rho_{\sigma\mu} g_{\rho\nu} + \Gamma^\rho_{\sigma\nu} g_{\mu\rho} = 0. \tag{4.112}$$

This is exactly the same relation as in (4.44) (with $\sigma \leftrightarrow \lambda$). Therefore, following exactly the same steps as in (4.44)-(4.49), we can determine the Christoffel symbol to be

$$\Gamma^\sigma_{\mu\nu} = -\frac{1}{2}\, g^{\sigma\lambda} \left(\partial_\mu g_{\nu\lambda} + \partial_\nu g_{\lambda\mu} - \partial_\lambda g_{\mu\nu} \right), \tag{4.113}$$

which is equivalent to saying that the Christoffel symbol can be determined from the metric compatibility condition.

◀

▶ **Example (Covariant derivative of a scalar density).** Let us consider the determinant of a second rank covariant tensor

$$A = \det A_{\mu\nu}, \tag{4.114}$$

which appears to be a scalar. However, under a general transformation A in (4.114) transforms as

$$A \to A' = \det A'_{\mu\nu} = \det \frac{\partial x^\alpha}{\partial x'^\mu} \frac{\partial x^\beta}{\partial x'^\nu} A_{\alpha\beta}$$
$$= \left(\det \frac{\partial x}{\partial x'} \right)^2 \det A_{\mu\nu} = \left| \frac{\partial x'}{\partial x} \right|^{-2} A \neq A. \tag{4.115}$$

Thus we see that $A = \det A_{\mu\nu}$ is not a scalar, rather it is a scalar density of weight $w = -2$ (see (3.31)).

We can construct a scalar quantity from this determinant by multiplying (appropriate power of) the determinant of the metric tensor of the manifold, in this case as

$$\bar{A} = (-g)^{-1} A, \tag{4.116}$$

where $g = \det g_{\mu\nu}$ and $\bar{A}$ would transform like a scalar under a general coordinate transformation. Since $\bar{A}$ is a genuine scalar, it follows from (4.83) that

$$\partial_\mu \bar{A} = D_\mu \bar{A} = D_\mu \left((-g)^{-1} A \right) = -g^{-1} D_\mu A,$$
$$\text{or,} \quad -g^{-1} D_\mu A = \partial_\mu \bar{A} = \partial_\mu \left((-g)^{-1} A \right),$$
$$\text{or,} \quad D_\mu A = g \partial_\mu \left(g^{-1} A \right). \tag{4.117}$$

Here we have used the metric compatibility condition (4.105) in the intermediate steps. This shows how we can define the covariant derivative of the determinant of any second rank covariant tensor (or for that matter any tensor density) through the use of powers of g. In particular, we note that when $A_{\mu\nu} = g_{\mu\nu}$, (4.117) leads to

$$D_\mu g = g\partial_\mu \left(g^{-1}g\right) = 0, \tag{4.118}$$

consistent with metric compatibility (4.105).

◀

4.5 Covariant derivative of covariant and mixed tensors

For the present, however, we note that since the metric is covariantly flat, raising and lowering of tensor indices commute with covariant differentiation. That is, for a covariant tensor of rank n, we have

$$
\begin{aligned}
D_\sigma T_{\mu_1\mu_2\ldots\mu_n} &= D_\sigma \left(g_{\mu_1\nu_1}g_{\mu_2\nu_2}\cdots g_{\mu_n\nu_n}T^{\nu_1\nu_2\ldots\nu_n}\right) \\
&= g_{\mu_1\nu_1}g_{\mu_2\nu_2}\cdots g_{\mu_n\nu_n}D_\sigma T^{\nu_1\nu_2\ldots\nu_n}.
\end{aligned}
\tag{4.119}
$$

This rule is of particular interest since we can now write down the covariant derivative of a covariant vector in the following way

$$
\begin{aligned}
D_\sigma A_\mu &= D_\sigma \left(g_{\mu\nu}A^\nu\right) = g_{\mu\nu}D_\sigma A^\nu \\
&= g_{\mu\nu}\left(\partial_\sigma A^\nu - \Gamma^\nu_{\sigma\lambda}A^\lambda\right) \\
&= \partial_\sigma \left(g_{\mu\nu}A^\nu\right) - \left(\partial_\sigma g_{\mu\nu}\right)A^\nu - g_{\mu\nu}\Gamma^\nu_{\sigma\lambda}A^\lambda \\
&= \partial_\sigma A_\mu - \left(\partial_\sigma g_{\mu\lambda}\right)A^\lambda - g_{\mu\nu}\Gamma^\nu_{\sigma\lambda}A^\lambda.
\end{aligned}
\tag{4.120}
$$

To simplify this let us use the form of the Christoffel symbol in (4.49) which leads to

$$\left(\partial_\sigma g_{\mu\lambda} + g_{\mu\nu}\Gamma^\nu_{\sigma\lambda}\right) A^\lambda$$

$$= \left[\partial_\sigma g_{\mu\lambda} - g_{\mu\nu} \times \frac{1}{2}\, g^{\nu\rho} \left(\partial_\sigma g_{\lambda\rho} + \partial_\lambda g_{\rho\sigma} - \partial_\rho g_{\sigma\lambda}\right)\right] A^\lambda$$

$$= \left[\partial_\sigma g_{\mu\lambda} - \frac{1}{2}\, \delta^\rho_\mu \left(\partial_\sigma g_{\lambda\rho} + \partial_\lambda g_{\rho\sigma} - \partial_\rho g_{\sigma\lambda}\right)\right] A^\lambda$$

$$= \left[\partial_\sigma g_{\mu\lambda} - \frac{1}{2}\, \left(\partial_\sigma g_{\lambda\mu} + \partial_\lambda g_{\mu\sigma} - \partial_\mu g_{\sigma\lambda}\right)\right] A^\lambda$$

$$= \frac{1}{2}\, \left(\partial_\sigma g_{\mu\lambda} + \partial_\mu g_{\lambda\sigma} - \partial_\lambda g_{\sigma\mu}\right) A^\lambda$$

$$= \frac{1}{2}\, \left(\partial_\sigma g_{\mu\lambda} + \partial_\mu g_{\lambda\sigma} - \partial_\lambda g_{\sigma\mu}\right) g^{\lambda\rho} A_\rho$$

$$= -\Gamma^\rho_{\sigma\mu} A_\rho = -\Gamma^\lambda_{\sigma\mu} A_\lambda. \tag{4.121}$$

As a result, we can write the covariant derivative in (4.120) as

$$D_\sigma A_\mu = \partial_\sigma A_\mu + \Gamma^\lambda_{\sigma\mu} A_\lambda. \tag{4.122}$$

We can derive this also from the fact that for the scalar product of two vectors (which is a scalar) we have

$$D_\sigma\left(A_\mu B^\mu\right) = \partial_\sigma\left(A_\mu B^\mu\right), \tag{4.123}$$

which leads to (using the Leibniz rule (4.100))

$$\left(D_\sigma A_\mu\right) B^\mu + A_\mu \left(D_\sigma B^\mu\right) = \partial_\sigma\left(A_\mu B^\mu\right),$$

$$\text{or,} \quad \left(D_\sigma A_\mu\right) B^\mu + A_\mu \left(\partial_\sigma B^\mu - \Gamma^\mu_{\sigma\nu} B^\nu\right) = \partial_\sigma\left(A_\mu B^\mu\right),$$

$$\text{or,} \quad \left(D_\sigma A_\mu\right) B^\mu = \left(\partial_\sigma A_\mu + \Gamma^\lambda_{\sigma\mu} A_\lambda\right) B^\mu,$$

$$\text{or,} \quad D_\sigma A_\mu = \partial_\sigma A_\mu + \Gamma^\lambda_{\sigma\mu} A_\lambda. \tag{4.124}$$

That this behaves covariantly under a coordinate transformation can be seen as follows.

$$D_\sigma A_\mu(x) \to D'_\sigma A'_\mu(x') = \partial'_\sigma A'_\mu + \Gamma'^\lambda_{\sigma\mu} A'_\lambda$$

$$= \frac{\partial x^{\sigma_1}}{\partial x'^\sigma} \partial_{\sigma_1} \left(\frac{\partial x^{\mu_1}}{\partial x'^\mu} A_{\mu_1} \right)$$

$$+ \left(\frac{\partial x'^\lambda}{\partial x^{\lambda_1}} \frac{\partial x^{\sigma_1}}{\partial x'^\sigma} \frac{\partial x^{\mu_1}}{\partial x'^\mu} \Gamma^{\lambda_1}_{\sigma_1\mu_1} + \frac{\partial^2 x'^\lambda}{\partial x^{\sigma_1}\partial x^{\mu_1}} \frac{\partial x^{\sigma_1}}{\partial x'^\sigma} \frac{\partial x^{\mu_1}}{\partial x'^\mu} \right) \frac{\partial x^{\lambda_2}}{\partial x'^\lambda} A_{\lambda_2}$$

$$= \frac{\partial x^{\sigma_1}}{\partial x'^\sigma} \frac{\partial x^{\mu_1}}{\partial x'^\mu} \partial_{\sigma_1} A_{\mu_1} + \frac{\partial x^{\sigma_1}}{\partial x'^\sigma} \partial_{\sigma_1} \left(\frac{\partial x^{\mu_1}}{\partial x'^\mu} \right) A_{\mu_1}$$

$$+ \delta^{\lambda_2}_{\lambda_1} \frac{\partial x^{\sigma_1}}{\partial x'^\sigma} \frac{\partial x^{\mu_1}}{\partial x'^\mu} \Gamma^{\lambda_1}_{\sigma_1\mu_1} A_{\lambda_2} + \frac{\partial^2 x'^\lambda}{\partial x^{\sigma_1}\partial x^{\mu_1}} \frac{\partial x^{\lambda_2}}{\partial x'^\lambda} \frac{\partial x^{\mu_1}}{\partial x'^\mu} \frac{\partial x^{\sigma_1}}{\partial x'^\sigma} A_{\lambda_2}.$$

$$(4.125)$$

To simplify this expression we use the relation

$$\frac{\partial x'^\lambda}{\partial x^{\mu_1}} \frac{\partial x^{\lambda_2}}{\partial x'^\lambda} = \delta^{\lambda_2}_{\mu_1}, \qquad (4.126)$$

which leads to

$$\partial_{\sigma_1} \left(\frac{\partial x'^\lambda}{\partial x^{\mu_1}} \frac{\partial x^{\lambda_2}}{\partial x'^\lambda} \right) = 0,$$

or, $$\frac{\partial^2 x'^\lambda}{\partial x^{\sigma_1}\partial x^{\mu_1}} \frac{\partial x^{\lambda_2}}{\partial x'^\lambda} + \frac{\partial x'^\lambda}{\partial x^{\mu_1}} \partial_{\sigma_1} \left(\frac{\partial x^{\lambda_2}}{\partial x'^\lambda} \right) = 0,$$

or, $$\frac{\partial^2 x'^\lambda}{\partial x^{\sigma_1}\partial x^{\mu_1}} \frac{\partial x^{\lambda_2}}{\partial x'^\lambda} = -\partial_{\sigma_1} \left(\frac{\partial x^{\lambda_2}}{\partial x'^\lambda} \right) \frac{\partial x'^\lambda}{\partial x^{\mu_1}}. \qquad (4.127)$$

Using this relation in (4.125) we obtain

$$D_\sigma A_\mu(x) \to D'_\sigma A'_\mu(x')$$

$$= \frac{\partial x^{\sigma_1}}{\partial x'^\sigma} \frac{\partial x^{\mu_1}}{\partial x'^\mu} \partial_{\sigma_1} A_{\mu_1} + \frac{\partial x^{\sigma_1}}{\partial x'^\sigma} \partial_{\sigma_1} \left(\frac{\partial x^{\mu_1}}{\partial x'^\mu} \right) A_{\mu_1}$$

$$+ \frac{\partial x^{\sigma_1}}{\partial x'^\sigma} \frac{\partial x^{\mu_1}}{\partial x'^\mu} \Gamma^{\lambda_1}_{\sigma_1\mu_1} A_{\lambda_1} - \partial_{\sigma_1} \left(\frac{\partial x^{\lambda_2}}{\partial x'^\lambda} \right) \frac{\partial x'^\lambda}{\partial x^{\mu_1}} \frac{\partial x^{\mu_1}}{\partial x'^\mu} \frac{\partial x^{\sigma_1}}{\partial x'^\sigma} A_{\lambda_2}$$

$$= \frac{\partial x^{\sigma_1}}{\partial x'^{\sigma}} \frac{\partial x^{\mu_1}}{\partial x'^{\mu}} \left(\partial_{\sigma_1} A_{\mu_1} + \Gamma^{\lambda_1}_{\sigma_1 \mu_1} A_{\lambda_1} \right)$$

$$+ \frac{\partial x^{\sigma_1}}{\partial x'^{\sigma}} \partial_{\sigma_1} \left(\frac{\partial x^{\mu_1}}{\partial x'^{\mu}} \right) A_{\mu_1} - \frac{\partial x^{\sigma_1}}{\partial x'^{\sigma}} \partial_{\sigma_1} \left(\frac{\partial x^{\lambda_2}}{\partial x'^{\lambda}} \right) \delta^{\lambda}_{\mu} A_{\lambda_2}$$

$$= \frac{\partial x^{\sigma_1}}{\partial x'^{\sigma}} \frac{\partial x^{\mu_1}}{\partial x'^{\mu}} D_{\sigma_1} A_{\mu_1}. \tag{4.128}$$

Similarly, we can show that

$$D^{\sigma} A_{\mu} = g^{\sigma \rho} D_{\rho} A_{\mu} = g^{\sigma \rho} \left(\partial_{\rho} A_{\mu} + \Gamma^{\lambda}_{\rho \mu} A_{\lambda} \right)$$

$$= \partial^{\sigma} A_{\mu} + g^{\sigma \rho} \Gamma^{\lambda}_{\rho \mu} A_{\lambda}, \tag{4.129}$$

also transforms covariantly under a general coordinate transformation. We can now write down the covariant derivative of a covariant tensor of rank n as

$$D_{\sigma} T_{\mu_1 \mu_2 \ldots \mu_n} = \partial_{\sigma} T_{\mu_1 \ldots \mu_n} + \Gamma^{\lambda}_{\sigma \mu_1} T_{\lambda \mu_2 \ldots \mu_n} + \Gamma^{\lambda}_{\sigma \mu_2} T_{\mu_1 \lambda \ldots \mu_n}$$

$$+ \cdots + \Gamma^{\lambda}_{\sigma \mu_n} T_{\mu_1 \ldots \mu_{n-1} \lambda}. \tag{4.130}$$

The covariant derivative for a mixed tensor can likewise be shown to correspond to

$$D_{\sigma} T^{\mu_1 \ldots \mu_m}{}_{\nu_1 \ldots \nu_n}$$

$$= \partial_{\sigma} T^{\mu_1 \ldots \mu_m}{}_{\nu_1 \ldots \nu_n} - \Gamma^{\mu_1}_{\sigma \lambda} T^{\lambda \ldots \mu_m}{}_{\nu_1 \ldots \nu_n} - \cdots - \Gamma^{\mu_m}_{\sigma \lambda} T^{\mu_1 \ldots \lambda}{}_{\nu_1 \ldots \nu_n}$$

$$+ \Gamma^{\lambda}_{\sigma \nu_1} T^{\mu_1 \ldots \mu_m}{}_{\lambda \ldots \nu_n} + \cdots + \Gamma^{\lambda}_{\sigma \nu_n} T^{\mu_1 \ldots \mu_m}{}_{\nu_1 \ldots \lambda}. \tag{4.131}$$

4.6 Electromagnetic analogy

We have seen that in dealing with gravitation and hence with curved manifolds, we have to introduce the notion of a connection in order to define the covariant derivative (of tensors). This is deeply connected with the fact that there is a local gauge symmetry associated with gravitation which we recognize as general covariance (invariance under a general coordinate transformation). A similar situation also arises in flat space-time in the study of gauge theories. For example, if $\psi(x)$ represents the quantum mechanical wave function of a charged particle, then we know that under a local Abelian gauge transformation, the wave function transforms as

$$\psi(x) \to \psi'(x) = e^{-ie\alpha(x)}\psi(x). \tag{4.132}$$

Under this gauge transformation we find that

$$\begin{aligned}
\partial_\mu \psi(x) &\to \partial_\mu \psi'(x) = \partial_\mu \left(e^{-ie\alpha(x)}\psi(x) \right) \\
&= -ie(\partial_\mu \alpha(x))e^{-ie\alpha(x)}\psi(x) + e^{-ie\alpha(x)}\partial_\mu \psi(x), \tag{4.133}
\end{aligned}$$

so that the expectation value of the momentum operator is no longer invariant under this transformation (the derivative of the wave function does not transform covariantly under the gauge transformation). On the other hand, if we require physical results to be invariant under such a gauge transformation, we find that invariance can be restored if we assume that there exists a connection $A_\mu(x)$ (vector potential) which transforms under the gauge transformation as

$$A_\mu(x) \to A'_\mu(x) = A_\mu + \partial_\mu \alpha(x), \tag{4.134}$$

and if we generalize the ordinary derivative to a covariant derivative of the form (recall minimal coupling)

$$D_\mu \psi(x) = (\partial_\mu + ieA_\mu(x)) \psi(x). \tag{4.135}$$

In this case, under the gauge transformations (4.132) and (4.134)

$$D_\mu \psi(x) \quad \rightarrow \quad D'_\mu \psi'(x) = \left(\partial_\mu + ie A'_\mu(x) \right) \psi'(x)$$

$$= \partial_\mu \psi'(x) + ie A'_\mu(x) \psi'(x)$$

$$= -ie(\partial_\mu \alpha(x)) e^{-ie\alpha(x)} \psi(x) + e^{-ie\alpha(x)} \partial_\mu \psi(x)$$

$$+ ie(A_\mu(x) + \partial_\mu \alpha(x)) e^{-ie\alpha(x)} \psi(x)$$

$$= e^{-ie\alpha(x)} \left(\partial_\mu + ie A_\mu(x) \right) \psi(x)$$

$$= e^{-ie\alpha(x)} D_\mu \psi(x). \tag{4.136}$$

Namely, the covariant derivative of the wave function transforms covariantly under the Abelian gauge transformation. We note here that the connection $A_\mu(x)$, in this case, is nothing other than the electromagnetic potential (1.83) or the Maxwell gauge field and the presence of the gauge field or the connection is intimately connected with the local gauge symmetry associated with electromagnetism. Let us also note from (4.134) that the connection (gauge field) does not transform covariantly under the gauge transformation because of the inhomogeneous term much like the transformation of the affine connection in (4.22).

Similarly in the case of gravitation (curved manifold) we note that the presence of the connection signals the presence of a local gauge symmetry in the theory which corresponds to the general coordinate invariance of the system. Furthermore, this also suggests that Einstein theory, which describes gravitational forces, must have the nature of a gauge theory and we will study this later in the course.

4.7 Gradient, divergence and curl

Although the derivatives in the case of a curved manifold have to be generalized to covariant derivatives, there are special cases where they take simple forms. We have already seen in (4.97) that for a scalar function

$$D_\mu \phi(x) = \partial_\mu \phi(x), \tag{4.137}$$

so that the gradient of a scalar function remains unchanged in a curved manifold. Furthermore, we note that the covariant derivative of a contravariant vector is defined as (see (4.87))

$$D_\mu A^\nu = \partial_\mu A^\nu - \Gamma^\nu_{\mu\sigma} A^\sigma. \tag{4.138}$$

Therefore, we can define the (covariant) divergence of a contravariant vector as

$$D_\mu A^\mu = A^\mu_{;\mu} = \partial_\mu A^\mu - \Gamma^\mu_{\mu\sigma} A^\sigma, \tag{4.139}$$

where a semi-colon conventionally denotes covariant differentiation. To simplify this expression let us recall that

$$
\begin{aligned}
\Gamma^\mu_{\mu\sigma} &= -\frac{1}{2} g^{\mu\rho} \left(\partial_\mu g_{\sigma\rho} + \partial_\sigma g_{\rho\mu} - \partial_\rho g_{\mu\sigma} \right) \\
&= -\frac{1}{2} \left(\partial^\rho g_{\sigma\rho} + g^{\mu\rho} \partial_\sigma g_{\rho\mu} - \partial^\mu g_{\mu\sigma} \right) \\
&= -\frac{1}{2} g^{\mu\rho} \partial_\sigma g_{\rho\mu}.
\end{aligned}
\tag{4.140}
$$

Let us now use the following simplifying relation. If A denotes any (square) matrix, then we note that we can represent its determinant as

$$\det A = e^{\operatorname{Tr} \ln A}. \tag{4.141}$$

This is obvious if the matrix A is diagonal, for in this case

$$\det A = \prod_i \lambda_i, \tag{4.142}$$

where the diagonal elements λ_i represent the eigenvalues of the matrix and we have

$$e^{\operatorname{Tr} \ln A} = e^{\sum_i \ln \lambda_i} = \prod_i \lambda_i = \det A, \tag{4.143}$$

which establishes the equivalence in (4.141). If A is not diagonal, on the other hand, we can write

$$A = SA_D S^{-1}, \tag{4.144}$$

where A_D is the diagonalized form of A and S is the similarity transformation that takes A to its diagonal form. In this case, we have

$$
\begin{aligned}
e^{\operatorname{Tr} \ln A} &= e^{\operatorname{Tr} \ln\left(SA_D S^{-1}\right)} = e^{\operatorname{Tr}(\ln S + \ln A_D - \ln S)} \\
&= e^{\operatorname{Tr} \ln A_D} = \det A_D = \det SA_D S^{-1} = \det A. \quad (4.145)
\end{aligned}
$$

This proves the formula (4.141) in general. If we further assume that the matrix A depends on space-time coordinates we can take the derivative of the determinant of the matrix with respect to coordinates to obtain

$$
\begin{aligned}
\partial_\sigma(\det A) &= \partial_\sigma e^{\operatorname{Tr} \ln A} = \partial_\sigma(\operatorname{Tr} \ln A) e^{\operatorname{Tr} \ln A} \\
&= \operatorname{Tr}(\partial_\sigma \ln A) \det A \\
&= \operatorname{Tr}\left(A^{-1}\partial_\sigma A\right) \det A. \tag{4.146}
\end{aligned}
$$

Let us apply the relation in (4.146) to the determinant of the metric tensor (metric tensor can be thought of as a 4×4 matrix in four dimensions), which leads to

$$\partial_\sigma g = \partial_\sigma \det g_{\mu\nu} = \left(g^{\mu\nu}\partial_\sigma g_{\nu\mu}\right) g, \tag{4.147}$$

where we have identified $g = \det g_{\mu\nu}$. Therefore, we have

$$
\begin{aligned}
\partial_\sigma \sqrt{-g} &= \frac{1}{2\sqrt{-g}}\left(-\partial_\sigma g\right) = \frac{(-g)}{2\sqrt{-g}}\, g^{\mu\nu}\partial_\sigma g_{\nu\mu} \\
&= \frac{\sqrt{-g}}{2}\, g^{\mu\nu}\partial_\sigma g_{\nu\mu}, \tag{4.148}
\end{aligned}
$$

so that comparing with (4.140) we can write

$$\Gamma^\mu{}_{\mu\sigma} = -\frac{1}{2}\, g^{\mu\rho}\partial_\sigma g_{\rho\mu} = -\frac{1}{\sqrt{-g}}\, \partial_\sigma\sqrt{-g}. \tag{4.149}$$

The divergence of a contravariant vector now follows to be (see (4.139))

$$\begin{aligned}
D_\mu A^\mu &= \partial_\mu A^\mu - \Gamma^\mu{}_{\mu\sigma} A^\sigma \\
&= \partial_\mu A^\mu + \frac{1}{\sqrt{-g}}\, (\partial_\sigma\sqrt{-g})A^\sigma \\
&= \frac{1}{\sqrt{-g}}\, \partial_\mu(\sqrt{-g}\, A^\mu).
\end{aligned} \tag{4.150}$$

This leads to an invariant formulation of Gauss' theorem in a curved manifold. Namely, if the vector field $A^\mu(x)$ vanishes at infinity, then

$$\begin{aligned}
\int \sqrt{-g}\, \mathrm{d}^4x\, D_\mu A^\mu &= \int \mathrm{d}^4x\, \sqrt{-g}\, \frac{1}{\sqrt{-g}}\, \partial_\mu(\sqrt{-g}A^\mu) \\
&= \int \mathrm{d}^4x\, \partial_\mu(\sqrt{-g}\, A^\mu) = 0.
\end{aligned} \tag{4.151}$$

A particularly simple application of the expression for the covariant divergence in (4.150) arises when the vector field itself corresponds to the gradient of a scalar function. In other words,

$$D_\mu D^\mu \phi(x) = D_\mu(\partial^\mu\phi(x)),$$

or, $$\quad \Box\phi(x) = \frac{1}{\sqrt{-g}}\, \partial_\mu(\sqrt{-g}\partial^\mu\phi(x)), \tag{4.152}$$

where $\Box$ stands for the invariant D'Alembertian in a curved manifold. We note here that much like the covariant derivative, the exact form of the D'Alembertian depends on the space of functions on which it acts. However, from (4.152) we see that acting on the space of scalar functions, the invariant D'Alembertian can be identified with the operator

$$\Box = \frac{1}{\sqrt{-g}} \, \partial_\mu \sqrt{-g} \, \partial^\mu = \frac{1}{\sqrt{-g}} \, \partial_\mu \sqrt{-g} \, g^{\mu\nu} \partial_\nu. \tag{4.153}$$

Let us apply this result to the simple case of three dimensional Euclidean space in spherical coordinates with the line element given by

$$ds^2 = dr^2 + r^2(d\theta^2 + \sin^2\theta d\phi^2). \tag{4.154}$$

In this case, we can read out the nontrivial components of the covariant metric tensor to be

$$g_{rr} = 1, \qquad g_{\theta\theta} = r^2, \qquad g_{\phi\phi} = r^2 \sin^2\theta, \tag{4.155}$$

and the inverse (contravariant) metric tensor is given by

$$g^{rr} = 1, \qquad g^{\theta\theta} = \frac{1}{r^2}, \qquad g^{\phi\phi} = \frac{1}{r^2 \sin^2\theta}, \tag{4.156}$$

which leads to

$$g = \det g_{ij} = g_{rr} g_{\theta\theta} g_{\phi\phi} = r^4 \sin^2\theta. \tag{4.157}$$

Thus using (4.152) we see that, in spherical coordinates, the Laplacian (analog of the D'Alembertian in three dimensions) acting on scalar functions becomes

$$
\begin{aligned}
\boldsymbol{\nabla}^2 \Phi \;=\;& \frac{1}{\sqrt{g}} \, \partial_i \left(\sqrt{g} g^{ij} \partial_j \Phi \right) = \frac{1}{r^2 \sin\theta} \, \partial_i \left(r^2 \sin\theta g^{ij} \partial_j \Phi \right) \\[2mm]
=\;& \frac{1}{r^2 \sin\theta} \left[\partial_r \left(r^2 \sin\theta g^{rr} \partial_r \right) + \partial_\theta \left(r^2 \sin\theta g^{\theta\theta} \partial_\theta \right) \right. \\[2mm]
& \left. + \partial_\phi \left(r^2 \sin\theta g^{\phi\phi} \partial_\phi \right) \right] \Phi \\[2mm]
=\;& \frac{1}{r^2 \sin\theta} \left[\partial_r (r^2 \sin\theta \partial_r) + \partial_\theta \left(r^2 \sin\theta \frac{1}{r^2} \, \partial_\theta \right) \right.
\end{aligned}
$$

$$+ \partial_\phi \left(r^2 \sin\theta \; \frac{1}{r^2 \sin^2\theta} \; \partial_\phi \right) \right] \Phi$$

$$= \frac{1}{r^2} \left[\frac{\partial}{\partial r} \, r^2 \frac{\partial}{\partial r} + \frac{1}{\sin\theta} \frac{\partial}{\partial\theta} \, \sin\theta \, \frac{\partial}{\partial\theta} + \frac{1}{\sin^2\theta} \frac{\partial^2}{\partial\phi^2} \right] \Phi,$$

$$(4.158)$$

so that, in this case, the form of the Laplacian (acting on scalar functions) in spherical coordinates has the familiar form

$$\nabla^2 = \frac{1}{r^2} \left[\frac{\partial}{\partial r} \, r^2 \frac{\partial}{\partial r} + \frac{1}{\sin\theta} \frac{\partial}{\partial\theta} \sin\theta \frac{\partial}{\partial\theta} + \frac{1}{\sin^2\theta} \frac{\partial^2}{\partial\phi^2} \right]. \qquad (4.159)$$

It is worth noting that the divergence of some of the tensors also takes a particularly simpler form. For example, we note that for a second rank tensor

$$\begin{aligned} D_\mu T^{\mu\nu} &= \partial_\mu T^{\mu\nu} - \Gamma^\mu_{\mu\sigma} T^{\sigma\nu} - \Gamma^\nu_{\mu\sigma} T^{\mu\sigma} \\ &= \partial_\mu T^{\mu\nu} + \frac{1}{\sqrt{-g}} \left(\partial_\sigma \sqrt{-g} \right) T^{\sigma\nu} - \Gamma^\nu_{\mu\sigma} T^{\mu\sigma} \\ &= \frac{1}{\sqrt{-g}} \, \partial_\mu (\sqrt{-g} \, T^{\mu\nu}) - \Gamma^\nu_{\mu\sigma} T^{\mu\sigma}. \end{aligned} \qquad (4.160)$$

In particular, if the tensor is anti-symmetric, namely,

$$T^{\mu\nu} = -T^{\nu\mu}, \qquad (4.161)$$

then the second term in (4.160) vanishes (because of the symmetry of the Christoffel symbol in the lower indices) and we can write (for $T^{\mu\nu} = -T^{\nu\mu}$)

$$D_\mu T^{\mu\nu} = \frac{1}{\sqrt{-g}} \, \partial_\mu (\sqrt{-g} T^{\mu\nu}). \qquad (4.162)$$

This result generalizes to any tensor that is completely anti-symmetric in all its indices.

The covariant curl of a covariant vector takes a particularly simple form. For example, we know that (see (4.122))

$$D_\mu A_\nu = \partial_\mu A_\nu + \Gamma^\lambda_{\mu\nu} A_\lambda, \tag{4.163}$$

from which it follows that

$$\begin{aligned}
D_\mu A_\nu - D_\nu A_\mu &= A_{\nu;\mu} - A_{\mu;\nu} \\
&= \partial_\mu A_\nu + \Gamma^\lambda_{\mu\nu} A_\lambda - \partial_\nu A_\mu - \Gamma^\lambda_{\nu\mu} A_\lambda \\
&= \partial_\mu A_\nu - \partial_\nu A_\mu, \tag{4.164}
\end{aligned}$$

where we have used the symmetry of the Christoffel symbol (4.50). This shows that the curl of a covariant vector remains unchanged in the presence of gravitation (or in a curved manifold). However, it is worth emphasizing here that this follows only because we have assumed the connection to be torsion free (the connection is symmetric). In the presence of torsion, this relation would modify since the connection will have an anti-symmetric part as well. In the present case we note that the curl of a covariant vector is a second rank anti-symmetric tensor,

$$A_{\mu\nu} = D_\mu A_\nu - D_\nu A_\mu = \partial_\mu A_\nu - \partial_\nu A_\mu = -A_{\nu\mu}, \tag{4.165}$$

which does not involve the Christoffel symbol at all.

Let us next note that for a covariant second rank tensor which is anti-symmetric $A_{\mu\nu} = -A_{\nu\mu}$ (see (4.130)),

$$\begin{aligned}
A_{\mu\nu;\lambda} = D_\lambda A_{\mu\nu} &= \partial_\lambda A_{\mu\nu} + \Gamma^\sigma_{\lambda\mu} A_{\sigma\nu} + \Gamma^\sigma_{\lambda\nu} A_{\mu\sigma} \\
&= \partial_\lambda A_{\mu\nu} - \Gamma^\sigma_{\lambda\mu} A_{\nu\sigma} + \Gamma^\sigma_{\nu\lambda} A_{\mu\sigma}, \tag{4.166}
\end{aligned}$$

where we have used the symmetry property of the Christoffel symbol as well as the anti-symmetry of $A_{\mu\nu}$ and have introduced the conventional notation of representing a covariant derivative by a semi-colon. Taking cyclic permutations of this relation we obtain

$$\begin{aligned}
D_\mu A_{\nu\lambda} &= \partial_\mu A_{\nu\lambda} - \Gamma^\sigma_{\mu\nu} A_{\lambda\sigma} + \Gamma^\sigma_{\lambda\mu} A_{\nu\sigma}, \\
D_\nu A_{\lambda\mu} &= \partial_\nu A_{\lambda\mu} - \Gamma^\sigma_{\nu\lambda} A_{\mu\sigma} + \Gamma^\sigma_{\mu\nu} A_{\lambda\sigma}. \tag{4.167}
\end{aligned}$$

Taking the sum these three relations we obtain

$$D_\lambda A_{\mu\nu} + D_\mu A_{\nu\lambda} + D_\nu A_{\lambda\mu}$$

$$= \partial_\lambda A_{\mu\nu} - \Gamma^\sigma_{\lambda\mu} A_{\nu\sigma} + \Gamma^\sigma_{\nu\lambda} A_{\mu\sigma} + \partial_\mu A_{\nu\lambda} - \Gamma^\sigma_{\mu\nu} A_{\lambda\sigma} + \Gamma^\sigma_{\lambda\mu} A_{\nu\sigma}$$

$$+ \partial_\nu A_{\lambda\mu} - \Gamma^\sigma_{\nu\lambda} A_{\mu\sigma} + \Gamma^\sigma_{\mu\nu} A_{\lambda\sigma}$$

$$= \partial_\lambda A_{\mu\nu} + \partial_\mu A_{\nu\lambda} + \partial_\nu A_{\lambda\mu}. \tag{4.168}$$

That is, in this case all of the effects of gravitation drop out. This is particularly interesting since it implies that a relation such as the Bianchi identity in the case of electromagnetic theory remains unchanged in the presence of gravitation (or in a curved manifold). Namely,

$$D_\lambda F_{\mu\nu} + D_\mu F_{\nu\lambda} + D_\nu F_{\lambda\mu} = \partial_\lambda F_{\mu\nu} + \partial_\mu F_{\nu\lambda} + \partial_\nu F_{\lambda\mu} = 0, \tag{4.169}$$

which is the Bianchi identity in flat space.

▶ **Example (Nijenhuis torsion tensor).** As we understand now, the ordinary derivative acting on tensors in a curved manifold does not transform, in general, like a tensor under a general coordinate transformation. We need to use the covariant derivative for this purpose. However, we have also seen in (4.164) and (4.168) that under special circumstances, the connection (Christoffel symbol) in the covariant derivative drops out and the combination of ordinary derivatives indeed leads to a true tensor. Nijenhuis torsion tensor is one such tensor in a curved manifold.

Let S^μ_ν be a mixed tensor of rank 2 with one covariant and one contravariant index. Using the covariant derivative (4.131), we can now define

$$N^\lambda_{\mu\nu} = S^\sigma_\mu D_\sigma S^\lambda_\nu - S^\sigma_\nu D_\sigma S^\lambda_\mu - S^\lambda_\sigma (D_\mu S^\sigma_\nu - D_\nu S^\sigma_\mu) = -N^\lambda_{\nu\mu}, \tag{4.170}$$

which by construction is a mixed tensor of rank 3 with one contravariant and two covariant indices.

If we write out the tensor (4.170) in detail, we obtain

$$N^\lambda_{\mu\nu} = S^\sigma_\mu (\partial_\sigma S^\lambda_\nu + \Gamma^\rho_{\sigma\nu} S^\lambda_\rho - \Gamma^\lambda_{\sigma\rho} S^\rho_\nu) - S^\sigma_\nu (\partial_\sigma S^\lambda_\mu + \Gamma^\rho_{\sigma\mu} S^\lambda_\rho - \Gamma^\lambda_{\sigma\rho} S^\rho_\mu)$$

$$- S^\lambda_\sigma (\partial_\mu S^\sigma_\nu + \Gamma^\rho_{\mu\nu} S^\sigma_\rho - \Gamma^\sigma_{\mu\rho} S^\rho_\nu) + S^\lambda_\sigma (\partial_\nu S^\sigma_\mu + \Gamma^\rho_{\nu\mu} S^\sigma_\rho - \Gamma^\sigma_{\nu\rho} S^\rho_\mu). \tag{4.171}$$

We note that if we use the symmetry property of the Christoffel symbol (4.50), all the terms involving the connection cancel pairwise and we have

$$N^\lambda_{\mu\nu} = -N^\lambda_{\nu\mu} = S^\sigma_\mu \partial_\sigma S^\lambda_\nu - S^\sigma_\nu \partial_\sigma S^\lambda_\mu - S^\lambda_\sigma (\partial_\mu S^\sigma_\nu - \partial_\nu S^\sigma_\mu). \tag{4.172}$$

Even though the tensor (4.172) involves only ordinary derivatives, it is a true tensor. This is known as the Nijenhuis torsion tensor associated with the mixed tensor S_ν^μ and plays an important role in the study of complex manifolds (as well as in integrable systems).

◄

Therefore, we see that if $A_{\mu\nu}$ is a covariant anti-symmetric tensor of second rank, namely,

$$A_{\mu\nu} = -A_{\nu\mu}, \tag{4.173}$$

then we can construct an anti-symmetric tensor of third rank from $A_{\mu\nu}$ independent of the Christoffel symbol (this would correspond to taking the curl of the second rank tensor), as

$$
\begin{aligned}
A_{\mu\nu\lambda} &= D_\lambda A_{\mu\nu} + D_\mu A_{\nu\lambda} + D_\nu A_{\lambda\mu} \\
&= \partial_\lambda A_{\mu\nu} + \partial_\mu A_{\nu\lambda} + \partial_\nu A_{\lambda\mu}.
\end{aligned} \tag{4.174}
$$

These results suggest the following general fact. Given a covariant rank n anti-symmetric tensor, we can form a unique covariant rank $n + 1$ anti-symmetric tensor which does not involve the Christoffel symbol (by taking its curl). The general result can be quickly proved with p-forms which we will not get into. Here let us study this only for a third rank tensor. If $A_{\mu\nu\lambda}$ is a completely anti-symmetric tensor, then

$$
\begin{aligned}
D_\rho A_{\mu\nu\lambda} &= \partial_\rho A_{\mu\nu\lambda} + \Gamma_{\rho\mu}^\sigma A_{\sigma\nu\lambda} + \Gamma_{\rho\nu}^\sigma A_{\mu\sigma\lambda} + \Gamma_{\rho\lambda}^\sigma A_{\mu\nu\sigma} \\
&= \partial_\rho A_{\mu\nu\lambda} + \Gamma_{\rho\mu}^\sigma A_{\nu\lambda\sigma} + \Gamma_{\rho\nu}^\sigma A_{\lambda\mu\sigma} + \Gamma_{\rho\lambda}^\sigma A_{\mu\nu\sigma}.
\end{aligned} \tag{4.175}
$$

Taking the cyclic permutations of this relation we obtain

$$
\begin{aligned}
D_\mu A_{\nu\lambda\rho} &= \partial_\mu A_{\nu\lambda\rho} + \Gamma_{\mu\nu}^\sigma A_{\lambda\rho\sigma} + \Gamma_{\mu\lambda}^\sigma A_{\rho\nu\sigma} + \Gamma_{\mu\rho}^\sigma A_{\nu\lambda\sigma}, \\
D_\nu A_{\lambda\rho\mu} &= \partial_\nu A_{\lambda\rho\mu} + \Gamma_{\nu\lambda}^\sigma A_{\rho\mu\sigma} + \Gamma_{\nu\rho}^\sigma A_{\mu\lambda\sigma} + \Gamma_{\nu\mu}^\sigma A_{\lambda\rho\sigma}, \\
D_\lambda A_{\rho\mu\nu} &= \partial_\lambda A_{\rho\mu\nu} + \Gamma_{\lambda\rho}^\sigma A_{\mu\nu\sigma} + \Gamma_{\lambda\mu}^\sigma A_{\nu\rho\sigma} + \Gamma_{\lambda\nu}^\sigma A_{\rho\mu\sigma}.
\end{aligned} \tag{4.176}
$$

It follows now that the totally anti-symmetric combination

$$A_{\mu\nu\lambda\rho} = D_\rho A_{\mu\nu\lambda} - D_\mu A_{\nu\lambda\rho} + D_\nu A_{\lambda\rho\mu} - D_\lambda A_{\rho\mu\nu}$$

$$= \partial_\rho A_{\mu\nu\lambda} - \partial_\mu A_{\nu\lambda\rho} + \partial_\nu A_{\lambda\rho\mu} - \partial_\lambda A_{\rho\mu\nu}, \qquad (4.177)$$

which is, of course, a covariant anti-symmetric tensor of rank 4 independent of the Christoffel symbol.

Some comments are in order here. First of all we note that we can increase the rank of a covariant anti-symmetric tensor by one simply by taking its curl. If we try to apply this procedure twice, then it yields a tensor which vanishes identically. For example, we note that given a scalar function, we can construct a covariant vector by applying the covariant derivative (curl)

$$A_\mu = D_\mu \phi = \partial_\mu \phi. \qquad (4.178)$$

If we try to construct an anti-symmetric second rank tensor from this by taking its curl, then we find

$$D_\mu A_\nu - D_\nu A_\mu = \partial_\mu A_\nu - \partial_\nu A_\mu$$

$$= \partial_\mu \partial_\nu \phi - \partial_\nu \partial_\mu \phi = 0. \qquad (4.179)$$

Let us next consider a covariant vector A_μ and construct the second rank anti-symmetric tensor by taking its curl as (see (4.164))

$$A_{\mu\nu} = D_\mu A_\nu - D_\nu A_\mu = \partial_\mu A_\nu - \partial_\nu A_\mu. \qquad (4.180)$$

If we now try to construct an anti-symmetric tensor of rank 3 by taking curl once more, we have (the square bracket represents anti-symmetrization)

$$A_{[\mu\nu\lambda]} = D_\lambda A_{\mu\nu} + D_\mu A_{\nu\lambda} + D_\nu A_{\lambda\mu}$$

$$= \partial_\lambda A_{\mu\nu} + \partial_\mu A_{\nu\lambda} + \partial_\nu A_{\lambda\mu}$$

$$= \partial_\lambda(\partial_\mu A_\nu - \partial_\nu A_\mu) + \partial_\mu(\partial_\nu A_\lambda - \partial_\lambda A_\nu) + \partial_\nu(\partial_\lambda A_\mu - \partial_\mu A_\lambda)$$

$$= 0, \qquad (4.181)$$

where we have used (4.168). We can show that this result is true in general and the general result follows in a simple manner with the use of p-forms that we will not go into. Furthermore, let us note that in n dimensions, the highest rank nontrivial anti-symmetric tensor would have rank n. This is easy to see because if there exists an anti-symmetric tensor of rank $n + 1$, then two of its indices have to be the same in which case it will vanish identically. The same is true for any higher rank tensor as well.

Let us also note that in n dimensions any anti-symmetric tensor of rank n must be proportional to the generalization of the Levi-Civita tensor to this space. This can also be seen simply because if

$$T_{[\mu_1...\mu_n]},   \tag{4.182}$$

is the anti-symmetric tensor (anti-symmetrization is denoted by the square bracket), then the combination

$$\epsilon^{\mu_1...\mu_n} T_{[\mu_1...\mu_n]},   \tag{4.183}$$

is a scalar function (density). Therefore, we can always write

$$T_{[\mu_1...\mu_n]} = \epsilon_{\mu_1...\mu_n} \, \phi(x),   \tag{4.184}$$

where $\phi(x)$ denotes a scalar density.

Let us define the following terminology. An anti-symmetric tensor with the property

$$D_{[\lambda} A_{\mu\nu...]} = 0,   \tag{4.185}$$

is called closed. (Namely, the covariant curl of the tensor vanishes.) Furthermore, an anti-symmetric tensor $A_{[\mu\nu...]}$ is called exact if it can be written in terms of a lower rank (anti-symmetric) tensor as

$$A_{[\mu\nu...]} = D_{[\mu} T_{[\nu...]]} = \partial_{[\mu} T_{[\nu...]]}.   \tag{4.186}$$

Thus for example, in the case of three dimensional electrostatic in Euclidean space, as we know, one of Maxwell's equations takes the form

$$\nabla \times \mathbf{E} = 0. \tag{4.187}$$

Therefore, the electric field can be thought of as a closed vector and since we know that it can be written as the gradient of a scalar potential

$$\mathbf{E} = -\nabla \phi, \tag{4.188}$$

it is also exact.

Another example follows from the definition of the electromagnetic field strength tensor itself (see (1.83) and (4.164))

$$F_{\mu\nu} = D_\mu A_\nu - D_\nu A_\mu = \partial_\mu A_\nu - \partial_\nu A_\mu, \tag{4.189}$$

where A_μ denotes the vector potential. Thus we see that one can think of the electromagnetic field strength as an exact tensor. Furthermore, we know that the field strength tensor satisfies Bianchi identity (see (4.168))

$$\begin{aligned} D_\lambda F_{\mu\nu} + D_\mu F_{\nu\lambda} + D_\nu F_{\lambda\mu} &= \partial_\lambda F_{\mu\nu} + \partial_\mu F_{\nu\lambda} + \partial_\nu F_{\lambda\mu} \\ &= \partial_\lambda(\partial_\mu A_\nu - \partial_\nu A_\mu) + \partial_\mu(\partial_\nu A_\lambda - \partial_\lambda A_\nu) + \partial_\nu(\partial_\lambda A_\mu - \partial_\mu A_\lambda) \\ &= 0, \end{aligned} \tag{4.190}$$

so that it is also closed. These observations lead to the following two theorems.

Theorem 1: *Every exact tensor is closed.*

It follows from the definition of an exact tensor in (4.186) that

$$T_{[\mu_1 \ldots \mu_n]} = \partial_{[\mu_1} \widetilde{T}_{[\mu_2 \ldots \mu_n]]}. \tag{4.191}$$

Therefore, since derivatives commute (their product is symmetric) we have

$$D_{[\mu_{n+1}}T_{[\mu_1...\mu_n]]} = \partial_{[\mu_{n+1}}T_{[\mu_1...\mu_n]]} = \partial_{[\mu_{n+1}}\partial_{[\mu_1}\widetilde{T}_{[\mu_2...\mu_n]]]} = 0,$$
$$(4.192)$$

which proves the theorem. (This is the same result as we had discussed earlier, namely, that taking the curl twice gives zero.)

Theorem 2: *Every closed tensor is at least locally exact, i.e., it admits a tensor potential locally.*

We will not prove this theorem here but simply study its consequences in four dimensions.

Theorem: *In four dimensions a second rank anti-symmetric tensor being closed is equivalent to its dual having zero divergence.*

Let $F_{\mu\nu}$ be a second rank anti-symmetric tensor in four dimensions. Then it is closed if (see (4.185))

$$D_\lambda F_{\mu\nu} + D_\mu F_{\nu\lambda} + D_\nu F_{\lambda\mu} = \partial_\lambda F_{\mu\nu} + \partial_\mu F_{\nu\lambda} + \partial_\nu F_{\lambda\mu} = 0. \quad (4.193)$$

Furthermore, the divergence free nature of the dual tensor (see (1.81)) is defined as (any factor of $\sqrt{-g}$ commutes with the covariant derivative and, therefore, can only be a multiplicative factor)

$$\begin{aligned}
D_\mu\widetilde{F}^{\mu\nu} &= D_\mu\left(\frac{1}{2}\,\epsilon^{\mu\nu\lambda\rho}F_{\lambda\rho}\right) = \frac{1}{2}\,\epsilon^{\mu\nu\lambda\rho}D_\mu F_{\lambda\rho} \\
&= \frac{1}{2}\times\frac{1}{3}\epsilon^{\mu\nu\lambda\rho}\left(D_\mu F_{\lambda\rho} + D_\lambda F_{\rho\mu} + D_\rho F_{\mu\lambda}\right) = 0. \quad (4.194)
\end{aligned}$$

This is known as the Bianchi identity and as we had noted earlier in chapter 1, the two relations (4.193) and (4.194) are equivalent.

CHAPTER 5

Geodesic equation

5.1 Covariant differentiation along a curve

Sometimes a vector field or a tensor field is defined only on a certain trajectory rather than on the entire space-time manifold. Thus for example, the four momentum or the stress tensor or the spin vector of a particle can be defined only along the trajectory of the particle. In this case we have to develop the notion of covariant differentiation along the curve.

From what we have seen already in the last chapter, the covariant derivative measures the change in a vector or a tensor in an absolute sense. Let us apply this notion to a vector ξ^μ defined along a curve $x^\mu(\tau)$ so that we can write

$$\xi^\mu(x(\tau)) = \xi^\mu(\tau), \tag{5.1}$$

and at an infinitesimally separated point on the curve we have

$$\xi^\mu(\tau + d\tau) = \xi^\mu(\tau) + d\tau \, \frac{d\xi^\mu(\tau)}{d\tau} + O(d\tau^2). \tag{5.2}$$

On the other hand, we have defined the parallel transport of a vector $\xi^\mu(\tau)$ to the point $\tau + d\tau$ as given by (see (4.15))

$$\xi^{*\,\mu}(\tau + d\tau) = \xi^\mu(\tau) + \Gamma^\mu_{\nu\lambda} \, \frac{dx^\nu}{d\tau} \, \xi^\lambda(\tau) d\tau. \tag{5.3}$$

From (5.2) and (5.3) we obtain

133

$$\xi^\mu(\tau + d\tau) - \xi^{*\,\mu}(\tau + d\tau)$$

$$= \xi^\mu(\tau) + d\tau \frac{d\xi^\mu(\tau)}{d\tau} - \xi^\mu(\tau) - \Gamma^\mu_{\nu\lambda} \frac{dx^\nu}{d\tau}\xi^\lambda(\tau)d\tau + O(d\tau^2)$$

$$= d\tau \frac{d\xi^\mu(\tau)}{d\tau} - \Gamma^\mu_{\nu\lambda} \frac{dx^\nu}{d\tau}\,\xi^\lambda(\tau)d\tau + O(d\tau^2)$$

$$= d\tau \left(\frac{d\xi^\mu(\tau)}{d\tau} - \Gamma^\mu_{\nu\lambda} \frac{dx^\nu}{d\tau}\,\xi^\lambda(\tau) \right) + O(d\tau^2)$$

$$= d\tau \frac{D\xi^\mu(\tau)}{D\tau} + O(d\tau^2), \tag{5.4}$$

which defines the covariant derivative of the vector ξ^μ along a curve to be (see also the discussion around (4.90))

$$\frac{D\xi^\mu}{D\tau} = \frac{d\xi^\mu}{d\tau} - \Gamma^\mu_{\nu\lambda} \frac{dx^\nu}{d\tau}\,\xi^\lambda$$

$$= \frac{dx^\nu}{d\tau} \frac{\partial \xi^\mu}{\partial x^\nu} - \Gamma^\mu_{\nu\lambda} \frac{dx^\nu}{d\tau}\,\xi^\lambda$$

$$= \frac{dx^\nu}{d\tau} \left(\partial_\nu \xi^\mu - \Gamma^\mu_{\nu\lambda}\xi^\lambda \right)$$

$$= \frac{dx^\nu}{d\tau}\, D_\nu \xi^\mu. \tag{5.5}$$

Comparing with (4.12) we recognize that the parallel transport of a vector along a curve can also be written as

$$\frac{D\xi^\mu}{D\tau} = \frac{dx^\nu}{d\tau}\, D_\nu \xi^\mu = 0. \tag{5.6}$$

Similarly, the covariant derivative of a contravariant tensor field along a curve can be expressed as

$$\frac{DT^{\mu_1\cdots\mu_n}}{D\tau} = \frac{dx^\nu}{d\tau}\, D_\nu T^{\mu_1\cdots\mu_n}. \tag{5.7}$$

We note from (5.7) that

$$\frac{\mathrm{D}g^{\mu\nu}}{\mathrm{D}\tau} = \frac{\mathrm{d}x^\lambda}{\mathrm{d}\tau} D_\lambda g^{\mu\nu} = 0, \tag{5.8}$$

which follows from the fact that the metric is flat under covariant differentiation (see (4.103)). Since $g_{\mu\nu}$ denotes the inverse of the contravariant metric tensor $g^{\mu\nu}$, it follows that

$$\frac{\mathrm{D}g_{\mu\nu}}{\mathrm{D}\tau} = \frac{\mathrm{d}x^\lambda}{\mathrm{d}\tau} D_\lambda g_{\mu\nu} = 0. \tag{5.9}$$

Using these, we can now derive

$$\frac{\mathrm{D}\xi_\mu}{\mathrm{D}\tau} = \frac{\mathrm{D}}{\mathrm{D}\tau}\left(g_{\mu\nu}\xi^\nu\right) = g_{\mu\nu}\frac{\mathrm{D}\xi^\nu}{\mathrm{D}\tau}, \tag{5.10}$$

Similarly, for an arbitrary covariant tensor we have

$$\begin{aligned}
\frac{\mathrm{D}}{\mathrm{D}\tau}\left(T_{\mu_1\dots\mu_n}\right) &= \frac{\mathrm{D}}{\mathrm{D}\tau}\left(g_{\mu_1\nu_1}\cdots g_{\mu_n\nu_n}T^{\nu_1\dots\nu_n}\right)\\
&= g_{\mu_1\nu_1}\cdots g_{\mu_n\nu_n}\frac{\mathrm{D}}{\mathrm{D}\tau}\left(T^{\nu_1\dots\nu_n}\right),
\end{aligned} \tag{5.11}$$

and so on.

5.2 Curvature from derivatives

So far we have talked about the metric and the connection as representing the effects of gravitation. However, we have also said on several occasions that gravitation produces curvature in the space-time manifold. But, we are yet to define and identify the curvature. We will do so now in two different ways before studying its properties in detail in chapter **7**.

First, let us go back to the example of electromagnetism in section **4.6**. We have noted earlier that invariance under the $U(1)$ gauge transformation requires that the ordinary derivative be replaced by the covariant derivative defined as (see (4.135))

$$D_\mu\psi = (\partial_\mu + ieA_\mu)\psi. \tag{5.12}$$

On the other hand, the first property that we notice from the definition of the covariant derivative is that unlike ordinary derivatives, the covariant derivatives do not commute. Namely,

$$
\begin{aligned}
[D_\mu, D_\nu]\psi &= (D_\mu D_\nu - D_\nu D_\mu)\psi \\
&= [(\partial_\mu + ieA_\mu)(\partial_\nu + ieA_\nu) - (\partial_\nu + ieA_\nu)(\partial_\mu + ieA_\mu)]\psi \\
&= ie\partial_\mu(A_\nu\psi) + ieA_\mu\partial_\nu\psi - ie\partial_\nu(A_\mu\psi) - ieA_\nu\partial_\mu\psi \\
&= ie(\partial_\mu A_\nu)\psi + ieA_\nu\partial_\mu\psi + ieA_\mu\partial_\nu\psi - ie(\partial_\nu A_\mu)\psi \\
&\quad -ieA_\mu\partial_\nu\psi - ieA_\nu\partial_\mu\psi \\
&= ie(\partial_\mu A_\nu - \partial_\nu A_\mu)\psi = ieF_{\mu\nu}\psi, \qquad (5.13)
\end{aligned}
$$

where $F_{\mu\nu}$ represents the electromagnetic field strength tensor. The non-commutativity of the covariant derivatives arisesbecause of curvature in the space (in this case the Hilbert space) and we identify the observable curvature of the space with the field strength tensor $F_{\mu\nu}$.

Similarly, we can derive the curvature in the case of a gravitational field (or a curved manifold) by taking the commutator of two covariant derivatives. We have defined the covariant derivative of a contravariant vector as (see (4.87))

$$
D_\mu A^\rho = \partial_\mu A^\rho - \Gamma^\rho_{\mu\sigma} A^\sigma. \qquad (5.14)
$$

It follows from this as well as from the definition of the covariant derivative for mixed tensors that (see (4.131))

$$
\begin{aligned}
D_\nu D_\mu A^\rho &= \partial_\nu (D_\mu A^\rho) + \Gamma^\lambda_{\nu\mu} D_\lambda A^\rho - \Gamma^\rho_{\nu\lambda} D_\mu A^\lambda \\
&= \partial_\nu (\partial_\mu A^\rho - \Gamma^\rho_{\mu\sigma} A^\sigma) + \Gamma^\lambda_{\nu\mu} (\partial_\lambda A^\rho - \Gamma^\rho_{\lambda\sigma} A^\sigma) \\
&\quad -\Gamma^\rho_{\nu\lambda}(\partial_\mu A^\lambda - \Gamma^\lambda_{\mu\sigma} A^\sigma) \\
&= \partial_\nu\partial_\mu A^\rho - (\partial_\nu\Gamma^\rho_{\mu\sigma})A^\sigma - \Gamma^\rho_{\mu\sigma}\partial_\nu A^\sigma \\
&\quad + \Gamma^\lambda_{\nu\mu}(\partial_\lambda A^\rho - \Gamma^\rho_{\lambda\sigma} A^\sigma) - \Gamma^\rho_{\nu\lambda}(\partial_\mu A^\lambda - \Gamma^\lambda_{\mu\sigma} A^\sigma). \quad (5.15)
\end{aligned}
$$

Interchanging $\mu \leftrightarrow \nu$, we obtain from this

$$D_\mu D_\nu A^\rho = \partial_\mu \partial_\nu A^\rho - (\partial_\mu \Gamma^\rho_{\nu\sigma}) A^\sigma - \Gamma^\rho_{\nu\sigma} \partial_\mu A^\sigma$$
$$+ \; \Gamma^\lambda_{\mu\nu} (\partial_\lambda A^\rho - \Gamma^\rho_{\lambda\sigma} A^\sigma) - \Gamma^\rho_{\mu\lambda} (\partial_\nu A^\lambda - \Gamma^\lambda_{\nu\sigma} A^\sigma), \quad (5.16)$$

and these results lead to the commutator of the covariant derivatives acting on a contravariant vector as

$$[D_\mu, D_\nu] A^\rho = (D_\mu D_\nu A^\rho - D_\nu D_\mu A^\rho)$$
$$= (\partial_\nu \Gamma^\rho_{\mu\sigma} - \partial_\mu \Gamma^\rho_{\nu\sigma}) A^\sigma + (\Gamma^\rho_{\mu\lambda} \Gamma^\lambda_{\nu\sigma} - \Gamma^\rho_{\nu\lambda} \Gamma^\lambda_{\mu\sigma}) A^\sigma$$
$$= (\partial_\nu \Gamma^\rho_{\mu\sigma} - \partial_\mu \Gamma^\rho_{\nu\sigma} + \Gamma^\rho_{\mu\lambda} \Gamma^\lambda_{\nu\sigma} - \Gamma^\rho_{\nu\lambda} \Gamma^\lambda_{\mu\sigma}) A^\sigma$$
$$= R^\rho{}_{\sigma\mu\nu} A^\sigma, \quad (5.17)$$

where we have identified

$$R^\rho{}_{\sigma\mu\nu} = \partial_\nu \Gamma^\rho_{\mu\sigma} - \partial_\mu \Gamma^\rho_{\nu\sigma} + \Gamma^\rho_{\mu\lambda} \Gamma^\lambda_{\nu\sigma} - \Gamma^\rho_{\nu\lambda} \Gamma^\lambda_{\mu\sigma}. \quad (5.18)$$

This is known as the Riemann-Christoffel curvature tensor of the manifold. Similarly, for a covariant vector we can show that

$$[D_\mu, D_\nu] A_\rho = D_\mu (D_\nu A_\rho) - D_\nu (D_\mu A_\rho)$$
$$= \partial_\mu (D_\nu A_\rho) + \Gamma^\lambda_{\mu\nu} D_\lambda A_\rho + \Gamma^\lambda_{\mu\rho} D_\nu A_\lambda$$
$$- \partial_\nu (D_\mu A_\rho) - \Gamma^\lambda_{\nu\mu} D_\lambda A_\rho - \Gamma^\lambda_{\nu\rho} D_\mu A_\lambda$$
$$= \partial_\mu \left(\partial_\nu A_\rho + \Gamma^\sigma_{\nu\rho} A_\sigma \right) + \Gamma^\lambda_{\mu\rho} \left(\partial_\nu A_\lambda + \Gamma^\sigma_{\nu\lambda} A_\sigma \right)$$
$$- \partial_\nu \left(\partial_\mu A_\rho + \Gamma^\sigma_{\mu\rho} A_\sigma \right) - \Gamma^\lambda_{\nu\rho} \left(\partial_\mu A_\lambda + \Gamma^\sigma_{\mu\lambda} A_\sigma \right)$$
$$= - \left[\partial_\nu \Gamma^\sigma_{\mu\rho} - \partial_\mu \Gamma^\sigma_{\nu\rho} + \Gamma^\lambda_{\nu\rho} \Gamma^\sigma_{\mu\lambda} - \Gamma^\lambda_{\mu\rho} \Gamma^\sigma_{\nu\lambda} \right] A_\sigma$$
$$= -R^\sigma{}_{\rho\mu\nu} A_\sigma. \quad (5.19)$$

Furthermore, note that since the curvature tensor is a function of the connection and its derivative, if the space were flat the curvature tensor would vanish and then the covariant derivatives would reduce to ordinary derivatives and they would commute. (Since curvature is a tensor, if it vanishes in one coordinate system, say the Cartesian coordinate system, then it will vanish in any curvilinear coordinate system.)

▶ **Example (Commutator of covariant derivatives on a tensor).** Let us study how the commutator of two covariant derivatives acts on tensors, in particular on a second rank tensor $T^{\mu\nu}$. From the definition of the covariant derivatives in (4.98) and (4.131) we have

$$D_\lambda D_\rho T^{\mu\nu} = \partial_\lambda (D_\rho T^{\mu\nu}) + \Gamma^\sigma_{\lambda\rho} D_\sigma T^{\mu\nu} - \Gamma^\mu_{\lambda\sigma} D_\rho T^{\sigma\nu} - \Gamma^\mu_{\lambda\sigma} D_\rho T^{\sigma\nu},$$

$$D_\rho D_\lambda T^{\mu\nu} = \partial_\rho (D_\lambda T^{\mu\nu}) + \Gamma^\sigma_{\rho\lambda} D_\sigma T^{\mu\nu} - \Gamma^\mu_{\rho\sigma} D_\lambda T^{\sigma\nu} - \Gamma^\mu_{\rho\sigma} D_\lambda T^{\sigma\nu}. \qquad (5.20)$$

Therefore, we obtain

$$
\begin{aligned}
[D_\lambda,\, D_\rho] T^{\mu\nu} &= \partial_\lambda (D_\rho T^{\mu\nu}) + \Gamma^\sigma_{\lambda\rho} D_\sigma T^{\mu\nu} - \Gamma^\mu_{\lambda\sigma} D_\rho T^{\sigma\nu} - \Gamma^\mu_{\lambda\sigma} D_\rho T^{\sigma\nu} \\
&\quad - \partial_\rho (D_\lambda T^{\mu\nu}) + \Gamma^\sigma_{\rho\lambda} D_\sigma T^{\mu\nu} - \Gamma^\mu_{\rho\sigma} D_\lambda T^{\sigma\nu} - \Gamma^\mu_{\rho\sigma} D_\lambda T^{\sigma\nu} \\
&= \partial_\lambda (\partial_\rho T^{\mu\nu} - \Gamma^\mu_{\rho\sigma} T^{\sigma\nu} - \Gamma^\nu_{\rho\sigma} T^{\mu\sigma}) \\
&\quad - \Gamma^\mu_{\lambda\sigma} (\partial_\rho T^{\sigma\nu} - \Gamma^\sigma_{\rho\eta} T^{\eta\nu} - \Gamma^\nu_{\rho\eta} T^{\eta\sigma}) \\
&\quad - \Gamma^\nu_{\lambda\sigma} (\partial_\rho T^{\mu\sigma} - \Gamma^\mu_{\lambda\sigma} T^{\sigma\nu} - \Gamma^\nu_{\lambda\sigma} T^{\mu\sigma}) - (\rho \leftrightarrow \lambda) \\
&= (\partial_\rho \Gamma^\mu_{\lambda\sigma} - \partial_\lambda \Gamma^\mu_{\rho\sigma} + \Gamma^\mu_{\lambda\eta} \Gamma^\eta_{\rho\sigma} - \Gamma^\mu_{\rho\eta} \Gamma^\eta_{\lambda\sigma}) T^{\sigma\nu} \\
&\quad + (\partial_\rho \Gamma^\nu_{\lambda\sigma} - \partial_\lambda \Gamma^\nu_{\rho\sigma} + \Gamma^\nu_{\lambda\eta} \Gamma^\eta_{\rho\sigma} - \Gamma^\nu_{\rho\eta} \Gamma^\eta_{\lambda\sigma}) T^{\mu\sigma} \\
&= R^\mu_{\sigma\lambda\rho} T^{\sigma\nu} + R^\nu_{\sigma\lambda\rho} T^{\mu\sigma}. \qquad (5.21)
\end{aligned}
$$

We see that when the commutator of two covariant derivatives acts on a tensor, it leads to a sum of terms involving the Riemann-Christoffel curvature tensor. ◀

▶ **Example (Curvature for a two dimensional manifold).** Let us consider the two dimensional space defined by the line element

$$d\tau^2 = \frac{1}{t^2}(dt^2 - dx^2). \qquad (5.22)$$

The nonzero components of the metric tensor follow from (5.22) to have the forms

$$g_{tt} = \frac{1}{t^2}, \quad g_{xx} = -\frac{1}{t^2}, \quad g^{tt} = t^2, \quad g^{xx} = -t^2. \qquad (5.23)$$

The metric tensors depend only on the coordinate t and, therefore, the nontrivial components of the connection (4.49)

$$\Gamma^{\sigma}_{\mu\nu} = -\frac{1}{2}\, g^{\sigma\lambda}\left(\partial_\mu g_{\nu\lambda} + \partial_\nu g_{\lambda\mu} - \partial_\lambda g_{\mu\nu}\right), \quad \mu,\nu,\sigma = t,x, \qquad (5.24)$$

can be determined to have the forms

$$\Gamma^t_{tt} = \frac{1}{t}, \quad \Gamma^t_{xx} = \frac{1}{t}, \quad \Gamma^x_{tx} = \Gamma^x_{xt} = \frac{1}{t}, \qquad (5.25)$$

so that the geodesic equations are given by

$$\ddot{t} - \Gamma^t_{tt}\dot{t}^2 - \Gamma^t_{xx}\dot{x}^2 = 0,$$

$$\text{or,}\quad \ddot{t} - \frac{1}{t}\dot{t}^2 - \frac{1}{t}\dot{x}^2 = 0,$$

$$\ddot{x} - \Gamma^t_{tx}\dot{t}\dot{x} - \Gamma^x_{xt}\dot{x}\dot{t} = 0,$$

$$\text{or,}\quad \ddot{x} - \frac{2}{t}\dot{t}\dot{x} = 0. \qquad (5.26)$$

In two dimensions there is only one nontrivial, independent component of the Riemann-Christoffel curvature tensor (5.18) which we denote as $R^t{}_{xtx}$ and it has the form

$$R^t{}_{xtx} = \partial_x \Gamma^t_{xt} - \partial_t \Gamma^t_{xx} + \Gamma^t_{t\sigma}\Gamma^\sigma_{xx} - \Gamma^t_{\sigma x}\Gamma^\sigma_{tx}$$

$$= -\partial_t \Gamma^t_{xx} + \Gamma^t_{tt}\Gamma^t_{xx} - \Gamma^t_{xx}\Gamma^x_{tx}$$

$$= \frac{1}{t^2} + \frac{1}{t}\frac{1}{t} - \frac{1}{t}\frac{1}{t} = \frac{1}{t^2}. \qquad (5.27)$$

It follows now that

$$R_{txtx} = g_{tt}R^t{}_{xtx} = \frac{1}{t^4}. \qquad (5.28)$$

◀

5.3 Parallel transport along a closed curve

Let us next discuss the notion of curvature from a second point of view. To differentiate between flat space and a curved space, let us study the example of parallel transport of a vector along a closed path. First, let us consider a closed curve in flat space as shown in Fig. 5.1. If we take any vector and parallel transport it along the curve keeping it always parallel to itself, then when we come back

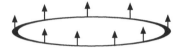

Figure 5.1: Parallel transport of a vector along a closed path in flat space.

to the starting point the final vector would coincide with the initial vector. This is the characteristic of a flat space.

Let us next consider the surface of a sphere. If we move a vector from the north pole Q along the closed curve QPRQ always maintaining the vector locally parallel (always facing the south pole) as shown in Fig. 5.2, then when we reach the starting point (the north pole), the final vector would be pointing in a different direction from the initial direction. This is a characteristic of curved spaces, namely, when a vector is parallel transported along a closed curve it does not come back to itself. The angle between the initial and the final vector depends on the closed path. Therefore, the change in the vector must be related to the curvature as well as the closed path along which it is transported.

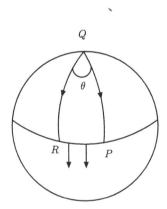

Figure 5.2: Parallel transport along a closed path on the surface of a sphere.

From the definition of parallel transport in (4.14) we know that when a vector is parallel transported infinitesimally in a curved space the change in its components is given by

$$d\xi^{\mu}(x) = \Gamma^{\mu}_{\nu\lambda}(x)dx^{\nu}\xi^{\lambda}(x). \tag{5.29}$$

However, since the value of the connection is different at different points in a curved space-time manifold, the change in a vector parallel transported from point A to point B along path 1 would be different from the change along path 2 (see Fig. 5.3). The other way of saying this is that a vector parallel transported along a closed curve would not come back to itself. Therefore, the change in a vector around a closed path would be a measure of the curvature of the manifold.

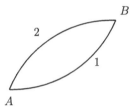

Figure 5.3: Parallel transporting of a vector from A to B along two paths.

▶ **Example (Parallel transport on the surface of the unit sphere).** Let us consider a vector $A^{\alpha} = (A^{\theta}, A^{\phi})$ on the surface of the unit sphere with components $(1, 0)$ at the coordinates $(\theta = \theta_0, \phi = 0)$, namely,

$$A^{\alpha}(\theta = \theta_0, \phi = 0) = (A^{\theta}(\theta_0, 0), A^{\phi}(\theta_0, 0)) = (1, 0). \tag{5.30}$$

We would like to calculate the components of this vector when it is parallel transported along a circle at $\theta = \theta_0$ and, in particular, when it is parallel transported back to the starting point.

The line element on the surface of the unit sphere is given by

$$d\tau^2 = d\theta^2 + \sin^2\theta d\phi^2, \tag{5.31}$$

This is the same as the line element studied in (4.58) with $r = 1, dr = 0$ and, therefore, we have

$$g_{\theta\theta} = 1, \quad g_{\phi\phi} = \sin^2\theta, \tag{5.32}$$

and we can carry over the nontrivial components of the connection from (4.62) ($\Gamma^r_{\mu\nu}$ vanishes in this case since $g_{rr} = 0$)

$$\Gamma^\theta_{\phi\phi} = \sin\theta\cos\theta, \quad \Gamma^\phi_{\theta\phi} = \Gamma^\phi_{\phi\theta} = -\cot\theta. \tag{5.33}$$

At this point there are several ways to solve the problem. Let us indicate only two methods.

Method 1: The equation for the parallel transport of the vector A^α has the form (see (5.6))

$$\frac{dx^\beta}{d\tau}\left(\partial_\beta A^\alpha - \Gamma^\alpha_{\beta\gamma}A^\gamma\right) = 0, \quad \alpha, \beta, \gamma = \theta, \phi. \tag{5.34}$$

Furthermore, since we are parallel transporting along a circle where ϕ is changing but $\theta = \theta_0$ is fixed, the relevant equation to study is

$$\partial_\phi A^\alpha - \Gamma^\alpha_{\phi\gamma}(\theta = \theta_0)A^\gamma = 0, \tag{5.35}$$

which in components (with the use of (5.33)) becomes

$$\partial_\phi A^\theta - \Gamma^\theta_{\phi\phi}(\theta = \theta_0)A^\phi = 0,$$

$$\text{or,} \quad \partial_\phi A^\theta = \sin\theta_0\cos\theta_0\, A^\phi, \tag{5.36}$$

$$\partial_\phi A^\phi - \Gamma^\phi_{\phi\theta}(\theta = \theta_0)A^\theta = 0,$$

$$\text{or,} \quad \partial_\phi A^\phi = -\cot\theta_0\, A^\theta. \tag{5.37}$$

The two coupled equations (5.36) and (5.37) lead to

$$\partial^2_\phi A^\phi = -\cot\theta_0\partial_\phi A^\theta = -\cot\theta_0\left(\sin\theta_0\cos\theta_0 A^\phi\right)$$

$$= -\cos^2\theta_0 A^\phi, \tag{5.38}$$

whose general solution has the form

$$A^\phi(\theta_0, \phi) = C\,\sin(\phi\cos\theta_0) + D\,\cos(\phi\cos\theta_0). \tag{5.39}$$

On the other hand, the "initial" condition (5.30)

$$A^\phi(\theta_0, 0) = 0, \tag{5.40}$$

determines $D = 0$ so that we can write

$$A^\phi(\theta_0, \phi) = C \, \sin(\phi \cos \theta_0). \tag{5.41}$$

Substituting (5.41) into (5.36) we obtain

$$\partial_\phi A^\theta(\theta_0, \phi) = \sin \theta_0 \cos \theta_0 A^\phi(\theta_0, \phi) = C \, \sin \theta_0 \cos \theta_0 \sin(\phi \cos \theta_0),$$

or, $\quad A^\theta(\theta_0, \phi) = -C \, \sin \theta_0 \cos(\phi \cos \theta_0). \tag{5.42}$

Imposing the "initial" condition (5.30)

$$A^\theta(\theta_0, 0) = 1, \tag{5.43}$$

determines

$$C = -\csc \theta_0, \tag{5.44}$$

so that we can write the solutions to be

$$A^\theta(\theta_0, \phi) = \cos(\phi \cos \theta_0), \quad A^\phi(\theta_0, \phi) = -\csc \theta_0 \, \sin(\phi \cos \theta_0). \tag{5.45}$$

We note that the length of the "initial" vector is given by

$$g_{\theta\theta} A^\theta(\theta_0, 0) A^\theta(\theta_0, 0) + g_{\phi\phi} A^\phi(\theta_0, 0) A^\phi(\theta_0, 0) = 1. \tag{5.46}$$

The length of the parallel transported vector at any subsequent point on the circle is given by

$$g_{\theta\theta} A^\theta(\theta_0, \phi) A^\theta(\theta_0, \phi) + g_{\phi\phi} A^\phi(\theta_0, \phi) A^\phi(\theta_0, \phi)$$
$$= \cos^2(\phi \cos \theta_0) + \sin^2 \theta_0 \csc^2 \theta_0 \, \sin^2(\phi \cos \theta_0) = 1, \tag{5.47}$$

which shows that the length of the vector remains invariant under parallel transport. However, the vector gets rotated (the components change) as we move along the circle. In particular when we return to the starting point, namely, when $\phi = 2\pi$, the vector has the components

$$A^\alpha = (\cos(2\pi \cos \theta_0), -\csc \theta_0 \, \sin(2\pi \cos \theta_0)), \tag{5.48}$$

which is different from the "initial" vector (5.30) unless $\theta_0 = 0$, or $\frac{\pi}{2}$, namely, at the pole or at the equator .

Method 2: We have already determined the formula for parallel transport by an infinitesimal amount in spherical coordinates in (4.63). In the present problem, however, we have $dr = d\theta = 0, r = 1, A^r = 0$ so that the change in the vector under an infinitesimal parallel transport along the circle $\theta = \theta_0$ is given by

$$dA^\theta = \Gamma^\theta_{\phi\phi}(\theta_0)d\phi A^\phi = \sin\theta_0\cos\theta_0 A^\phi d\phi,$$

$$dA^\phi = \Gamma^\phi_{\theta\phi}(\theta_0)d\phi A^\theta = -\cot\theta_0 A^\theta d\phi. \tag{5.49}$$

Let us take the total angle traversed along the circle at $\theta = \theta_0$ to be ϕ and divide it into N equal infinitesimal parts so that

$$N d\phi = \phi, \tag{5.50}$$

where N is assumed to be large and we will take the limit $N \to \infty, d\phi \to 0$ such that (5.50) holds. Namely, we have divided the parallel transport by a finite amount into a series of successive infinitesimal parallel transports of equal amount. From (5.50) we recognize that at every order of the parallel transport, the components of the transported vector will be related as $(n < N)$

$$\begin{pmatrix} A^\theta_{n+1} \\ A^\phi_{n+1} \end{pmatrix} = \left[1 + \begin{pmatrix} 0 & \sin\theta_0\cos\theta_0 d\phi \\ -\cot\theta_0 d\phi & 0 \end{pmatrix} \right] \begin{pmatrix} A^\theta_n \\ A^\phi_n \end{pmatrix}. \tag{5.51}$$

We can write the components of the vector after N successive infinitesimal parallel transports compactly as

$$A^\alpha_N = \prod_{n=1}^{N} (1 + \Delta_n) A^\alpha_0, \tag{5.52}$$

where A^α_0 denotes the "initial" vector (5.30) (in column form) and we have identified

$$\Delta_n = \begin{pmatrix} 0 & \sin\theta_0\cos\theta_0 d\phi \\ -\cot\theta_0 d\phi & 0 \end{pmatrix} = \frac{\phi\cos\theta_0}{N} \begin{pmatrix} 0 & \sin\theta_0 \\ -\csc\theta_0 & 0 \end{pmatrix}$$

$$= \frac{\Delta}{N}, \tag{5.53}$$

where we have defined, for later use,

$$\Delta = \phi\cos\theta_0 \begin{pmatrix} 0 & \sin\theta_0 \\ -\csc\theta_0 & 0 \end{pmatrix}. \tag{5.54}$$

Let us recall here the important identity that

$$\lim_{N\to\infty} \prod_{n=1}^{N} (1 + \frac{\Delta}{N}) = e^\Delta, \tag{5.55}$$

so that in the limit $N \to \infty$ the transformation (5.52) takes the form

$$A^\alpha(\phi) = e^\Delta A_0^\alpha. \tag{5.56}$$

Since the matrix in (5.54) squares to $\Delta^2 = -(\phi \cos\theta_0)^2 \mathbb{1}$ we have the general identities

$$\Delta^{2n} = (-1)^n (\phi \cos\theta_0)^{2n} \mathbb{1}, \quad \Delta^{2n+1} = (-1)^n (\phi \cos\theta_0)^{2n} \Delta, \tag{5.57}$$

which can be used to simplify the exponent in (5.56)

$$e^\Delta = \cos(\phi \cos\theta_0) \mathbb{1} + \frac{1}{\phi \cos\theta_0} \sin(\phi \cos\theta_0) \Delta. \tag{5.58}$$

Using (5.54) as well as (5.58) in (5.56), we finally have

$$\begin{pmatrix} A^\theta(\theta_0, \phi) \\ A^\phi(\theta_0, \phi) \end{pmatrix} = \begin{pmatrix} \cos(\phi \cos\theta_0) & \sin\theta_0 \sin(\phi \cos\theta_0) \\ [6pt] -\csc\theta_0 \sin(\phi \cos\theta_0) & \cos(\phi \cos\theta_0) \end{pmatrix} \begin{pmatrix} 1 \\ 0 \end{pmatrix}$$

$$= \begin{pmatrix} \cos(\phi \cos\theta_0) \\ [6pt] -\csc\theta_0 \ \sin(\phi \cos\theta_0) \end{pmatrix}, \tag{5.59}$$

which coincides with (5.45).

◀

To calculate the change in a vector (in general) when parallel transported around a closed curve, we note that any finite path can be decomposed into many infinitesimal closed loops. And hence the problem of studying the change in a vector parallel transported around a finite loop is equivalent to studying the change in going around an infinitesimal loop as shown in Fig. 5.4. If we use the rule for parallel displacement (see (4.12))

$$d\xi^\mu = \Gamma^\mu_{\nu\lambda} dx^\nu \xi^\lambda,$$

$$\text{or,} \quad \frac{d\xi^\mu}{d\tau} = \Gamma^\mu_{\nu\lambda} \frac{dx^\nu}{d\tau} \xi^\lambda, \tag{5.60}$$

then, the change in the vector in going around a closed infinitesimal loop can be written as

$$\Delta\xi^\mu = \oint d\xi^\mu = \oint d\tau \frac{d\xi^\mu}{d\tau} = \oint d\tau \, \Gamma^\mu_{\nu\lambda}(x) \frac{dx^\nu}{d\tau} \xi^\lambda(x). \tag{5.61}$$

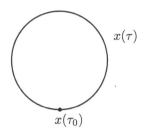

Figure 5.4: Parallel transport of a vector along an infinitesimal closed path.

Here we are assuming that the curve is parameterized by the proper time τ. Furthermore, the evaluation of the integral can be simplified by expanding quantities inside the integral around a fixed point on the curve, say, $x(\tau_0) = x_0$. Since the loop is infinitesimal, we have

$$\Gamma^\mu_{\nu\lambda}(x) = \Gamma^\mu_{\nu\lambda}(x_0) + (x - x_0)^\rho \partial_\rho \Gamma^\mu_{\nu\lambda}(x_0) + O\left((x - x_0)^2\right), \quad (5.62)$$

and similarly,

$$\begin{aligned}
\xi^\mu(x) &\simeq \xi^\mu(x_0) + \delta\xi^\mu(x_0) \\
&= \xi^\mu(x_0) + \Gamma^\mu_{\nu\lambda}(x_0)(x - x_0)^\nu \xi^\lambda(x_0). \quad (5.63)
\end{aligned}$$

Using these relations and keeping terms only up to linear power in $(x - x_0)$, the expression for the change in the vector in (5.61) becomes

$$\begin{aligned}
\Delta\xi^\mu &= \oint d\tau \left[\Gamma^\mu_{\nu\lambda}(x_0) + (x - x_0)^\gamma \partial_\gamma \Gamma^\mu_{\nu\lambda}(x_0) + \cdots\right] \\
&\quad \times \left[\xi^\lambda(x_0) + \Gamma^\lambda_{\rho\sigma}(x - x_0)^\rho \xi^\sigma(x_0)\right] \frac{dx^\nu}{d\tau} \\
&= \oint d\tau \left[\Gamma^\mu_{\nu\lambda}(x_0)\xi^\lambda(x_0)\frac{dx^\nu}{d\tau} + (x - x_0)^\rho \xi^\sigma(x_0)\left\{\partial_\rho \Gamma^\mu_{\nu\sigma}(x_0)\right.\right. \\
&\quad \left.\left. + \Gamma^\mu_{\nu\lambda}(x_0)\Gamma^\lambda_{\rho\sigma}(x_0)\right\}\frac{dx^\nu}{d\tau}\right]
\end{aligned}$$

$$= \Gamma^\mu_{\nu\lambda}(x_0)\xi^\lambda(x_0) \oint d\tau \, \frac{dx^\nu}{d\tau}$$

$$+\xi^\sigma(x_0)\left\{\partial_\rho\Gamma^\mu_{\nu\sigma}(x_0) + \Gamma^\mu_{\nu\lambda}(x_0)\Gamma^\lambda_{\rho\sigma}(x_0)\right\}$$

$$\times \oint d\tau (x-x_0)^\rho \frac{dx^\nu}{d\tau}.$$

$$(5.64)$$

Let us note that for a closed curve

$$\oint d\tau \, \frac{dx^\nu}{d\tau} = \oint dx^\nu = 0, \tag{5.65}$$

since the coordinate comes back to the same point. The expression for the change in the vector in (5.64), therefore, simplifies to

$$\Delta\xi^\mu = \xi^\sigma(x_0)\left\{\partial_\rho\Gamma^\mu_{\nu\sigma}(x_0) + \Gamma^\mu_{\nu\lambda}(x_0)\Gamma^\lambda_{\rho\sigma}(x_0)\right\} \oint d\tau \, x^\rho \frac{dx^\nu}{d\tau}. \tag{5.66}$$

Furthermore, let us note that

$$\oint d\tau \, x^\rho \frac{dx^\nu}{d\tau} = \oint d\tau \left[\frac{d}{d\tau}(x^\rho x^\nu) - \frac{dx^\rho}{d\tau} x^\nu\right]$$

$$= -\oint d\tau \, x^\nu \frac{dx^\rho}{d\tau}. \tag{5.67}$$

Namely, the integral is anti-symmetric in the indices ρ and ν so that we can manifestly antisymmetrize the coefficient of the integral in the indices ρ, ν and write

$$\Delta\xi^\mu = \frac{1}{2}\,\xi^\sigma(x_0)\left\{\partial_\rho\Gamma^\mu_{\nu\sigma}(x_0) - \partial_\nu\Gamma^\mu_{\rho\sigma}(x_0)\right.$$

$$\left. +\Gamma^\mu_{\nu\lambda}(x_0)\Gamma^\lambda_{\rho\sigma}(x_0) - \Gamma^\mu_{\rho\lambda}(x_0)\Gamma^\lambda_{\nu\sigma}(x_0)\right\} \oint d\tau \, x^\rho \frac{dx^\nu}{d\tau}$$

$$= \frac{1}{2}\,\xi^\sigma(x_0)R^\mu{}_{\sigma\nu\rho}(x_0) \oint d\tau \, x^\rho \frac{dx^\nu}{d\tau}, \tag{5.68}$$

where $R^{\mu}{}_{\sigma\nu\rho}$ is the Riemann-Christoffel curvature tensor defined earlier in (5.18).

The remaining integral is easily recognized as the area of the infinitesimal loop. This can be seen simply by recalling that in three dimensional Euclidean space the area enclosed by a curve is given by

$$\mathbf{A} \;=\; -\frac{1}{2}\oint d\mathbf{r}\times\mathbf{r},$$

$$\text{or,}\quad A_i \;=\; -\frac{1}{2}\oint \epsilon_{ijk}dx_j x_k. \tag{5.69}$$

Thus, we note from (5.68) that the change in a vector parallel transported along a closed infinitesimal curve at any point is proportional to the curvature tensor at that point as well as the area enclosed by the curve. Thus we see that the vector would not change when parallel transported along an infinitesimal closed curve around a point only if the curvature tensor vanishes at that point. Furthermore since a finite closed curve can be thought of as a sum of many infinitesimal closed curves, when parallel transported along a finite closed curve, a vector would not change only if the curvature tensor vanishes identically at every point in that region.

If the curvature vanishes identically in any finite region of spacetime, then, of course, any vector can be parallel transported along any closed curve in that region without any change. This simply means that in that region parallel transporting a vector from point A to B (see Fig. 5.3) is independent of the path along which the vector is parallel transported. Thus in such a case, given a vector at a point, namely, $\xi^{\mu}(x_0)$, we can obtain its value at any other point $\xi^{\mu}(x)$ in the region uniquely simply because parallel transport does not depend on the path. Furthermore, in such a case if we choose a curve $x^{\mu}(\tau)$, then along this curve

$$\frac{d\xi^{\mu}}{d\tau} = \frac{dx^{\nu}}{d\tau}\,\frac{\partial\xi^{\mu}}{\partial x^{\nu}}. \tag{5.70}$$

On the other hand, the law of parallel transport gives

$$\frac{d\xi^{\mu}}{d\tau} = \Gamma^{\mu}{}_{\nu\lambda}\,\frac{dx^{\nu}}{d\tau}\,\xi^{\lambda}, \tag{5.71}$$

so that combining the two relations we obtain

$$\frac{dx^\nu}{d\tau} \partial_\nu \xi^\mu = \Gamma^\mu_{\nu\lambda} \frac{dx^\nu}{d\tau} \xi^\lambda,$$

$$\text{or,} \quad \frac{dx^\nu}{d\tau} \left(\partial_\nu \xi^\mu - \Gamma^\mu_{\nu\lambda} \xi^\lambda \right) = \frac{dx^\nu}{d\tau} D_\nu \xi^\mu = 0. \tag{5.72}$$

Since this is true for any $x^\mu(\tau)$ we conclude that for this to be true the covariant derivative of the vector must vanish in that region, namely,

$$D_\nu \xi^\mu = 0. \tag{5.73}$$

Conversely, if we can find a vector field whose covariant derivative vanishes in some region of space-time, the curvature must also vanish in that region. This follows because if

$$D_\nu \xi^\mu = 0, \tag{5.74}$$

then the vector can be parallel transported along an arbitrary closed curve without any change. This implies

$$\Delta \xi^\mu = \frac{1}{2} R^\mu_{\sigma\nu\rho} \xi^\sigma \oint d\tau \, x^\rho \frac{dx^\nu}{d\tau} = 0,$$

$$\text{or,} \quad R^\mu_{\sigma\nu\rho} \xi^\sigma = 0,$$

$$\text{or,} \quad R^\mu_{\sigma\nu\rho} = 0. \tag{5.75}$$

This can also be obtained from the fact that if the covariant derivative of a vector vanishes, then so will the commutator of two covariant derivatives and, therefore, the curvature.

5.4 Geodesic equation

There are several ways to derive the geodesic equations in a curved manifold. Let us first derive it from the requirement of general covariance.

Let us ask how we can generalize the equations of motion for a particle in flat space to a curved manifold (in the presence of gravitation). The simplest equation to consider is, of course, the equation of motion of a free particle. In the locally flat Cartesian coordinate system we know that the equation is given by

$$\frac{\mathrm{d}^2 x^\mu}{\mathrm{d}\tau^2} = 0, \tag{5.76}$$

where $\mathrm{d}\tau^2 = \eta_{\mu\nu}\mathrm{d}x^\mu\mathrm{d}x^\nu$ is the proper time. If we now go to a general coordinate frame defined by

$$x^\mu(\tau) \to x'^\mu(x(\tau)), \tag{5.77}$$

then

$$\frac{\mathrm{d}x^\mu}{\mathrm{d}\tau} \to \frac{\mathrm{d}x'^\mu}{\mathrm{d}\tau} = \frac{\partial x'^\mu}{\partial x^\nu}\frac{\mathrm{d}x^\nu}{\mathrm{d}\tau}, \tag{5.78}$$

where it is understood that $\frac{\partial x'^\mu}{\partial x^\nu}$ is evaluated at $x(\tau)$. It now follows that

$$
\begin{aligned}
\frac{\mathrm{d}^2 x^\mu}{\mathrm{d}\tau^2} \;\to\; & \frac{\mathrm{d}}{\mathrm{d}\tau}\left(\frac{\mathrm{d}x'^\mu}{\mathrm{d}\tau}\right) \\
= \; & \frac{\mathrm{d}}{\mathrm{d}\tau}\left(\frac{\partial x'^\mu}{\partial x^\nu}\frac{\mathrm{d}x^\nu}{\mathrm{d}\tau}\right) \\
= \; & \frac{\partial x'^\mu}{\partial x^\nu}\frac{\mathrm{d}^2 x^\nu}{\mathrm{d}\tau^2} + \frac{\partial^2 x'^\mu}{\partial x^\lambda \partial x^\nu}\frac{\mathrm{d}x^\lambda}{\mathrm{d}\tau}\frac{\mathrm{d}x^\nu}{\mathrm{d}\tau}.
\end{aligned}
\tag{5.79}
$$

In other words, the naive generalization of the free particle equation (5.76) to a curved space does not transform covariantly under a general coordinate transformation and, consequently, it cannot represent the true equation of motion if general covariance is to hold. On the other hand, we know that the affine connection changes under a general coordinate transformation as (see (4.22))

$$\Gamma^{\mu}_{\nu\lambda} \;\to\; \Gamma'^{\mu}_{\nu\lambda}$$

$$= \frac{\partial x'^{\mu}}{\partial x^{\mu_1}} \frac{\partial x^{\nu_1}}{\partial x'^{\nu}} \frac{\partial x^{\lambda_1}}{\partial x'^{\lambda}} \, \Gamma^{\mu_1}_{\nu_1\lambda_1} + \frac{\partial^2 x'^{\mu}}{\partial x^{\nu_1} \partial x^{\lambda_1}} \frac{\partial x^{\nu_1}}{\partial x'^{\nu}} \frac{\partial x^{\lambda_1}}{\partial x'^{\lambda}}, \quad (5.80)$$

so that

$$\Gamma^{\mu}_{\nu\lambda} \frac{\mathrm{d}x^{\nu}}{\mathrm{d}\tau} \frac{\mathrm{d}x^{\lambda}}{\mathrm{d}\tau} \to \Gamma'^{\mu}_{\nu\lambda} \frac{\mathrm{d}x'^{\nu}}{\mathrm{d}\tau} \frac{\mathrm{d}x'^{\lambda}}{\mathrm{d}\tau}$$

$$= \left[\frac{\partial x'^{\mu}}{\partial x^{\mu_1}} \frac{\partial x^{\nu_1}}{\partial x'^{\nu}} \frac{\partial x^{\lambda_1}}{\partial x'^{\lambda}} \, \Gamma^{\mu_1}_{\nu_1\lambda_1} + \frac{\partial^2 x'^{\mu}}{\partial x^{\nu_1} \partial x^{\lambda_1}} \frac{\partial x^{\nu_1}}{\partial x'^{\nu}} \frac{\partial x^{\lambda_1}}{\partial x'^{\lambda}} \right]$$

$$\times \frac{\partial x'^{\nu}}{\partial x^{\nu_2}} \frac{\mathrm{d}x^{\nu_2}}{\mathrm{d}\tau} \frac{\partial x'^{\lambda}}{\partial x^{\lambda_2}} \frac{\mathrm{d}x^{\lambda_2}}{\mathrm{d}\tau}$$

$$= \delta^{\nu_1}_{\nu_2} \delta^{\lambda_1}_{\lambda_2} \left[\frac{\partial x'^{\mu}}{\partial x^{\mu_1}} \, \Gamma^{\mu_1}_{\nu_1\lambda_1} + \frac{\partial^2 x'^{\mu}}{\partial x^{\nu_1} \partial x^{\lambda_1}} \right] \frac{\mathrm{d}x^{\nu_2}}{\mathrm{d}\tau} \frac{\mathrm{d}x^{\lambda_2}}{\mathrm{d}\tau}$$

$$= \left[\frac{\partial x'^{\mu}}{\partial x^{\mu_1}} \, \Gamma^{\mu_1}_{\nu_1\lambda_1} + \frac{\partial^2 x'^{\mu}}{\partial x^{\nu_1} \partial x^{\lambda_1}} \right] \frac{\mathrm{d}x^{\nu_1}}{\mathrm{d}\tau} \frac{\mathrm{d}x^{\lambda_1}}{\mathrm{d}\tau}. \quad (5.81)$$

As a result, we see that the combination

$$\frac{\mathrm{d}^2 x^{\mu}}{\mathrm{d}\tau^2} - \Gamma^{\mu}_{\nu\lambda} \frac{\mathrm{d}x^{\nu}}{\mathrm{d}\tau} \frac{\mathrm{d}x^{\lambda}}{\mathrm{d}\tau} \to \frac{\mathrm{d}^2 x'^{\mu}}{\mathrm{d}\tau^2} - \Gamma'^{\mu}_{\nu\lambda} \frac{\mathrm{d}x'^{\nu}}{\mathrm{d}\tau} \frac{\mathrm{d}x'^{\lambda}}{\mathrm{d}\tau}$$

$$= \frac{\partial x'^{\mu}}{\partial x^{\sigma}} \left[\frac{\mathrm{d}^2 x^{\sigma}}{\mathrm{d}\tau^2} - \Gamma^{\sigma}_{\nu\lambda} \frac{\mathrm{d}x^{\nu}}{\mathrm{d}\tau} \frac{\mathrm{d}x^{\lambda}}{\mathrm{d}\tau} \right], \quad (5.82)$$

transforms like a contravariant vector under a general coordinate transformation. In a locally flat Cartesian coordinate system, the connection vanishes and hence this simply reduces to $\frac{\mathrm{d}^2 x^{\mu}}{\mathrm{d}\tau^2}$. Therefore, from considerations of general covariance alone we conclude that the equation of motion for a free particle in a gravitational field (curved manifold) is given by

$$\frac{\mathrm{d}^2 x^{\mu}}{\mathrm{d}\tau^2} - \Gamma^{\mu}_{\nu\lambda} \frac{\mathrm{d}x^{\nu}}{\mathrm{d}\tau} \frac{\mathrm{d}x^{\lambda}}{\mathrm{d}\tau} = 0. \quad (5.83)$$

This equation is known as the geodesic equation and determines the trajectory of a free particle in a curved manifold. We recall here that the four velocity of a particle is defined as (see (2.5))

$$u^\mu = \frac{dx^\mu}{d\tau}, \tag{5.84}$$

and behaves like a contravariant vector under a general coordinate transformation (see (5.78)). The free particle equation (5.76) can be written in flat space in terms of the proper velocity u^μ as

$$\frac{du^\mu}{d\tau} = 0. \tag{5.85}$$

As we have seen in (5.79), this is not covariant in a curved space and we recall that covariantizing this equation simply would correspond to replacing $\frac{d}{d\tau}$ by the appropriate covariant derivative (see (5.5)). Thus the covariant equation for a free particle in a curved manifold is given by

$$\frac{Du^\mu}{D\tau} = 0,$$

$$\text{or,} \quad \frac{du^\mu}{d\tau} - \Gamma^\mu_{\nu\lambda} \frac{dx^\nu}{d\tau} u^\lambda = 0. \tag{5.86}$$

If we now substitute the form of u^μ in (5.84) into (5.86), we obtain the geodesic equation (5.83) or the equation for the free particle as

$$\frac{d^2 x^\mu}{d\tau^2} - \Gamma^\mu_{\nu\lambda} \frac{dx^\nu}{d\tau} \frac{dx^\lambda}{d\tau} = 0. \tag{5.87}$$

▶ **Example (Affine parameter).** We note that the geodesic equation (5.83) describes the equation of motion for a free particle in a curved manifold

$$\frac{d^2 x^\mu}{d\tau^2} - \Gamma^\mu_{\nu\lambda} \frac{dx^\nu}{d\tau} \frac{dx^\lambda}{d\tau} = 0, \tag{5.88}$$

where τ is any parameter labeling the trajectory which we can choose to be the proper time as well.

Let us choose a new parameter

$$s = s(\tau), \tag{5.89}$$

to parameterize the trajectory. The geodesic equation (5.88), in this new variable, takes the form

$$\frac{\mathrm{d}}{\mathrm{d}\tau}\left(\frac{\mathrm{d}s}{\mathrm{d}\tau}\frac{\mathrm{d}x^{\mu}}{\mathrm{d}s}\right) - \Gamma^{\mu}_{\nu\lambda}\left(\frac{\mathrm{d}s}{\mathrm{d}\tau}\frac{\mathrm{d}x^{\nu}}{\mathrm{d}s}\right)\left(\frac{\mathrm{d}s}{\mathrm{d}\tau}\frac{\mathrm{d}x^{\lambda}}{\mathrm{d}s}\right) = 0,$$

$$\text{or,} \quad \frac{\mathrm{d}s^2}{\mathrm{d}\tau^2}\frac{\mathrm{d}x^{\mu}}{\mathrm{d}s} + \left(\frac{\mathrm{d}s}{\mathrm{d}\tau}\right)^2\frac{\mathrm{d}^2x^{\mu}}{\mathrm{d}s^2} - \left(\frac{\mathrm{d}s}{\mathrm{d}\tau}\right)^2\Gamma^{\mu}_{\nu\lambda}\frac{\mathrm{d}x^{\nu}}{\mathrm{d}s}\frac{\mathrm{d}x^{\lambda}}{\mathrm{d}s} = 0,$$

$$\text{or,} \quad \frac{\mathrm{d}^2x^{\mu}}{\mathrm{d}s^2} - \Gamma^{\mu}_{\nu\lambda}\frac{\mathrm{d}x^{\nu}}{\mathrm{d}s}\frac{\mathrm{d}x^{\lambda}}{\mathrm{d}s} = -\frac{\frac{\mathrm{d}^2s}{\mathrm{d}\tau^2}}{\left(\frac{\mathrm{d}s}{\mathrm{d}\tau}\right)^2}\frac{\mathrm{d}x^{\mu}}{\mathrm{d}s}. \tag{5.90}$$

Thus, the form of the geodesic equation (5.88) will remain invariant in the new parameterization if

$$\frac{\mathrm{d}^2s}{\mathrm{d}\tau^2} = 0,$$

$$\text{or,} \quad s(\tau) = \alpha\tau + \beta, \tag{5.91}$$

where α and β are constants. This is a linear transformation relating the two parameters – also known as an affine transformation. Parameters which leave the form of the geodesic equation (5.83) or (5.88) invariant are also known as affine parameters (related through affine transformations).

◀

The geodesic equation can alternatively be derived as the straightest path in a curved manifold as follows. From the study of parallel transport we have seen that under parallel transport the change in a vector is given by (see (4.12))

$$\frac{\mathrm{d}\xi^{\mu}}{\mathrm{d}\tau} = \Gamma^{\mu}_{\nu\lambda}\frac{\mathrm{d}x^{\nu}}{\mathrm{d}\tau}\xi^{\lambda}. \tag{5.92}$$

Thus any vector field which satisfies the differential equation

$$\frac{\mathrm{d}\xi^{\mu}}{\mathrm{d}\tau} = \Gamma^{\mu}_{\nu\lambda}\frac{\mathrm{d}x^{\nu}}{\mathrm{d}\tau}\xi^{\lambda},$$

$$\text{or,} \quad \frac{\mathrm{d}\xi^{\mu}}{\mathrm{d}\tau} - \Gamma^{\mu}_{\nu\lambda}\frac{\mathrm{d}x^{\nu}}{\mathrm{d}\tau}\xi^{\lambda} = 0, \tag{5.93}$$

is parallel to itself along a curve $x^\mu(\tau)$. We recall from the study of Euclidean geometry that a straight line carries the tangent along itself. So we can generalize this definition to a curved manifold by defining a "straight line" in a curved manifold as one which carries its tangent vector parallel to itself along the curve. Defining ξ^μ to be the tangent vector to the curve we have (this is the proper velocity)

$$\xi^\mu = \frac{dx^\mu}{d\tau}. \tag{5.94}$$

Thus the tangent vector would be carried parallel to itself only if

$$\frac{d\xi^\mu}{d\tau} - \Gamma^\mu_{\nu\lambda} \frac{dx^\nu}{d\tau} \xi^\lambda = 0,$$

$$\text{or,} \quad \frac{d^2x^\mu}{d\tau^2} - \Gamma^\mu_{\nu\lambda} \frac{dx^\nu}{d\tau} \frac{dx^\lambda}{d\tau} = 0. \tag{5.95}$$

This is precisely the geodesic equation (5.83) and we conclude that the geodesic curve carries its tangent parallel to itself and hence is the straightest curve in a curved manifold.

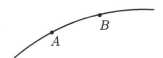

Figure 5.5: Motion from A to B along a given trajectory.

▶ **Example (Solution of the geodesic equation).** Let us consider the geodesic equation in the plane in polar coordinates where the line element is given by

$$d\tau^2 = dr^2 + r^2 d\theta^2. \tag{5.96}$$

This is the same space as in (4.51) and, therefore, we can carry over the metric and the nontrivial components of the connection from (4.52) and (4.55) which have the forms

$$g_{rr} = 1, \quad g_{\theta\theta} = r^2, \quad g^{rr} = 1, \quad g^{\theta\theta} = \frac{1}{r^2},$$

$$\Gamma^r_{\theta\theta} = r, \quad \Gamma^\theta_{r\theta} = \Gamma^\theta_{\theta r} = -\frac{1}{r}. \tag{5.97}$$

The geodesic equations (5.83) are given by

$$\frac{\mathrm{d}^2 x^\mu}{\mathrm{d}\tau^2} - \Gamma^\mu_{\nu\lambda} \frac{\mathrm{d}x^\nu}{\mathrm{d}\tau} \frac{\mathrm{d}x^\lambda}{\mathrm{d}\tau} = 0, \quad \mu.\nu, \lambda = r, \theta. \tag{5.98}$$

In the present case, there are two equations. For $\mu = r$, we have

$$\frac{\mathrm{d}^2 r}{\mathrm{d}\tau^2} - \Gamma^r_{\theta\theta} \frac{\mathrm{d}\theta}{\mathrm{d}\tau} \frac{\mathrm{d}\theta}{\mathrm{d}\tau} = 0,$$

$$\text{or,} \quad \frac{\mathrm{d}^2 r}{\mathrm{d}\tau^2} - r \left(\frac{\mathrm{d}\theta}{\mathrm{d}\tau} \right)^2 = 0, \tag{5.99}$$

while for $\mu = \theta$, eq. (5.98) leads to

$$\frac{\mathrm{d}^2 \theta}{\mathrm{d}\tau^2} - 2\Gamma^\theta_{\theta r} \frac{\mathrm{d}\theta}{\mathrm{d}\tau} \frac{\mathrm{d}r}{\mathrm{d}\tau} = 0,$$

$$\text{or,} \quad \frac{\mathrm{d}^2 \theta}{\mathrm{d}\tau^2} + \frac{2}{r} \frac{\mathrm{d}\theta}{\mathrm{d}\tau} \frac{\mathrm{d}r}{\mathrm{d}\tau} = 0. \tag{5.100}$$

Equation (5.100) can be integrated as follows

$$\frac{\mathrm{d}^2 \theta}{\mathrm{d}\tau^2} + \frac{2}{r} \frac{\mathrm{d}\theta}{\mathrm{d}\tau} \frac{\mathrm{d}r}{\mathrm{d}\tau} = 0,$$

$$\text{or,} \quad \frac{1}{\frac{\mathrm{d}\theta}{\mathrm{d}\tau}} \frac{\mathrm{d}^2 \theta}{\mathrm{d}\tau^2} + \frac{2}{r} \frac{\mathrm{d}r}{\mathrm{d}\tau} = 0,$$

$$\text{or,} \quad \frac{\mathrm{d}}{\mathrm{d}\tau} \ln \left(r^2 \frac{\mathrm{d}\theta}{\mathrm{d}\tau} \right) = 0,$$

$$\text{or,} \quad r^2 \frac{\mathrm{d}\theta}{\mathrm{d}\tau} = \ell = \text{constant}, \tag{5.101}$$

where we recognize ℓ to correspond to the orbital angular momentum of the particle. Note that (5.100) and (5.101) lead to

$$\frac{\mathrm{d}}{\mathrm{d}\tau} \left(r^2 \left(\frac{\mathrm{d}\theta}{\mathrm{d}\tau} \right)^2 \right) = 2r^2 \frac{\mathrm{d}\theta}{\mathrm{d}\tau} \frac{\mathrm{d}^2 \theta}{\mathrm{d}\tau^2} + 2r \frac{\mathrm{d}r}{\mathrm{d}\tau} \left(\frac{\mathrm{d}\theta}{\mathrm{d}\tau} \right)^2$$

$$= -4r \frac{\mathrm{d}r}{\mathrm{d}\tau} \left(\frac{\mathrm{d}\theta}{\mathrm{d}\tau} \right)^2 + 2r \frac{\mathrm{d}r}{\mathrm{d}\tau} \left(\frac{\mathrm{d}\theta}{\mathrm{d}\tau} \right)^2$$

$$= -2r \frac{\mathrm{d}r}{\mathrm{d}\tau} \left(\frac{\mathrm{d}\theta}{\mathrm{d}\tau} \right)^2. \tag{5.102}$$

Using (5.102) in (5.99) we obtain

$$\frac{d^2 r}{d\tau^2} - r \left(\frac{d\theta}{d\tau} \right)^2 = 0,$$

$$\text{or,} \quad 2\frac{dr}{d\tau}\frac{d^2 r}{d\tau^2} - 2r\frac{dr}{d\tau} \left(\frac{d\theta}{d\tau} \right)^2 = 0,$$

$$\text{or,} \quad \frac{d}{d\tau} \left(\frac{dr}{d\tau} \right)^2 + \frac{d}{d\tau} \left(r^2 \left(\frac{d\theta}{d\tau} \right)^2 \right) = 0,$$

$$\text{or,} \quad \left(\frac{dr}{d\tau} \right)^2 + r^2 \left(\frac{d\theta}{d\tau} \right)^2 = 1. \tag{5.103}$$

Here the constant of integration is chosen to be compatible with the line element (5.96).

Since $r = r(\tau), \theta = \theta(\tau)$, we can also write r in the parametric form

$$r = r(\theta), \quad \frac{dr}{d\tau} = \frac{dr}{d\theta}\frac{d\theta}{d\tau},$$

$$\frac{d^2 r}{d\tau^2} = \left(\frac{d^2 r}{d\theta^2} - \frac{2}{r} \left(\frac{dr}{d\theta} \right)^2 \right) \left(\frac{d\theta}{d\tau} \right)^2, \tag{5.104}$$

where we have used (5.100). Using this in (5.99) we obtain

$$\left(\frac{d^2 r}{d\theta^2} - \frac{2}{r} \left(\frac{dr}{d\theta} \right)^2 - r \right) \left(\frac{d\theta}{d\tau} \right)^2 = 0,$$

$$\text{or,} \quad \frac{d^2 r}{d\theta^2} - \frac{2}{r} \left(\frac{dr}{d\theta} \right)^2 - r = 0. \tag{5.105}$$

The geodesic equation (5.105) describes a free particle motion and, therefore, in Euclidean coordinates we expect the general solution to be of the form

$$y = ax + b, \tag{5.106}$$

where a, b are constants depending on the initial conditions. In polar coordinates, (5.106) can be expressed as

$$r(\theta) = \frac{b}{\sin\theta - a\cos\theta}, \tag{5.107}$$

where we have used $x = r\cos\theta, y = r\sin\theta$. Indeed it can be checked easily that

$$\frac{dr}{d\theta} = -\frac{b(\cos\theta + a\sin\theta)}{(\sin\theta - a\cos\theta)^2},$$

$$\frac{d^2 r}{d\theta^2} = \frac{b}{\sin\theta - a\cos\theta} + \frac{2b(\cos\theta + a\sin\theta)^2}{(\sin\theta - a\cos\theta)^3}$$

$$= r + \frac{2}{r}\left(\frac{dr}{d\theta}\right)^2, \tag{5.108}$$

so that (5.108) gives the general solution of the geodesic equation (5.105). This shows that the geodesic describes the straightest path in the manifold.

◄

The geodesic equation can also be derived as the equation for the shortest (extremal) path between two points in a curved manifold as follows. Let τ_{AB} be the (proper) time taken by a particle in going from point A to B in a gravitational field (see Fig. 5.5), namely,

$$\tau_{AB} = \int_A^B d\tau, \tag{5.109}$$

where $(c = 1)$

$$ds^2 = d\tau^2 = g_{\mu\nu}dx^\mu dx^\nu. \tag{5.110}$$

Let us assume that the curve x^μ along which the particle moves is parameterized by an invariant parameter λ. Thus we can write

$$d\tau = \left(g_{\mu\nu}\frac{dx^\mu}{d\lambda}\frac{dx^\nu}{d\lambda}\right)^{\frac{1}{2}}d\lambda, \tag{5.111}$$

so that

$$\tau_{AB} = \int_A^B d\tau = \int_A^B d\lambda \left(g_{\mu\nu}\frac{dx^\mu}{d\lambda}\frac{dx^\nu}{d\lambda}\right)^{\frac{1}{2}}. \tag{5.112}$$

This is like an action and hence we can derive the equation of for the shortest (extremal) path simply by extremizing the action. Let us change the path infinitesimally,

$$x^\mu(\lambda) \to x^\mu(\lambda) + \delta x^\mu(\lambda), \tag{5.113}$$

subject to the condition

$$\delta x^\mu(A) = \delta x^\mu(B) = 0, \tag{5.114}$$

so that the end points are held fixed during the variation. Furthermore, let us note that the present action (5.112) has a one parameter group of gauge invariance. Namely, under the reparameterization

$$\lambda \to t = t(\lambda), \tag{5.115}$$

we note that

$$\frac{dx^\mu}{d\lambda} \to \frac{dx^\mu}{dt} = \frac{dx^\mu}{d\lambda} \frac{d\lambda}{dt}, \tag{5.116}$$

so that we have

$$
\begin{aligned}
\tau_{AB} &= \int_A^B d\lambda \left(g_{\mu\nu} \frac{dx^\mu}{d\lambda} \frac{dx^\nu}{d\lambda} \right)^{\frac{1}{2}} \\
&\to \int_A^B dt \left(g_{\mu\nu} \frac{dx^\mu}{dt} \frac{dx^\nu}{dt} \right)^{\frac{1}{2}} \\
&= \int_A^B dt \left(g_{\mu\nu} \frac{dx^\mu}{d\lambda} \frac{d\lambda}{dt} \frac{dx^\nu}{d\lambda} \frac{d\lambda}{dt} \right)^{\frac{1}{2}} \\
&= \int_A^B dt \frac{d\lambda}{dt} \left(g_{\mu\nu} \frac{dx^\mu}{d\lambda} \frac{dx^\nu}{d\lambda} \right)^{\frac{1}{2}} \\
&= \int_A^B d\lambda \left(g_{\mu\nu} \frac{dx^\mu}{d\lambda} \frac{dx^\nu}{d\lambda} \right)^{\frac{1}{2}} = \tau_{AB}.
\end{aligned}
\tag{5.117}
$$

Namely, the action does not change under the reparameterization (5.115) of the curve. This is reminiscent of the gauge invariance of Maxwell's theory. Thus we can choose a gauge condition and we choose

$$\lambda = \tau. \tag{5.118}$$

With this choice of the gauge condition as well as the condition of fixed end points in (5.114) we obtain from (5.112)

$$
\begin{aligned}
\delta \tau_{AB} &= \int_A^B d\lambda \; \frac{1}{2 \left(g_{\mu\nu} \frac{dx^\mu}{d\lambda} \frac{dx^\nu}{d\lambda} \right)^{\frac{1}{2}}} \\
&\qquad \times \left(\delta g_{\mu\nu} \frac{dx^\mu}{d\lambda} \frac{dx^\nu}{d\lambda} + 2 g_{\mu\nu} \frac{dx^\mu}{d\lambda} \frac{d\delta x^\nu}{d\lambda} \right) \\
&= \int_A^B d\lambda \; \frac{1}{\frac{d\tau}{d\lambda}} \left(g_{\mu\nu} \frac{dx^\mu}{d\lambda} \frac{d\delta x^\nu}{d\lambda} + \frac{1}{2} \partial_\sigma g_{\mu\nu} \delta x^\sigma \frac{dx^\mu}{d\lambda} \frac{dx^\nu}{d\lambda} \right) \\
&= \int_A^B d\tau \left(g_{\mu\nu} \frac{dx^\mu}{d\tau} \frac{d\delta x^\nu}{d\tau} + \frac{1}{2} \partial_\sigma g_{\mu\nu} \frac{dx^\mu}{d\tau} \frac{dx^\nu}{d\tau} \delta x^\sigma \right), \tag{5.119}
\end{aligned}
$$

where we have used (5.111) and (5.118). Integrating the first term by parts and remembering that the end points do not change under the variation (5.114), we obtain

$$
\begin{aligned}
\delta \tau_{AB} &= \left. g_{\mu\nu} \frac{dx^\mu}{d\tau} \delta x^\nu \right|_A^B + \int_A^B d\tau \left[-\delta x^\nu \frac{d}{d\tau} \left(g_{\mu\nu} \frac{dx^\mu}{d\tau} \right) \right. \\
&\qquad \left. + \frac{1}{2} \partial_\sigma g_{\mu\nu} \frac{dx^\mu}{d\tau} \frac{dx^\nu}{d\tau} \delta x^\sigma \right] \\
&= \int_A^B d\tau \left[-\delta x^\nu \left(\partial_\sigma g_{\mu\nu} \frac{dx^\sigma}{d\tau} \frac{dx^\mu}{d\tau} + g_{\mu\nu} \frac{d^2 x^\mu}{d\tau^2} \right) \right. \\
&\qquad \left. + \frac{1}{2} \partial_\sigma g_{\mu\nu} \frac{dx^\mu}{d\tau} \frac{dx^\nu}{d\tau} \delta x^\sigma \right]
\end{aligned}
$$

$$
\begin{aligned}
= & -\int_A^B \mathrm{d}\tau\, \delta x^\sigma \left[\partial_\nu g_{\mu\sigma}\, \frac{\mathrm{d}x^\nu}{\mathrm{d}\tau}\, \frac{\mathrm{d}x^\mu}{\mathrm{d}\tau} + g_{\mu\sigma}\, \frac{\mathrm{d}^2 x^\mu}{\mathrm{d}\tau^2} \right. \\
& \left. \qquad\qquad\qquad - \frac{1}{2}\, \partial_\sigma g_{\mu\nu}\, \frac{\mathrm{d}x^\mu}{\mathrm{d}\tau}\, \frac{\mathrm{d}x^\nu}{\mathrm{d}\tau} \right] \\[2mm]
= & -\int_A^B \mathrm{d}\tau\, \delta x^\sigma \left[g_{\mu\sigma}\, \frac{\mathrm{d}^2 x^\mu}{\mathrm{d}\tau^2} \right. \\
& \left. \qquad + \frac{1}{2}\left(\partial_\nu g_{\mu\sigma} + \partial_\mu g_{\nu\sigma} - \partial_\sigma g_{\mu\nu} \right) \frac{\mathrm{d}x^\mu}{\mathrm{d}\tau}\, \frac{\mathrm{d}x^\nu}{\mathrm{d}\tau} \right]. \quad (5.120)
\end{aligned}
$$

If the action has to be stationary along the actual path (geodesic) of the particle then, we must have

$$
\delta\tau_{AB} = 0, \tag{5.121}
$$

and since the variation δx^σ is arbitrary this is possible only if the integrand in (5.120) vanishes identically. Namely, the action will be stationary only along the trajectory given by the equation

$$
g_{\mu\sigma}\, \frac{\mathrm{d}^2 x^\mu}{\mathrm{d}\tau^2} + \frac{1}{2}\left(\partial_\mu g_{\nu\sigma} + \partial_\nu g_{\sigma\mu} - \partial_\sigma g_{\mu\nu} \right) \frac{\mathrm{d}x^\mu}{\mathrm{d}\tau}\, \frac{\mathrm{d}x^\nu}{\mathrm{d}\tau} = 0,
$$

$$
\text{or,}\quad \frac{\mathrm{d}^2 x^\lambda}{\mathrm{d}\tau^2} + \frac{1}{2}\, g^{\lambda\sigma}\left(\partial_\mu g_{\nu\sigma} + \partial_\nu g_{\sigma\mu} - \partial_\sigma g_{\mu\nu} \right) \frac{\mathrm{d}x^\mu}{\mathrm{d}\tau}\, \frac{\mathrm{d}x^\nu}{\mathrm{d}\tau} = 0,
$$

$$
\text{or,}\quad \frac{\mathrm{d}^2 x^\lambda}{\mathrm{d}\tau^2} - \Gamma^\lambda_{\mu\nu}\, \frac{\mathrm{d}x^\mu}{\mathrm{d}\tau}\, \frac{\mathrm{d}x^\nu}{\mathrm{d}\tau} = 0. \tag{5.122}
$$

This is, of course, the geodesic equation (5.83) which shows that the trajectory of a freely falling particle (under the influence of gravitation) is given by its geodesic which is the shortest (longest) path between two points. Furthermore, note that although the equation seems to be a vector equation, the particle in reality has only three independent degrees of freedom. The other degree of freedom is expressed in terms of the independent degrees through the gauge condition (5.118)

$$
\lambda = \tau, \tag{5.123}
$$

which leads to (see (5.111))

$$\frac{\mathrm{d}\tau}{\mathrm{d}\lambda} = \left(g_{\mu\nu}\frac{\mathrm{d}x^\mu}{\mathrm{d}\lambda}\frac{\mathrm{d}x^\nu}{\mathrm{d}\lambda}\right)^{\frac{1}{2}} = 1,$$

or, $g_{\mu\nu}u^\mu u^\nu = 1.$ (5.124)

We have already seen such a condition in the case of the Lorentz covariant description of particle motion (see (2.8)) and we conclude that the particle truly has only three independent degrees of freedom as it should.

5.5 Derivation of geodesic equation from a Lagrangian

The discussion of the last section suggests that the motion of a free particle in a curved manifold can be given a Lagrangian description much like in flat space. As we have seen in (5.112), the action for the free particle can be written as

$$\tau_{AB} = \int_A^B \mathrm{d}\lambda \left(g_{\mu\nu}\frac{\mathrm{d}x^\mu}{\mathrm{d}\lambda}\frac{\mathrm{d}x^\nu}{\mathrm{d}\lambda}\right)^{\frac{1}{2}},$$ (5.125)

where (see also (5.111))

$$\frac{\mathrm{d}\tau}{\mathrm{d}\lambda} = \left(g_{\mu\nu}\frac{\mathrm{d}x^\mu}{\mathrm{d}\lambda}\frac{\mathrm{d}x^\nu}{\mathrm{d}\lambda}\right)^{\frac{1}{2}} = (g_{\mu\nu}(x)\dot{x}^\mu\dot{x}^\nu)^{\frac{1}{2}},$$ (5.126)

where a dot denotes a derivative with respect to the parameter λ. Thus we can think of the Lagrangian for the system to be

$$L = \left(g_{\mu\nu}(x)\frac{\mathrm{d}x^\mu}{\mathrm{d}\lambda}\frac{\mathrm{d}x^\nu}{\mathrm{d}\lambda}\right)^{\frac{1}{2}} = (g_{\mu\nu}(x)\dot{x}^\mu\dot{x}^\nu)^{\frac{1}{2}},$$ (5.127)

which is a function of the coordinate $x^\mu(\lambda)$ and the velocity $\dot{x}^\mu(\lambda)$ with

$$\frac{\partial L}{\partial \dot{x}^\mu} = \frac{1}{2\left(g_{\alpha\beta}\dot{x}^\alpha\dot{x}^\beta\right)^{\frac{1}{2}}} 2g_{\mu\nu}\dot{x}^\nu = \frac{1}{\left(g_{\alpha\beta}\dot{x}^\alpha\dot{x}^\beta\right)^{\frac{1}{2}}} g_{\mu\nu}\dot{x}^\nu,$$

$$\text{or,} \quad \frac{\partial L}{\partial \dot{x}^\mu} = \frac{\mathrm{d}\lambda}{\mathrm{d}\tau} g_{\mu\nu} \frac{\mathrm{d}x^\nu}{\mathrm{d}\lambda} = g_{\mu\nu} \frac{\mathrm{d}x^\nu}{\mathrm{d}\tau}, \tag{5.128}$$

where we have used (5.111).
Similarly, we also have

$$\frac{\partial L}{\partial x^\mu} = \frac{1}{2\left(g_{\alpha\beta}\dot{x}^\alpha\dot{x}^\beta\right)^{\frac{1}{2}}} \partial_\mu g_{\sigma\nu}\dot{x}^\sigma\dot{x}^\nu = \frac{1}{2}\frac{\mathrm{d}\lambda}{\mathrm{d}\tau} \partial_\mu g_{\sigma\nu} \frac{\mathrm{d}x^\sigma}{\mathrm{d}\lambda} \frac{\mathrm{d}x^\nu}{\mathrm{d}\lambda}$$

$$= \frac{1}{2}\frac{\mathrm{d}\tau}{\mathrm{d}\lambda} \partial_\mu g_{\sigma\nu} \frac{\mathrm{d}x^\sigma}{\mathrm{d}\tau} \frac{\mathrm{d}x^\nu}{\mathrm{d}\tau}. \tag{5.129}$$

The Euler-Lagrange equation for the system can be written as

$$\frac{\mathrm{d}}{\mathrm{d}\lambda} \frac{\partial L}{\partial \dot{x}^\mu} - \frac{\partial L}{\partial x^\mu} = 0,$$

$$\text{or,} \quad \frac{\mathrm{d}\tau}{\mathrm{d}\lambda} \frac{\mathrm{d}}{\mathrm{d}\tau} \frac{\partial L}{\partial \dot{x}^\mu} - \frac{\partial L}{\partial x^\mu} = 0,$$

$$\text{or,} \quad \frac{\mathrm{d}\tau}{\mathrm{d}\lambda} \frac{\mathrm{d}}{\mathrm{d}\tau} \left(g_{\mu\nu} \frac{\mathrm{d}x^\nu}{\mathrm{d}\tau}\right) - \frac{1}{2}\frac{\mathrm{d}\tau}{\mathrm{d}\lambda} \partial_\mu g_{\sigma\nu} \frac{\mathrm{d}x^\sigma}{\mathrm{d}\tau} \frac{\mathrm{d}x^\nu}{\mathrm{d}\tau} = 0,$$

$$\text{or,} \quad \frac{\mathrm{d}\tau}{\mathrm{d}\lambda} \left[g_{\mu\nu} \frac{\mathrm{d}^2 x^\nu}{\mathrm{d}\tau^2} + \partial_\sigma g_{\mu\nu} \frac{\mathrm{d}x^\sigma}{\mathrm{d}\tau} \frac{\mathrm{d}x^\nu}{\mathrm{d}\tau} - \frac{1}{2}\partial_\mu g_{\sigma\nu} \frac{\mathrm{d}x^\sigma}{\mathrm{d}\tau} \frac{\mathrm{d}x^\nu}{\mathrm{d}\tau}\right] = 0,$$

$$\text{or,} \quad g_{\mu\nu} \frac{\mathrm{d}^2 x^\nu}{\mathrm{d}\tau^2} + \frac{1}{2}(\partial_\sigma g_{\mu\nu} + \partial_\nu g_{\mu\sigma} - \partial_\mu g_{\sigma\nu}) \frac{\mathrm{d}x^\sigma}{\mathrm{d}\tau} \frac{\mathrm{d}x^\nu}{\mathrm{d}\tau} = 0, \tag{5.130}$$

Multiplying with the inverse metric, this can be written as

$$\frac{\mathrm{d}^2 x^\lambda}{\mathrm{d}\tau^2} + \frac{1}{2} g^{\lambda\mu}(\partial_\sigma g_{\mu\nu} + \partial_\nu g_{\mu\sigma} - \partial_\mu g_{\sigma\nu}) \frac{\mathrm{d}x^\sigma}{\mathrm{d}\tau} \frac{\mathrm{d}x^\nu}{\mathrm{d}\tau} = 0,$$

$$\text{or,} \quad \frac{\mathrm{d}^2 x^\lambda}{\mathrm{d}\tau^2} - \Gamma^\lambda_{\sigma\nu} \frac{\mathrm{d}x^\sigma}{\mathrm{d}\tau} \frac{\mathrm{d}x^\nu}{\mathrm{d}\tau} = 0, \tag{5.131}$$

where we have used the identification (4.49) for the Christoffel symbol. This is, of course, the geodesic equation (5.83) that we have derived earlier from various points of view. As we will see in the next chapter, the Lagrangian description gives a simpler method for determining the components of the connection.

▶ **Example (Alternative action for the geodesic).** For a massive particle, the action that leads to the geodesic equations can be written as

$$
S = m \int_{\tau_1}^{\tau_2} d\tau \left(g_{\mu\nu} \frac{dx^\mu}{d\tau} \frac{dx^\nu}{d\tau} \right)^{\frac{1}{2}},
\tag{5.132}
$$

which coincides with (5.112) except for the multiplicative factor m (mass of the particle) which makes the action dimensionless. The Euler-Lagrange equation following from this gives

$$
\frac{d^2 x^\mu}{d\tau^2} - \Gamma^\mu_{\nu\lambda} \frac{dx^\nu}{d\tau} \frac{dx^\lambda}{d\tau} = 0,
\tag{5.133}
$$

which coincides with the geodesic equation. There are several things to note here. First of all the parameter τ can be any affine parameter labelling the trajectory and not necessarily the proper time unless a specific gauge choice is used (see (5.118)). Furthermore, the geodesic equation is the same for any massive particle (the mass drops out in the equation) reflecting the fact that gravitation acts the same way on all particles.

However, there are certain disadvantages in using the action in the form (5.132). For example, the action is meaningful only for time-like trajectories (because of the square root) for which we have

$$
g_{\mu\nu} \frac{dx^\mu}{d\tau} \frac{dx^\nu}{d\tau} > 0.
\tag{5.134}
$$

Furthermore, since mass is an overall multiplicative parameter in the action (5.132), it is not a good action for massless particles for which it identically vanishes. For these reasons we look for an alternative action which would lead to the geodesic equations as Euler-Lagrange equations and would also overcome the disadvantages in (5.132). Normally this is done through the introduction of auxiliary (non-dynamical) variables (fields).

Let us consider the action

$$
\tilde{S} = \int_{\tau_1}^{\tau_2} d\tau \, L = \frac{1}{2} \int_{\tau_1}^{\tau_2} d\tau \left(\frac{1}{\sqrt{F}} g_{\mu\nu} \frac{dx^\mu}{d\tau} \frac{dx^\nu}{d\tau} + \sqrt{F} m^2 \right),
\tag{5.135}
$$

where F is an auxiliary (non-dynamical) variable. We note that the action (5.135) is defined for arbitrary m including $m = 0$. Similarly, since there is no square root, the action is defined as well for

$$g_{\mu\nu} \frac{dx^\mu}{d\tau} \frac{dx^\nu}{d\tau} \leq 0,$$ (5.136)

in addition to time like trajectories.

Since F is an auxiliary variable (there is no $\frac{dF}{d\tau}$ term in the Lagrangian), the Euler-Lagrange equation for F is quite simple and has the form

$$\frac{\partial L}{\partial F} = 0,$$

or, $$-\frac{1}{2F^{3/2}} \, g_{\mu\nu} \frac{dx^\mu}{d\tau} \frac{dx^\nu}{d\tau} + \frac{1}{2F^{1/2}} \, m^2 = 0,$$

or, $$F = \frac{1}{m^2} \, g_{\mu\nu} \frac{dx^\mu}{d\tau} \frac{dx^\nu}{d\tau}.$$ (5.137)

This is a constraint equation (not a dynamical equation) and if this is used to eliminate F in the action (5.136) we recover the action (5.132)

$$\tilde{S} = \frac{1}{2} \int_{\tau_1}^{\tau_2} d\tau \left(\frac{m}{\sqrt{g_{\lambda\rho} \frac{dx^\lambda}{d\tau} \frac{dx^\rho}{d\tau}}} \, g_{\mu\nu} \frac{dx^\mu}{d\tau} \frac{dx^\nu}{d\tau} + \sqrt{g_{\mu\nu} \frac{dx^\mu}{d\tau} \frac{dx^\nu}{d\tau}} \, m \right)$$

$$= m \int_{\tau_1}^{\tau_2} d\tau \sqrt{g_{\mu\nu} \frac{dx^\mu}{d\tau} \frac{dx^\nu}{d\tau}} = S.$$ (5.138)

Therefore, we see that the alternative action (5.135) is equivalent to the action (5.132) if the constraint involving the auxiliary variable F is used. Therefore, the action (5.136) would also lead to the geodesic equation which can be explicitly verified from the Euler-Lagrange equation for x^μ together with (5.138). However, as we have emphasized this alternative action does not suffer from the criticisms which apply to (5.132).

The variable F can be identified with the induced one-dimensional metric, $g_{\tau\tau}$, on the trajectory. This alternative formulation of the action is particularly useful in theories like string theory.

◀

▶ **Example (Geodesic equation in polar coordinates).** The action for a particle moving on a plane in polar coordinates is given by

$$S = \int dt \, L = \frac{m}{2} \int dt \, (\dot{r}^2 + r^2 \dot{\theta}^2),$$ (5.139)

where dots denote derivatives with respect to t. The Euler-Lagrange equations lead to the dynamical equations

$$\ddot{r} - r\dot{\theta}^2 = 0,$$

$$r^2\ddot{\theta} + 2r\dot{r}\dot{\theta} = 0,$$

or, $\quad \ddot{\theta} + \dfrac{2}{r}\dot{r}\dot{\theta} = 0.$ \hfill (5.140)

Let us note that the geodesic equation (5.83) describing the motion of the particle for the present problem can be written as $(\mu, \nu, \lambda = r, \theta)$

$$\frac{\mathrm{d}^2 x^\mu}{\mathrm{d}\tau^2} - \Gamma^\mu_{\nu\lambda}\frac{\mathrm{d}x^\nu}{\mathrm{d}\tau}\frac{\mathrm{d}x^\lambda}{\mathrm{d}\tau} = 0,$$

or, $\quad \left(\dfrac{\mathrm{d}\tau}{\mathrm{d}t}\right)^2 \left(\dfrac{\mathrm{d}^2 x^\mu}{\mathrm{d}\tau^2} - \Gamma^\mu_{\nu\lambda}\dfrac{\mathrm{d}x^\nu}{\mathrm{d}\tau}\dfrac{\mathrm{d}x^\lambda}{\mathrm{d}\tau}\right) = 0,$

or, $\quad \ddot{x}^\mu - \Gamma^\mu_{\nu\lambda}\dot{x}^\nu\dot{x}^\lambda = 0.$ \hfill (5.141)

Comparing (5.140) with (5.141) leads immediately to the nontrivial components of the connection

$$\Gamma^r_{\theta\theta} = r,$$

$$\Gamma^\theta_{r\theta} = \Gamma^\theta_{\theta r} = -\frac{1}{r}.$$ \hfill (5.142)

These coincide with our earlier results derived in (4.55).

◀

CHAPTER 6

Applications of the geodesic equation

6.1 Geodesic as representing gravitational effect

We have seen that the geodesic equation leads to the extremal path in a curved manifold. However, it is not clear if it incorporates the true effect of gravitational force on a particle. More specifically, we may ask how we can choose the equation

$$\frac{\mathrm{d}^2 x^\mu}{\mathrm{d}\tau^2} - \Gamma^\mu_{\nu\lambda} \frac{\mathrm{d}x^\nu}{\mathrm{d}\tau} \frac{\mathrm{d}x^\lambda}{\mathrm{d}\tau} = 0, \tag{6.1}$$

over, say, the equation

$$\frac{\mathrm{d}^2 x^\mu}{\mathrm{d}\tau^2} - \Gamma^\mu_{\nu\lambda} \frac{\mathrm{d}x^\nu}{\mathrm{d}\tau} \frac{\mathrm{d}x^\lambda}{\mathrm{d}\tau} + \alpha R^\mu_{\nu\lambda\rho} S^\rho \frac{\mathrm{d}x^\nu}{\mathrm{d}\tau} \frac{\mathrm{d}x^\lambda}{\mathrm{d}\tau} = 0, \tag{6.2}$$

as representing the motion of a free particle in a curved manifold. Here α is a constant, $R^\mu_{\nu\lambda\rho}$ denotes the Riemann-Christoffel curvature tensor defined in (5.18) and S^ρ is the spin vector of the particle. General covariance arguments cannot distinguish between the two since both the equations transform covariantly like contravariant vectors under a general coordinate transformation. Furthermore, the principle of equivalence cannot distinguish between the two either since in a locally flat Cartesian coordinate system

$$\Gamma^\mu_{\nu\lambda} = 0, \qquad R^\mu_{\nu\lambda\sigma} = 0. \tag{6.3}$$

Therefore, both the equations (6.1) and (6.2) reduce to the free particle equation in a Lorentz (inertial) frame,

$$\frac{\mathrm{d}^2 x^\mu}{\mathrm{d}\tau^2} = 0. \tag{6.4}$$

The answer to this interesting question comes from the observation that whichever equation represents the true equation of motion must reduce to Newton's equation of motion in the nonrelativistic limit. Let us consider a particle falling freely under the influence of a weak, stationary gravitational field produced by a point mass M. Classically we know that the gravitational potential at a point $\mathbf{x}$ (with the mass M at the origin) can be written as

$$\phi(\mathbf{x}) = -\frac{G_\mathrm{N} M}{|\mathbf{x}|}, \tag{6.5}$$

where G_N denotes Newton's constant. Newton's equation of motion for a particle in this potential has the form

$$\frac{\mathrm{d}^2 \mathbf{x}}{\mathrm{d} t^2} = -\boldsymbol{\nabla} \phi(\mathbf{x}). \tag{6.6}$$

If we assume that the geodesic equation represents the true equation of motion incorporating the effects of gravitation, then we have

$$\frac{\mathrm{d}^2 x^\mu}{\mathrm{d} \tau^2} - \Gamma^\mu_{\nu\lambda} \frac{\mathrm{d} x^\nu}{\mathrm{d} \tau} \frac{\mathrm{d} x^\lambda}{\mathrm{d} \tau} = 0, \tag{6.7}$$

and it should reduce to (6.6) in the appropriate limit. Furthermore, if we assume the gravitational field to be weak, the metric would change only slightly from the Minkowski metric and we can decompose the metric as

$$g_{\mu\nu}(x) = \eta_{\mu\nu} + h_{\mu\nu}(x), \tag{6.8}$$

where $h_{\mu\nu}(x)$ is assumed to be small in magnitude. Consequently we can write the inverse metric as

$$g^{\mu\nu}(x) \simeq \eta^{\mu\nu} - h^{\mu\nu}, \tag{6.9}$$

so that

$$\begin{aligned} g^{\mu\nu}(x)g_{\nu\lambda}(x) \ &\simeq\ (\eta^{\mu\nu} - h^{\mu\nu})(\eta_{\nu\lambda} + h_{\nu\lambda}) \\ &\simeq\ \eta^{\mu\nu}\eta_{\nu\lambda} - \eta_{\nu\lambda}h^{\mu\nu} + \eta^{\mu\nu}h_{\nu\lambda} \\ &=\ \delta^{\mu}_{\lambda} - h^{\mu}{}_{\lambda} + h^{\mu}{}_{\lambda} \\ &=\ \delta^{\mu}_{\lambda}. \end{aligned} \tag{6.10}$$

We note that the indices of $h_{\mu\nu}, h^{\mu\nu}$ can be effectively raised (or lowered) with the Minkowski metric (since $h_{\mu\nu}$ is assumed to be a weak field).

Since the gravitational field is stationary, the metric must be independent of time and we have

$$\partial_0 g_{\mu\nu} = \partial_0 h_{\mu\nu} = 0. \tag{6.11}$$

Furthermore, if the gravitational field is weak and the particle is nonrelativistic, then we note that we can neglect terms of the type $\frac{dx^i}{d\tau}$ in (6.7) in comparison to $\frac{dx^0}{d\tau} = \frac{dt}{d\tau}$. (This is equivalent to assuming $|\mathbf{v}| \ll 1$ with $c = 1$.) With these assumptions the geodesic equation (6.7) in this limit reduces to

$$\frac{d^2 x^{\mu}}{d\tau^2} - \Gamma^{\mu}_{00} \frac{dt}{d\tau} \frac{dt}{d\tau} = 0. \tag{6.12}$$

Let us also recall the definition of the connections (see (4.49))

$$\Gamma^{\mu}_{\nu\lambda} = -\frac{1}{2}\, g^{\mu\rho}(\partial_{\nu} g_{\lambda\rho} + \partial_{\lambda} g_{\rho\nu} - \partial_{\rho} g_{\nu\lambda}), \tag{6.13}$$

from which it follows that

$$\Gamma^{\mu}_{00} = -\frac{1}{2}\, g^{\mu\rho}(\partial_0 g_{0\rho} + \partial_0 g_{\rho 0} - \partial_{\rho} g_{00}) = \frac{1}{2}\, g^{\mu\rho}\partial_{\rho} g_{00}, \tag{6.14}$$

where we have used the static nature of the metric tensor (6.11). Keeping only terms linear in the field $h_{\mu\nu}$, we have

$$\Gamma^{\mu}_{00} = \frac{1}{2}\, \eta^{\mu\rho}\partial_{\rho} h_{00}, \tag{6.15}$$

so that

$$\Gamma^0_{00} = 0, \quad \Gamma^i_{00} = \frac{1}{2}\,\partial^i h_{00}. \tag{6.16}$$

Substituting these into (6.12), we see that the geodesic equation now reduces to

$$\frac{\mathrm{d}^2 t}{\mathrm{d}\tau^2} = 0,$$

$$\frac{\mathrm{d}^2 \mathbf{x}}{\mathrm{d}\tau^2} + \frac{1}{2}\,\boldsymbol{\nabla} h_{00}(\mathbf{x})\left(\frac{\mathrm{d}t}{\mathrm{d}\tau}\right)^2 = 0. \tag{6.17}$$

The first equation simply says that $\frac{\mathrm{d}t}{\mathrm{d}\tau}$ is a constant. Therefore, we can divide the second equation by $\left(\frac{\mathrm{d}t}{\mathrm{d}\tau}\right)^2$ to obtain

$$\frac{\mathrm{d}^2 \mathbf{x}}{\mathrm{d}t^2} = -\frac{1}{2}\,\boldsymbol{\nabla} h_{00}(\mathbf{x}). \tag{6.18}$$

This has the right form as the classical Newtonian equation (6.6) and hence we conclude that the geodesic equation (5.83) represents the true gravitational effects. Furthermore comparing with the classical equation (6.6) we see that in the weakfield limit we can identify

$$h_{00}(\mathbf{x}) = 2\phi(\mathbf{x}) + \text{ constant} = -\frac{2G_{\mathrm{N}}M}{|\mathbf{x}|} + \text{ constant}. \tag{6.19}$$

The constant can be determined by imposing the asymptotic condition that infinitely far away from the gravitational source (the mass M) the metric must be Lorentzian (Minkowski) and hence

$$\lim_{|\mathbf{x}|\to\infty} h_{\mu\nu}(\mathbf{x}) \longrightarrow 0, \tag{6.20}$$

which determines the value of the constant to be zero. Consequently, we determine the metric to be diagonal with

$$h_{00}(\mathbf{x}) = 2\phi(\mathbf{x}) = -\frac{2G_N M}{|\mathbf{x}|},$$

$$\text{or,} \quad g_{00}(\mathbf{x}) = \eta_{00} + h_{00}(\mathbf{x}) = 1 + 2\phi(\mathbf{x}) = 1 - \frac{2G_N M}{|\mathbf{x}|},$$

$$g_{ij}(\mathbf{x}) = \eta_{ij}, \qquad g_{0i}(\mathbf{x}) = 0. \tag{6.21}$$

In this case, far away from the source the invariant line element reduces to

$$d\tau^2 = g_{\mu\nu} dx^\mu dx^\nu,$$

$$\text{or,} \quad \left(\frac{d\tau}{dt}\right)^2 = g_{\mu\nu} \frac{dx^\mu}{dt} \frac{dx^\nu}{dt}$$

$$= g_{00} + \eta_{ij} \frac{dx^i}{dt} \frac{dx^j}{dt}$$

$$= g_{00} - (\dot{\mathbf{x}})^2 = \left(1 - \frac{2GM}{|\mathbf{x}|}\right) - (\dot{\mathbf{x}})^2,$$

$$\text{or,} \quad \frac{d\tau}{dt} = \left(1 - (\dot{\mathbf{x}})^2 - \frac{2GM}{|\mathbf{x}|}\right)^{\frac{1}{2}}$$

$$\simeq 1 - \frac{1}{2}(\dot{\mathbf{x}})^2 - \frac{GM}{|\mathbf{x}|}$$

$$= 1 - \left(\frac{1}{2}(\dot{\mathbf{x}})^2 - \phi(\mathbf{x})\right)$$

$$= 1 - (T - V), \tag{6.22}$$

where we have used $|\dot{\mathbf{x}}| \ll 1$ and $|\mathbf{x}| \gg 1$. (T and V denote respectively the non-relativistic kinetic energy and the potential energy of the particle scaled by the mass of the particle.) Therefore, the action for the system can be written as (see (5.112))

$$\tau = \int_{\tau_1}^{\tau_2} d\tau = \int dt\, \frac{d\tau}{dt}$$

$$= \int dt\, [1 - (T - V)]$$

$$= \text{const} - \int dt\, (T - V)$$

$$= \text{const} - \int dt\, L, \tag{6.23}$$

which is familiar from classical mechanics. This leads to the dynamical equation from

$$-\delta\tau = \delta \int dt\, L = 0, \tag{6.24}$$

which we recognize as the minimum action principle.

6.2 Rotating coordinate system and the Coriolis force

The geodesic equation holds not only in a gravitational field but also in any curvilinear coordinate system. Therefore, we can again test the validity of the geodesic equation by expressing it in a uniformly rotating coordinate system in flat space and comparing with the classical expressions for the Coriolis and the centrifugal forces.

Let us recall that the earth provides an example of a uniformly rotating coordinate system. If we assume that the earth is rotating around the z-axis with a constant angular frequency ω, then this system has cylindrical symmetry as shown in Fig. 6.1. Therefore, we can write the invariant length in the inertial frame or the space fixed frame in cylindrical coordinates as

$$d\tau^2 = dt^2 - (dr^2 + r^2 d\phi^2 + dz^2). \tag{6.25}$$

We can go to the uniformly rotating coordinate system from this through the transformation

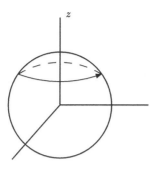

Figure 6.1: Rotating earth as a natural coordinate system with cylindrical symmetry.

$$t \;\to\; t, \qquad r \to r,$$
$$z \;\to\; z, \qquad \phi \to \phi + \omega t. \tag{6.26}$$

In the rotating coordinate frame the invariant length (6.25) takes the form

$$d\tau^2 = dt^2 - (dr^2 + r^2(d(\phi + \omega t))^2 + dz^2)$$
$$= dt^2 - (dr^2 + r^2 d\phi^2 + r^2\omega^2 dt^2 + 2r^2\omega dt d\phi + dz^2)$$
$$= (1 - r^2\omega^2)dt^2 - 2r^2\omega dt d\phi - (dr^2 + r^2 d\phi^2 + dz^2). \tag{6.27}$$

Therefore, we can simply read off the components of the metric tensor in this frame as

$$g_{tt} \;=\; (1 - r^2\omega^2), \qquad g_{rr} = -1,$$
$$g_{\phi\phi} \;=\; -r^2, \qquad g_{zz} = -1,$$
$$g_{t\phi} \;=\; g_{\phi t} = -r^2\omega. \tag{6.28}$$

It is worth noting here that the components of the metric tensor depend at best on the radial coordinate r. This also shows that if we restrict ourselves to time-like motion

$$g_{tt} = (1 - r^2\omega^2) > 0, \tag{6.29}$$

we must have $r\omega < 1$ and accordingly we assume that the frequency of rotation is small.

We note that we can invert the metric tensor in (6.28) and calculate all the components of the connection. However, (as we have already mentioned) a simpler way of determining the connection is to look at the dynamical equation from the Lagrangian (see (5.130)) and identify the coefficients of the double derivative terms, namely, terms of the kind $\frac{\mathrm{d}x^\mu}{\mathrm{d}\tau} \frac{\mathrm{d}x^\nu}{\mathrm{d}\tau}$ with the components of the connection (see (5.83)). To do this we write the geodesic equation in the alternate form (5.130) (by multiplying with the metric tensor) as

$$g_{\mu\nu} \frac{\mathrm{d}^2 x^\nu}{\mathrm{d}\tau^2} + \frac{1}{2} \left(\partial_\sigma g_{\mu\nu} + \partial_\nu g_{\mu\sigma} - \partial_\mu g_{\nu\sigma} \right) \frac{\mathrm{d}x^\sigma}{\mathrm{d}\tau} \frac{\mathrm{d}x^\nu}{\mathrm{d}\tau} = 0. \tag{6.30}$$

The r equation ($\mu = r$) has the form (a dot here denotes a derivative with respect to the proper time τ)

$$g_{rr} \frac{\mathrm{d}^2 r}{\mathrm{d}\tau^2} + \frac{1}{2} \left(\partial_\sigma g_{r\nu} + \partial_\nu g_{r\sigma} - \partial_r g_{\nu\sigma} \right) \frac{\mathrm{d}x^\sigma}{\mathrm{d}\tau} \frac{\mathrm{d}x^\nu}{\mathrm{d}\tau} = 0,$$

or, $\quad -\ddot{r} - \dfrac{1}{2} \partial_r g_{\nu\sigma} \dot{x}^\sigma \dot{x}^\nu = 0,$

or, $\quad \ddot{r} + \dfrac{1}{2} \left(\partial_r g_{tt} \dot{t}\dot{t} + \partial_r g_{\phi\phi} \dot{\phi}\dot{\phi} + 2\partial_r g_{t\phi} \dot{t}\dot{\phi} \right) = 0,$

or, $\quad \ddot{r} + \dfrac{1}{2} \left(-2r\omega^2 \dot{t}^2 - 2r\dot{\phi}^2 - 4r\omega \dot{t}\dot{\phi} \right) = 0,$

or, $\quad \ddot{r} - r\omega^2 \dot{t}^2 - r\dot{\phi}^2 - 2r\omega \dot{t}\dot{\phi} = 0. \tag{6.31}$

Therefore, comparing with the form of the geodesic equation (5.83) for $\mu = r$, we conclude that

$$\Gamma^r_{tt} = r\omega^2, \qquad \Gamma^r_{\phi\phi} = r, \qquad \Gamma^r_{t\phi} = \Gamma^r_{\phi t} = r\omega. \tag{6.32}$$

Similarly, the z-equation ($\mu = z$) is given by

$$g_{zz}\frac{\mathrm{d}^2 z}{\mathrm{d}\tau^2} + \frac{1}{2}\left(\partial_\sigma g_{z\nu} + \partial_\nu g_{z\sigma} - \partial_z g_{\nu\sigma}\right)\frac{\mathrm{d}x^\sigma}{\mathrm{d}\tau}\frac{\mathrm{d}x^\nu}{\mathrm{d}\tau} = 0,$$

or, $-\ddot{z} = 0,$

or, $\ddot{z} = 0,$ (6.33)

so that we have

$$\Gamma^z_{\mu\nu} = 0.$$ (6.34)

The ϕ equation ($\mu = \phi$) is similarly given by

$$g_{\phi\phi}\frac{\mathrm{d}^2\phi}{\mathrm{d}\tau^2} + g_{\phi t}\frac{\mathrm{d}^2 t}{\mathrm{d}\tau^2} + \frac{1}{2}(\partial_\sigma g_{\phi\nu} + \partial_\nu g_{\phi\sigma} - \partial_\phi g_{\nu\sigma})\frac{\mathrm{d}x^\sigma}{\mathrm{d}\tau}\frac{\mathrm{d}x^\nu}{\mathrm{d}\tau} = 0,$$

or, $-r^2\ddot{\phi} - r^2\omega\ddot{t} + \partial_\sigma g_{\phi\nu}\dot{x}^\sigma\dot{x}^\nu = 0,$

or, $r^2\ddot{\phi} + r^2\omega\ddot{t} - \partial_r g_{\phi\phi}\dot{r}\dot{\phi} - \partial_r g_{\phi t}\dot{r}\dot{t} = 0,$

or, $r^2\ddot{\phi} + r^2\omega\ddot{t} + 2r\dot{r}\dot{\phi} + 2r\omega\dot{r}\dot{t} = 0.$ (6.35)

Equation (6.35) can also be written as

$$\frac{\mathrm{d}}{\mathrm{d}\tau}\left(r^2\dot{\phi} + r^2\omega\dot{t}\right) = 0,$$ (6.36)

which shows that $r^2(\dot{\phi} + \omega\dot{t})$ is a constant of motion for the problem under study. Finally, the t-equation ($\mu = t$) has the form

$$g_{tt}\frac{\mathrm{d}^2 t}{\mathrm{d}\tau^2} + g_{t\phi}\frac{\mathrm{d}^2\phi}{\mathrm{d}\tau^2} + \frac{1}{2}\left(\partial_\sigma g_{t\nu} + \partial_\nu g_{t\sigma} - \partial_t g_{\nu\sigma}\right)\frac{\mathrm{d}x^\sigma}{\mathrm{d}\tau}\frac{\mathrm{d}x^\nu}{\mathrm{d}\tau} = 0,$$

or, $(1 - r^2\omega^2)\ddot{t} - r^2\omega\ddot{\phi} + \partial_\sigma g_{t\nu}\dot{x}^\sigma\dot{x}^\nu = 0,$

or, $(1 - r^2\omega^2)\ddot{t} - r^2\omega\ddot{\phi} + \partial_r g_{tt}\dot{r}\dot{t} + \partial_r g_{t\phi}\dot{r}\dot{\phi} = 0,$

or, $(1 - r^2\omega^2)\ddot{t} - r^2\omega\ddot{\phi} - 2r\omega^2\dot{r}\dot{t} - 2r\omega\dot{r}\dot{\phi} = 0.$ (6.37)

Equation (6.37) can also be written as

$$\frac{d}{d\tau} \left(\dot{t} - r^2\omega(\dot{\phi} + \omega\dot{t}) \right) = 0, \tag{6.38}$$

which shows that $\dot{t} - r^2\omega(\dot{\phi} + \omega\dot{t})$ is a conserved quantity. In fact, from these two conserved quantities, we conclude that $\dot{t}$ must be conserved.

That $\dot{t}$ is conserved can also be seen directly as follows. Multiplying (6.35) by ω and adding to (6.37) we obtain

$$(1 - r^2\omega^2)\ddot{t} - r^2\omega\ddot{\phi} - 2r\omega^2\dot{r}\dot{t} - 2r\omega\dot{r}\dot{\phi} + r^2\omega\ddot{\phi} + r^2\omega^2\ddot{t}$$

$$+2r\omega\dot{r}\dot{\phi} + 2r\omega^2\dot{r}\dot{t} = 0, \tag{6.39}$$

which simplifies to give

$$\ddot{t} = 0, \qquad \text{or,} \quad \dot{t} = \text{constant.} \tag{6.40}$$

This determines

$$\Gamma^t_{\mu\nu} = 0. \tag{6.41}$$

Similarly substituting (6.40) into (6.35), we obtain

$$r^2\ddot{\phi} + 2r\dot{r}\dot{\phi} + 2r\omega\dot{r}\dot{t} = 0,$$

$$\text{or,} \quad \ddot{\phi} + \frac{2}{r}\dot{r}\dot{\phi} + \frac{2}{r}\omega\dot{r}\dot{t} = 0, \tag{6.42}$$

which determines

$$\Gamma^\phi_{r\phi} = \Gamma^\phi_{\phi r} = -\frac{1}{r}, \qquad \Gamma^\phi_{rt} = \Gamma^\phi_{tr} = -\frac{\omega}{r}. \tag{6.43}$$

This completes the determination of all the components of the connection. Namely, the nontrivial components are given by

$$\Gamma^r_{tt} = r\omega^2, \qquad \Gamma^r_{\phi\phi} = r, \qquad \Gamma^r_{t\phi} = \Gamma^r_{\phi t} = r\omega,$$

$$\Gamma^\phi_{r\phi} = \Gamma^\phi_{\phi r} = -\frac{1}{r}, \qquad \Gamma^\phi_{rt} = \Gamma^\phi_{tr} = -\frac{\omega}{r}. \qquad (6.44)$$

To see the centrifugal and the Coriolis forces, let us note from (6.40) that

$$\dot{t} = \frac{dt}{d\tau} = \text{constant}. \qquad (6.45)$$

If we choose the scale of time suitably then the constant can be taken to be unity in which case we can identify

$$t = \tau. \qquad (6.46)$$

(Another way to say this is to note that we can divide the dynamical equations by factors of $\frac{dt}{d\tau}$.) Furthermore, in the rotating frame,

$$\dot{\phi} = 0, \qquad (6.47)$$

so that the r-equation (6.31) becomes (remember $\dot{t} = 1$)

$$\ddot{r} - r\omega^2 \dot{t}^2 = 0,$$

$$\text{or,} \qquad \frac{d^2 r}{dt^2} = r\omega^2, \qquad (6.48)$$

which gives the familiar centrifugal force.

Using (6.47), the Coriolis force is obtained from the ϕ-equation (6.42) and reads as (recall $\dot{t} = 1$)

$$r^2 \ddot{\phi} + 2r\dot{r}\dot{\phi} + 2r\omega\dot{r}\dot{t} = 0,$$

$$\text{or,} \qquad r^2 \frac{d^2\phi}{dt^2} + 2r\omega \frac{dr}{dt} = 0,$$

$$\text{or,} \qquad r \frac{d^2\phi}{dt^2} + 2\omega \frac{dr}{dt} = 0. \qquad (6.49)$$

We recognize the second term as the classical Coriolis force. (We note here that in a non-uniformly rotating frame, the force generalizes to what is known as the Euler force.) These examples give us further confidence that the geodesic equation is the true equation of motion for a free particle in a curved manifold as well as in a curvilinear system of coordinates.

▶ **Example (Calculation of connection from action).** The calculation of the components of the connection from the definition (4.49) is, in general, quite tedious and as we have mentioned earlier, the calculation can be simplified enormously by considering the action for the geodesic equation. The action for the geodesic equation has already been described in (5.112) (see also (5.132)) and an alternative form is given in (5.135). For purposes of calculation, however, there is still a simpler form of the action for a massive particle (we do not write the overall multiplicative mass term) given by

$$
S = \int_{\tau_A}^{\tau_B} \mathrm{d}\tau \, \frac{1}{2} \, g_{\mu\nu} \dot{x}^\mu \dot{x}^\nu = \int_{\tau_A}^{\tau_B} \mathrm{d}\tau \, L, \tag{6.50}
$$

where a dot denotes a derivative with respect to τ. The Euler-Lagrange equation for this action leads to

$$
\frac{\mathrm{d}}{\mathrm{d}\tau} \frac{\partial L}{\partial \dot{x}^\mu} - \frac{\partial L}{\partial x^\mu} = 0,
$$

$$
\text{or,} \quad \frac{\mathrm{d}}{\mathrm{d}\tau} \left(g_{\mu\nu} \dot{x}^\nu \right) - \frac{1}{2} (\partial_\mu g_{\sigma\nu}) \dot{x}^\sigma \dot{x}^\nu = 0,
$$

$$
\text{or,} \quad g_{\mu\nu} \ddot{x}^\nu + (\partial_\sigma g_{\mu\nu}) \dot{x}^\sigma \dot{x}^\nu - \frac{1}{2} (\partial_\mu g_{\sigma\nu}) \dot{x}^\sigma \dot{x}^\nu = 0,
$$

$$
\text{or,} \quad g_{\mu\nu} \ddot{x}^\nu + \frac{1}{2} \left(\partial_\sigma g_{\mu\nu} + \partial_\nu g_{\mu\sigma} - \partial_\mu g_{\sigma\nu} \right) \dot{x}^\sigma \dot{x}^\nu = 0,
$$

$$
\text{or,} \quad \ddot{x}^\mu + \frac{1}{2} g^{\mu\lambda} \left(\partial_\sigma g_{\lambda\nu} + \partial_\nu g_{\sigma\lambda} - \partial_\lambda g_{\sigma\nu} \right) \dot{x}^\sigma \dot{x}^\nu = 0,
$$

$$
\text{or,} \quad \ddot{x}^\mu - \Gamma^\mu_{\sigma\nu} \dot{x}^\sigma \dot{x}^\nu = 0, \tag{6.51}
$$

which we recognize to be the geodesic equation (5.83). The action (6.50), therefore, also leads to the Euler-Lagrange equation and is much simpler to manipulate.

Let us next consider a space described by the line element

$$
\mathrm{d}\tau^2 = e^{2\psi(r)} \, \mathrm{d}t^2 - e^{2\lambda(r)} \, \mathrm{d}r^2 - r^2 \left(\mathrm{d}\theta^2 + \sin^2\theta \mathrm{d}\phi^2 \right). \tag{6.52}
$$

The nonzero components of the metric tensor are obtained from (6.52) to be

$$
g_{tt} = e^{2\psi(r)}, \quad g_{rr} = -e^{2\lambda(r)}, \quad g_{\theta\theta} = -r^2, \quad g_{\phi\phi} = -r^2 \sin^2\theta. \tag{6.53}
$$

Consequently, the action for the geodesic (6.50) can be written as

$$S = \int\limits_{\tau_A}^{\tau_B} d\tau \, L = \int\limits_{\tau_A}^{\tau_B} d\tau \, \frac{1}{2} \left(e^{2\psi(r)} \dot{t}^2 - e^{2\lambda(r)} \dot{r}^2 - r^2 \dot{\theta}^2 - r^2 \sin^2\theta \, \dot{\phi}^2 \right), \qquad (6.54)$$

where a dot denotes a derivative with respect to τ. The Euler-Lagrange equations

$$\frac{d}{d\tau} \frac{\partial L}{\partial \dot{x}^\mu} - \frac{\partial L}{\partial x^\mu} = 0, \qquad (6.55)$$

following from (6.54) (for $\mu = t, r, \theta, \phi$, respectively) are given by

$$e^{2\psi(r)} \ddot{t} + 2e^{2\psi(r)} \psi'(r) \dot{r}\dot{t} = 0,$$

$$- e^{2\lambda(r)} \ddot{r} - e^{2\psi(r)} \psi'(r) \dot{t}^2 - e^{2\lambda(r)} \lambda'(r) \dot{r}^2 + r\dot{\theta}^2 + r \sin^2\theta \dot{\phi}^2 = 0,$$

$$- r^2 \ddot{\theta} - 2r\dot{r}\dot{\theta} + r^2 \sin\theta \cos\theta \dot{\phi}^2 = 0,$$

$$- r^2 \sin^2\theta \ddot{\phi} - 2r \sin^2\theta \dot{r}\dot{\phi} - 2r^2 \sin\theta \cos\theta \dot{\theta}\dot{\phi} = 0, \qquad (6.56)$$

where a prime denotes a derivative with respect to the radial coordinate r. We can simplify and rewrite the equations in (6.56) also as

$$\ddot{t} + 2\psi'(r) \dot{r}\dot{t} = 0,$$

$$\ddot{r} + e^{2\psi(r) - 2\lambda(r)} \psi'(r) \dot{t}^2 + \lambda'(r) \dot{r}^2 - re^{-2\lambda(r)} \dot{\theta}^2 - re^{-2\lambda(r)} \sin^2\theta \dot{\phi}^2 = 0,$$

$$\ddot{\theta} + 2\frac{\dot{r}}{r}\dot{\theta} - \sin\theta \cos\theta \dot{\phi}^2 = 0,$$

$$\ddot{\phi} + 2\frac{\dot{r}}{r}\dot{\phi} + 2\cot\theta \dot{\theta}\dot{\phi} = 0. \qquad (6.57)$$

Comparing these with the geodesic equations (5.83) or (6.51), we obtain the nonzero components of the connection to be

$$\Gamma^t_{rt} = \Gamma^t_{tr} = -\psi'(r), \qquad\qquad \Gamma^r_{tt} = -e^{2\psi(r) - 2\lambda(r)} \psi'(r),$$

$$\Gamma^r_{rr} = \lambda'(r), \qquad\qquad \Gamma^r_{\theta\theta} = re^{-2\lambda(r)},$$

$$\Gamma^r_{\phi\phi} = r \sin^2\theta e^{-2\lambda(r)}, \qquad\qquad \Gamma^\theta_{r\theta} = \Gamma^\theta_{\theta r} = \Gamma^\phi_{r\phi} = \Gamma^\phi_{\phi r} = -\frac{1}{r},$$

$$\Gamma^\theta_{\phi\phi} = \frac{1}{2} \sin 2\theta, \qquad\qquad \Gamma^\phi_{\theta\phi} = \Gamma^\phi_{\phi\theta} = -\cot\theta. \qquad (6.58)$$

This shows that it is a lot easier to derive the components of the connection from the action for the geodesic.

◀

6.3 Gravitational red shift

In flat space we are all familiar with the phenomenon of Doppler shift. Let us briefly recapitulate here the principle behind this before discussing the phenomenon of gravitational red shift. Let us consider the following one dimensional problem. A source S which emits a light wave of frequency ν_S moves with a velocity v with respect to an observer O. (For $v > 0$, the source moves away from the observer while for $v < 0$, it moves towards the observer.) If we denote by dt_S and dt_O the time intervals (between two events) as seen in the source and the observer frames respectively, then they are related as $(c = 1)$

$$dt_O = \gamma(v)dt_S, \qquad (6.59)$$

where (basically if the observer is at rest and the source is moving with a velocity v, then $d\tau = dt_O = \gamma(v)dt_S$)

$$\gamma(v) = (1 - v^2)^{-\frac{1}{2}}. \qquad (6.60)$$

If Δt_S denotes the time interval between two successive pulses in the frame of the source, then during this interval the source would have moved a distance

$$\Delta t_S\, v. \qquad (6.61)$$

Consequently, the time interval (distance) between successive crests of a light wave in the source frame is given by

$$\Delta \tilde{t}_S = \Delta t_S + \Delta t_S\, v = (1 + v)\Delta t_S. \qquad (6.62)$$

Using (6.59), we conclude that the time interval seen in the frame of the observer is given by

$$\begin{aligned} \Delta t_O &= \gamma(v)\Delta \tilde{t}_S = \gamma(v)(1 + v)\Delta t_S, \\ \text{or,} \quad \frac{\Delta t_O}{\Delta t_S} &= \gamma(v)(1 + v). \end{aligned} \qquad (6.63)$$

Consequently, if ν_S and ν_O denote respectively the frequencies of the photon as observed in the source and the observer frames, then from (6.63) we obtain

$$\frac{\nu_S}{\nu_O} = \gamma(v)(1+v) = \frac{1+v}{(1-v^2)^{\frac{1}{2}}} = \left(\frac{1+v}{1-v}\right)^{\frac{1}{2}}. \tag{6.64}$$

For small v, this can be approximated by

$$\frac{\nu_S}{\nu_O} \simeq 1+v, \qquad \frac{\lambda_O}{\lambda_S} \simeq 1+v, \tag{6.65}$$

which shows (see also (6.64)) that if $v > 0$,

$$\nu_S > \nu_O, \qquad \lambda_S < \lambda_O. \tag{6.66}$$

Namely, if the source is moving away from the observer, then the frequency of the light wave measured by the observer will be red shifted.

On the other hand, if $v < 0$, namely, if the source is moving towards the observer, then it follows from (6.64) or (6.65) that

$$\nu_S < \nu_O, \qquad \lambda_S > \lambda_S. \tag{6.67}$$

Namely, in this case, the frequency of the light wave measured by the observer will be blue shifted. This phenomenon is known as Doppler shift. A similar phenomenon is also observed in the presence of a gravitational field resulting in a red shift in the light wave (spectral lines) emitted by stars.

Let us assume that the gravitational field we are interested in is produced by a static stellar body (star) of mass M and is weak. For such a field we have already determined the metric to be diagonal with (see (6.21))

$$g_{00}(x) = 1 + 2\phi(\mathbf{x}) = 1 - \frac{2GM}{|\mathbf{x}|}, \qquad g_{ij} = \eta_{ij}. \tag{6.68}$$

Let us suppose that a photon (light wave) is emitted by the star and is moving away in this gravitational field as shown in Fig. 6.2. Let us assume that at point A, the light wave (photon) has a frequency ν_A and that when it falls to the point B the frequency has changed to ν_B. At point A, the total energy of the photon is

$$E_A = h\nu_A, \tag{6.69}$$

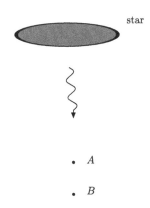

Figure 6.2: Light emitted by a distant star passing through two points A and B.

where we are assuming that the gravitational potential is zero at this point (reference point). Furthermore, we know that for purposes of gravitational interactions, a photon of frequency ν can be thought of as a particle of mass $m = h\nu$ (recall $c = 1$). Thus like any other particle, when the photon falls in the gravitational field it would gain a potential energy equivalent to $m|\Delta\phi|$. Therefore, we conclude that the total energy of the photon when it reaches the point B would be

$$E_B = h\nu_B + m|\Delta\phi|. \tag{6.70}$$

Energy conservation leads to

$$E_A = E_B,$$

$$\text{or,} \quad h\nu_A = h\nu_B + m|\Delta\phi| = h\nu_B + h\nu_B|\Delta\phi|,$$

$$\text{or,} \quad \frac{\nu_B}{\nu_A} = \frac{1}{1 + |\Delta\phi|} = \frac{1}{1 + \frac{G_N M}{r^2} \Delta r} \simeq 1 - \frac{G_N M}{r^2} \Delta r. \quad (6.71)$$

Here Δr is the radial separation between the two points and r is the radial distance of point B from the star. This shows that the frequency of the photon decreases as it falls in a gravitational field. We can calculate the ratio of the wavelengths from the relation (remember $c = 1$)

$$\lambda\nu = 1, \quad (6.72)$$

so that we have

$$\frac{\lambda_B}{\lambda_A} = \frac{\nu_A}{\nu_B} \simeq 1 + \frac{G_N M}{r^2} \Delta r. \quad (6.73)$$

In other words, the wavelength of the light wave received at B would be longer than that at A. That is, there would be a red shift in the photon frequency and wavelength. Let us emphasize here that in our calculation we have kept only linear changes. The photon mass also keeps on changing as it falls in the gravitational field. We have not taken such effects into consideration since they only induce higher order corrections.

We can also derive the red shift (6.71) or (6.73) purely from the principle of equivalence. According to the principle of equivalence the effect of gravitation can be completely rotated away locally by simply changing coordinate frames. Equivalently we can think of the observer at B as accelerating downwards with a constant acceleration g with respect to the observer at A (we note that the observer at A is attracted more strongly to the star than the observer at B). If at $t = 0$ both the observers are moving with the same velocity, then observer B would move away from A with time. Let observer A send a light signal at $t = 0$ towards the observer B as shown in Fig. 6.3. If the signal reaches B at time t, then ($c = 1$)

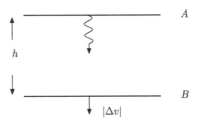

Figure 6.3: Gravitational red shift from the principle of equivalence.

$$t = h,   \tag{6.74}$$

the radial separation between the two observers. Here again we are neglecting higher order terms like retardation due to motion of the observers. Clearly, in this time interval, observer B would have gained a velocity with respect to A which is given by

$$|\Delta v| = gt = gh.   \tag{6.75}$$

Thus the light wave reaching B would be Doppler shifted and the wavelength would be given by (see (6.65))

$$\lambda_B = \lambda_A(1 + |\Delta v|),$$
$$\text{or,} \quad \frac{\lambda_B}{\lambda_A} = 1 + gh. \tag{6.76}$$

That is, the light wave would be red shifted. Furthermore, this is the same result as in (6.73) if we identify

$$g = \frac{G_N M}{r^2}, \qquad h = \Delta r.   \tag{6.77}$$

It is useful to keep the limitations of this method in mind. Principle of equivalence applies only locally and hence this method is not applicable when we are talking about widely separated points in space time.

We would next derive the gravitational red shift through yet another method, namely through the dilation of time. But to do that we have to discuss how measurements are done in a curved manifold or in a gravitational frame of reference. First of all we note that the invariant proper time interval in the presence of gravitation (in a curved manifold) is given by

$$d\tau = (g_{\mu\nu}(x)dx^{\mu}dx^{\nu})^{\frac{1}{2}}. \tag{6.78}$$

Unlike the Minkowski space, it is only the infinitesimal proper interval that makes any sense in a curved manifold. There is no global proper time. This can be seen from the fact that the time interval between two events A and B

$$(\Delta\tau)_{AB} = \int_{\tau_A}^{\tau_B} d\tau, \tag{6.79}$$

is path dependent. Therefore, although it is invariant under a coordinate change it is not very useful. (By the way, this kind of an approach gives a simple solution to the twin paradox which we will discuss shortly. Since the two persons have two different world lines, there is no reason why the associated time intervals would be equal.) In contrast, however, the coordinate system used by any observer gives a unique time interval between two events although it is not coordinate invariant, namely,

$$(\Delta t)_{AB} = \int_{\tau_A}^{\tau_B} dt = t_B - t_A, \tag{6.80}$$

is unique but not invariant. The proper time of a clock is, of course, related to the coordinate time by the relation

$$d\tau = (g_{00})^{\frac{1}{2}}dt, \tag{6.81}$$

when the clock is at rest.

Let us now extend these ideas to the case of a star which sends out N light waves of frequency ν_{star}. If the interval between pulses as measured by an observer on earth is dt, then the proper time interval between pulses at the star is given by (i.e., the time interval measured by a clock on the star)

$$d\tau_{\text{star}} = (g_{00}(x_{\text{star}}))^{\frac{1}{2}}\, dt. \tag{6.82}$$

Similarly on earth, the proper time interval between pulses is

$$d\tau_{\text{earth}} = (g_{00}(x_{\text{earth}}))^{\frac{1}{2}}\, dt. \tag{6.83}$$

Let us assume that the N waves are emitted in a proper time interval $\Delta\tau_{\text{star}}$. On earth N waves are received but in an interval $\Delta\tau_{\text{earth}}$. Thus, we have

$$N = \nu_{\text{star}}\Delta\tau_{\text{star}}, \qquad N = \nu_{\text{earth}}\Delta\tau_{\text{earth}}. \tag{6.84}$$

Clearly then

$$
\begin{aligned}
\frac{\nu_{\text{star}}}{\nu_{\text{earth}}} &= \frac{\Delta\tau_{\text{earth}}}{\Delta\tau_{\text{star}}} = \frac{(g_{00}(x_{\text{earth}}))^{\frac{1}{2}}}{(g_{00}(x_{\text{star}}))^{\frac{1}{2}}} \\
&= \left(\frac{1 + 2\phi_{\text{earth}}}{1 + 2\phi_{\text{star}}}\right)^{\frac{1}{2}} \\
&= 1 + (\phi_{\text{earth}} - \phi_{\text{star}}) \\
&= 1 + |\Delta\phi|,
\end{aligned}
\tag{6.85}
$$

where we have used (6.68) as well as the fact that the gravitational potential ϕ is negative and inversely proportional to distance (see (6.5)). This again shows that the frequency received on earth would be lower or correspondingly the wavelength on earth

$$\frac{\lambda_{\text{earth}}}{\lambda_{\text{star}}} = 1 + |\Delta\phi| > 1, \tag{6.86}$$

would be longer. Consequently we say that a light wave moving away from a gravitational source must be red shifted.

The phenomenon of gravitational red shift has been experimentally verified. If we assume that light waves emitted by a star (M_S, R_S) are received on earth (M_E, R_E), then (6.85) gives

$$\frac{\Delta\lambda}{\lambda} = |\Delta\phi| = G_N \left(\frac{M_S}{R_S} - \frac{M_E}{R_E}\right). \tag{6.87}$$

The red shift of hydrogen spectrum of the white dwarf star 40 Eridani B is measured and the value

$$\left(\frac{\Delta\lambda}{\lambda}\right)_{\text{EXPT}} = 7 \times 10^{-5}, \tag{6.88}$$

is obtained. Theoretically we can calculate the value to be

$$\left(\frac{\Delta\lambda}{\lambda}\right)_{\text{TH}} = 5.6 \times 10^{-5}, \tag{6.89}$$

where M_S, R_S are used as inputs. This is in quite good agreement. However, astronomical tests of red shift are hard since there is a lot of error due to other cosmological effects. For example, the expansion of the universe causes a Doppler shift, the overall metric of the universe is not so simple as we have used and so on.

Therefore, one looks for terrestrial measurements of gravitational red shift and since the shifts involved are so small we may think it is humanly impossible to verify them on earth. It is, however, possible because of the sensitivity gained through the Mössbauer effect. Such a measurement involves the study of the γ ray spectrum of a radioactive nucleus. In general it is hard to study the γ ray spectrum of a nucleus since the emission of the γ ray leads to a recoil of the nucleus which then Doppler shifts the spectrum. What Mössbauer had observed is that in certain crystals like Fe^{57}, the entire crystal rather than the nucleus alone, picks up the recoil and hence the process becomes virtually recoilless. In fact Fe^{57} shows a line at 14.4 keV with a fractional half width 10^{-12}. If a radioactive emitter, say of Fe^{57}, is placed near a nonradioactive absorber of the same material, then, of course, it would lead to resonant absorption. However, if the emitter is moved with a slight relative velocity, then the Doppler shift would change the frequency and hence the resonant absorption would be completely wiped out.

To measure the gravitational red shift, an emitter was placed at the bottom of a 21.6 m (72 ft) tower and an absorber at the top. Due to gravitational red shift there was no absorption. However, by moving the emitter with a relative velocity to compensate for the gravitational red shift, resonance was restored. And this led to the experimental value of

$$
\left. \left| \frac{\Delta\nu}{\nu} \right| \right|_{\text{EXPT}} = 2.56 \times 10^{-15}. \tag{6.90}
$$

On the other hand, theoretically we can calculate (see (6.85))

$$
\begin{aligned}
\left. \left| \frac{\Delta\nu}{\nu} \right| \right|_{\text{TH}} &= |\Delta\phi| = G_{\text{N}} M_E \left(\frac{1}{r_1} - \frac{1}{r_2} \right) \\
&= G_{\text{N}} M_E \frac{\Delta r}{R_E^2} = 2.46 \times 10^{-15}. \tag{6.91}
\end{aligned}
$$

Here we have made the approximation

$$
r_1 r_2 = R_E^2, \tag{6.92}
$$

and used $\Delta r = r_2 - r_1 = 21.6$ m (72 ft). This shows extremely good agreement between the theoretical calculation and the experimental measurement.

▶ **Example (Stereographic projection of unit sphere).** Let us consider a two dimensional space described by the line element

$$
d\tau^2 = \frac{4}{(1+r^2)^2} \, dr^2 + \frac{4r^2}{(1+r^2)^2} \, d\phi^2, \tag{6.93}
$$

where $r \in [0, \infty]$ and $\phi \in [0, 2\pi]$ with $\phi = 0$ and $\phi = 2\pi$ identified. The metric tensor follows from (6.93) to have the diagonal form

$$
g_{rr} = \frac{4}{(1+r^2)^2}, \qquad\qquad g_{\phi\phi} = \frac{4r^2}{(1+r^2)^2},
$$

$$
g^{rr} = \frac{1}{4} (1+r^2)^2, \qquad\qquad g^{\phi\phi} = \frac{1}{4r^2} (1+r^2)^2. \tag{6.94}
$$

The proper length (time) between the points A and B along a path is given by

$$\tau = \int_{\tau_A}^{\tau_B} d\tau = \int_{\tau_A}^{\tau_B} (g_{rr}dr^2 + g_{\phi\phi}d\phi^2)^{\frac{1}{2}}. \tag{6.95}$$

For a circle at $r = R_0$ we can identify the two points A and B to correspond to $\phi_A = 0$ and $\phi_B = 2\pi$ with $r = R_0$ fixed. In this case, (6.95) leads to

$$\tau = \int_0^{2\pi} d\phi \sqrt{g_{\phi\phi}(R_0)}$$

$$= \int_0^{2\pi} d\phi \, \frac{2R_0}{1 + R_0^2}$$

$$= \frac{4\pi R_0}{1 + R_0^2}. \tag{6.96}$$

Expression (6.96) gives the length of the circumference of the circle $r = R_0$ and shows that

$$\lim_{R_0 \to \infty} \tau \to 0, \tag{6.97}$$

so that the region described by $R_0 \to \infty$ actually corresponds to a point.

As we have seen, the action (6.50) (a dot denotes a derivative with respect to τ)

$$S = \int d\tau \, \frac{1}{2}(g_{rr}\dot{r}^2 + g_{\phi\phi}\dot{\phi}^2)$$

$$= \int d\tau \left(\frac{2\dot{r}^2}{(1 + r^2)^2} + \frac{2r^2\dot{\phi}^2}{(1 + r^2)^2} \right), \tag{6.98}$$

leads to the geodesic equations through the Euler-Lagrange equations. The Euler-Lagrange equations (see (6.55)) for r and ϕ, respectively give

$$\ddot{r} - \frac{2r}{1 + r^2}\dot{r}^2 - \frac{r(1 - r^2)}{1 + r^2}\dot{\phi}^2 = 0,$$

$$\frac{d}{d\tau}\left(\frac{4r^2\dot{\phi}}{(1 + r^2)^2} \right) = 0. \tag{6.99}$$

We conclude from the second equation in (6.99) that

$$\frac{4r^2\dot{\phi}}{(1 + r^2)^2} = \text{const.} \tag{6.100}$$

The region $r \to \infty$, as we have seen in (6.97), is mapped to a point and all other points seem to be regular. This is, in fact, nothing other than the unit sphere. To see this we note that if we define

$$r = \tan \frac{\theta}{2},$$
(6.101)

with $\theta \in [0, \pi]$ then,

$$dr = d\left(\tan \frac{\theta}{2}\right) = \frac{1}{2} \sec^2 \frac{\theta}{2} \, d\theta.$$
(6.102)

It follows now from (6.93), (6.101) and (6.102) that

$$
\begin{aligned}
d\tau^2 &= \frac{4}{(1 + r^2)^2} \, dr^2 + \frac{4r^2}{(1 + r^2)^2} \, d\phi^2 \\
&= \frac{1}{\sec^4 \frac{\theta}{2}} \sec^4 \frac{\theta}{2} \, d\theta^2 + \frac{4 \tan^2 \frac{\theta}{2}}{\sec^4 \frac{\theta}{2}} \, d\phi^2 \\
&= d\theta^2 + \sin^2 \theta \, d\phi^2.
\end{aligned}
$$
(6.103)

This is nothing other than the line element on a unit sphere. We can view the original coordinate as representing the stereographic projection of the unit sphere with the azimuth angle ϕ.

◀

6.4 Twin paradox and general covariance

In special relativity we know of the following paradox involving a pair of identical twins. If one of the twins stays on the earth while the other travels in a space craft then for the twin on earth, the partner is in a moving frame and hence his biological clock would slow down. Therefore, when he comes back to earth, he should be much younger than his twin partner on earth. On the other hand, in the rest frame of the twin on the space craft, the partner on earth is moving and hence his biological clock should slow down. Consequently, upon return the twin travelling in the space craft should find the earth partner to be much younger.

This, therefore, poses a problem and is referred to as the twin paradox. We can see now how the resolution of this puzzle (paradox) comes simply from considerations of general coordinate transformation. To simplify the problem let us assume that there exists only one space and one time dimension. Let us further assume that in the rest frame of A, the position of B is given by

$$x_B = f(t). \tag{6.104}$$

Then in the rest frame of B, the position of A is given by

$$X = x - x_B = x - f(t). \tag{6.105}$$

Let us further assume that both observers use the same coordinate time, i.e.,

$$t = T. \tag{6.106}$$

Then the invariant length interval ($c = 1$) in the frame of A has the general form

$$d\tau^2 = g_{xx}dx^2 + 2g_{xt}dxdt + g_{tt}dt^2, \tag{6.107}$$

whereas in the frame of B we can write

$$d\tau^2 = g_{XX}dX^2 + 2g_{XT}dXdT + g_{TT}dT^2. \tag{6.108}$$

From the transformation relations (6.105) and (6.106) between the two coordinate frames we note that

$$
\begin{aligned}
d\tau^2 &= g_{XX}dX^2 + 2g_{XT}dXdT + g_{TT}dT^2 \\
&= g_{XX}(dx - \dot{f}dt)^2 + 2g_{XT}(dx - \dot{f}dt)dt + g_{TT}dt^2 \\
&= g_{XX}dx^2 + (2g_{XT} - 2g_{XX}\dot{f})dxdt + (g_{TT} - 2g_{XT}\dot{f} + g_{XX}\dot{f}^2)dt^2 \\
&= g_{xx}dx^2 + 2g_{xt}dxdt + g_{tt}dt^2. \tag{6.109}
\end{aligned}
$$

This leads to the relations between the metric components in the two coordinate frames as

$$g_{xx} = g_{XX},$$

$$g_{xt} = g_{XT} - g_{XX}\dot{f},$$

$$g_{tt} = g_{TT} - 2g_{XT}\dot{f} + g_{XX}\dot{f}^2, \tag{6.110}$$

which can also be obtained from the transformation of the covariant metric tensor (3.17) under the coordinate transformations (6.105) and (6.106). Conversely, we can invert the metric components in (6.110) and write

$$g_{XX} = g_{xx},$$

$$g_{XT} = g_{xt} + g_{xx}\dot{f},$$

$$g_{TT} = g_{tt} + 2(g_{xt} + g_{xx}\dot{f})\dot{f} - g_{xx}\dot{f}^2$$

$$= g_{tt} + 2g_{xt}\dot{f} + g_{xx}\dot{f}^2. \tag{6.111}$$

Let us now assume that the take off time of the space craft is labeled by $t_1 = T_1$ and that the time of return by $t_2 = T_2$. Thus the age of B in the rest frame of A is given by

$$\tau_A(B) = \int_{t_1}^{t_2} \mathrm{d}\tau$$

$$= \int_{t_1}^{t_2} \mathrm{d}t \left(g_{tt} + 2g_{xt}\frac{\mathrm{d}x_B}{\mathrm{d}t} + g_{xx}\left(\frac{\mathrm{d}x_B}{\mathrm{d}t}\right)^2 \right)^{\frac{1}{2}}$$

$$= \int_{t_1}^{t_2} \mathrm{d}t \, (g_{tt} + 2g_{xt}\dot{f} + g_{xx}\dot{f}^2)^{\frac{1}{2}}$$

$$= \int_{T_1}^{T_2} \mathrm{d}T \, \sqrt{g_{TT}}. \tag{6.112}$$

On the other hand, the age of B in his own (rest) frame is given by

$$\tau_B(B) = \int_{T_1}^{T_2} \mathrm{d}T \, \sqrt{g_{TT}}, \tag{6.113}$$

which shows that

$$\tau_A(B) = \tau_B(B). \tag{6.114}$$

In other words, the age of a person is independent of the coordinate system. It is also clear that the age difference of the two persons is unique and is given by

$$\tau_A(B) - \tau_A(A) = \int_{t_1}^{t_2} \mathrm{d}t \, (\sqrt{g_{TT}} - \sqrt{g_{tt}}) = \tau_B(B) - \tau_B(A), \tag{6.115}$$

so that there is really no paradox involving (the frames of) the twins.

6.5 Other equations in the presence of gravitation

We have already seen that the free particle equation in the presence of gravitation is given by

$$m \, \frac{\mathrm{d}^2 x^\mu}{\mathrm{d}\tau^2} - m \, \Gamma^\mu_{\nu\lambda} \, \frac{\mathrm{d}x^\nu}{\mathrm{d}\tau} \, \frac{\mathrm{d}x^\lambda}{\mathrm{d}\tau} = 0. \tag{6.116}$$

In addition, if the particle experiences an external (nongravitational) force f^μ, then the equation of motion would modify to

$$m \, \frac{\mathrm{d}^2 x^\mu}{\mathrm{d}\tau^2} - m \, \Gamma^\mu_{\nu\lambda} \, \frac{\mathrm{d}x^\nu}{\mathrm{d}\tau} \, \frac{\mathrm{d}x^\lambda}{\mathrm{d}\tau} = f^\mu, \tag{6.117}$$

where f^μ represents the external force four vector. As in the case of flat space-time (see (2.9)), the force four vector (acceleration) has to be orthogonal to the proper velocity four vector, namely,

$$u_\mu f^\mu = g_{\mu\nu} u^\mu f^\nu = 0. \tag{6.118}$$

There are various ways of seeing this, but probably the simplest is to note that the equation (6.117) can be written in terms of the proper velocity as

$$m \frac{\mathrm{D}u^\mu}{\mathrm{D}\tau} = f^\mu, \tag{6.119}$$

which leads to

$$mg_{\mu\nu}u^\mu \frac{\mathrm{D}u^\nu}{\mathrm{D}\tau} = g_{\mu\nu}u^\mu f^\nu,$$

$$\text{or,} \quad \frac{m}{2} \frac{\mathrm{D}(g_{\mu\nu}u^\mu u^\nu)}{\mathrm{D}\tau} = g_{\mu\nu}u^\mu f^\nu,$$

$$\text{or,} \quad \frac{m}{2} \frac{\mathrm{d}(g_{\mu\nu}u^\mu u^\nu)}{\mathrm{d}\tau} = 0 = g_{\mu\nu}u^\mu f^\nu. \tag{6.120}$$

Here, in the intermediate steps, we have used the metric compatibility condition (5.8) as well as the normalization of the proper velocity (2.8) in the curved manifold (5.124), namely,

$$g_{\mu\nu}u^\mu u^\nu = 1. \tag{6.121}$$

The electromagnetic theory or Maxwell's theory (with sources) can also be extended to curved space in the following way. We know that in flat space, Maxwell's equations are given by

$$\partial_\mu F^{\mu\nu} = J^\nu, \qquad \partial_{[\mu}F_{\nu\lambda]} = 0. \tag{6.122}$$

In the presence of gravitation the first equation can be written in the covariant form

$$D_\mu F^{\mu\nu} = J^\nu, \tag{6.123}$$

where (the connections do not vanish for the contravariant field strength tensor unlike in the case of the covariant tensor in (4.189))

$$F^{\mu\nu} = D^\mu A^\nu - D^\nu A^\mu = -F^{\nu\mu}. \tag{6.124}$$

Furthermore, we know that (see (4.162)) we can write

$$D_\mu F^{\mu\nu} = \frac{1}{\sqrt{-g}} \, \partial_\mu(\sqrt{-g} \, F^{\mu\nu}), \tag{6.125}$$

so that the first equation of Maxwell becomes

$$D_\mu F^{\mu\nu} = J^\nu,$$

$$\text{or,} \quad \frac{1}{\sqrt{-g}} \, \partial_\mu(\sqrt{-g} \, F^{\mu\nu}) = J^\nu,$$

$$\text{or,} \quad \partial_\mu(\sqrt{-g} \, F^{\mu\nu}) = \sqrt{-g} \, J^\nu. \tag{6.126}$$

The second equation (the Bianchi identity) takes the covariant form (see (4.190))

$$D_{[\mu}F_{\nu\lambda]} = \partial_{[\mu}F_{\nu\lambda]} = 0. \tag{6.127}$$

Thus, Maxwell's equations (with sources) in a curved manifold (or in the presence of a gravitational field) can be written as

$$\partial_\mu(\sqrt{-g} \, F^{\mu\nu}) = \sqrt{-g} \, J^\nu, \qquad \partial_{[\mu}F_{\nu\lambda]} = 0. \tag{6.128}$$

Furthermore, from the first equation we see that

$$\partial_\nu\partial_\mu(\sqrt{-g} \, F^{\mu\nu}) = \partial_\nu(\sqrt{-g} \, J^\nu) = 0, \tag{6.129}$$

which follows from the anti-symmetry of the field strength tensor. Therefore, we see that, in the presence of gravitation, current conservation takes the covariant form (see (4.150) and compare with (2.75))

$$\partial_\mu(\sqrt{-g} \, J^\mu) = \sqrt{-g} \, D_\mu J^\mu = 0. \tag{6.130}$$

It is now obvious that we can define a global charge associated with the system which is conserved. Namely, if we define

$$Q(t) = \int d^3x \ \sqrt{-g} \ J^0, \tag{6.131}$$

then, it follows that

$$\frac{dQ}{dt} = \int d^3x \ \partial_0(\sqrt{-g} \ J^0) = - \int d^3x \ \partial_i(\sqrt{-g} \ J^i) = 0, \tag{6.132}$$

where we have used (6.129) (or (6.130)) and have also assumed that the current vanishes asymptotically.

Let us now examine the stress tensor of a particle. We have already seen that if a particle is subjected to an external force field then we have to define a total energy momentum (stress) tensor as (see (2.128))

$$T^{\mu\nu}_{\text{total}} = T^{\mu\nu}_{\text{matter}} + T^{\mu\nu}_{\text{field}}, \tag{6.133}$$

and it is the total stress tensor which is conserved in flat space, namely,

$$\partial_\mu T^{\mu\nu}_{\text{total}} = 0. \tag{6.134}$$

In the presence of gravitation this equation should be covariantized which leads to

$$D_\mu T^{\mu\nu}_{\text{total}} = 0,$$

$$\text{or,} \quad \partial_\mu T^{\mu\nu}_{\text{total}} - \Gamma^\mu_{\mu\rho} T^{\rho\nu}_{\text{total}} - \Gamma^\nu_{\mu\rho} T^{\mu\rho}_{\text{total}} = 0,$$

$$\text{or,} \quad \partial_\mu T^{\mu\nu}_{\text{total}} + \frac{1}{\sqrt{-g}} \ \partial_\rho(\sqrt{-g}) T^{\rho\nu}_{\text{total}} = \Gamma^\nu_{\mu\rho} T^{\mu\rho}_{\text{total}},$$

$$\text{or,} \quad \frac{1}{\sqrt{-g}} \ \partial_\mu \left(\sqrt{-g} \ T^{\mu\nu}_{\text{total}} \right) = \Gamma^\nu_{\mu\rho} T^{\mu\rho}_{\text{total}}. \tag{6.135}$$

Thus we see that unlike the charge current, the total stress tensor for the matter and the external fields does not satisfy a conservation law. Consequently, we cannot define a global momentum from $\sqrt{-g}\,T^{0\mu}_{\text{total}}$ which would be a constant in time. In other words, if we define

$$P^{\mu}_{\text{total}} = \int \mathrm{d}^3x \left(\sqrt{-g}\, T^{0\mu}_{\text{total}} \right), \tag{6.136}$$

then it follows that

$$\frac{\mathrm{d}P^{\mu}_{\text{total}}}{\mathrm{d}t} \neq 0. \tag{6.137}$$

This difference from flat space is understandable since the total momentum that would be conserved has to include the momentum associated with the gravitational field as well. However, we do not yet know how to construct the energy momentum associated with the gravitational field. This is because so far we only know how to extend quantities from flat space to a curved manifold. However, in a locally flat Cartesian coordinate frame the gravitational energy would vanish since it would be constructed out of derivatives of the metric tensor which is constant in the local frame. This, therefore, leads to difficulties in defining a global conserved momentum for a gravitational system. However, let us note here that for an isolated (closed) system it is always possible to define a global momentum which is conserved. We would come back to this question later.

As another example of a dynamical equation, let us consider the relativistic Klein-Gordon equation for a scalar field in flat space-time

$$\left(\Box + m^2 \right) \phi(x) = \left(\partial_\mu \partial^\mu + m^2 \right) \phi(x) = 0. \tag{6.138}$$

In the presence of gravitation this generalizes to

$$\left(D_\mu D^\mu + m^2 \right) \phi(x) = 0,$$

$$\text{or,} \quad D_\mu \partial^\mu \phi(x) + m^2 \phi(x) = 0,$$

$$\text{or,} \quad \frac{1}{\sqrt{-g}}\, \partial_\mu \left(\sqrt{-g}\, \partial^\mu \phi(x) \right) + m^2 \phi(x) = 0,$$

$$\text{or,} \quad \partial_\mu \left(\sqrt{-g}\, \partial^\mu \phi(x) \right) + \sqrt{-g}\, m^2 \phi(x) = 0. \tag{6.139}$$

This discussion shows how any equation in flat space-time can be generalized to a curved manifold. (Generalization of equations involving fermion fields needs the concept of tetrads or vierbeins that we will not go into.)

CHAPTER 7

Curvature tensor and Einstein's equation

7.1 Curvilinear coordinates versus gravitational field

A natural question we can ask at this point is how we can distinguish between a curvilinear coordinate system (in a flat space) and a gravitational field. Quantitatively we may ask if we make a coordinate change such that

$$\eta_{\mu\nu} = \frac{\partial \xi^\lambda}{\partial x^\mu} \frac{\partial \xi^\rho}{\partial x^\nu} \, g_{\lambda\rho}(x), \tag{7.1}$$

and $g_{\mu\nu}(x)$ has a nontrivial form, how can we distinguish this metric from a genuine gravitational metric. The answer lies in the fact that in such a case, given $g_{\mu\nu}(x)$ we can always find a coordinate frame in which the metric becomes the flat Cartesian metric. Correspondingly in that frame the curvature tensor will vanish identically. Namely, in that frame

$$R^\mu{}_{\nu\lambda\rho} = 0, \tag{7.2}$$

and since this is a tensor equation it must be true in any coordinate frame even if the metric has a nontrivial form. Consequently the manifold must be flat. On the other hand, the Riemann-Christoffel curvature tensor will not vanish in the presence of a genuine gravitational field.

7.2 Definition of an inertial coordinate frame

We have seen that in any coordinate frame the metric must satisfy the metric compatibility condition (4.103)

$$D_\mu g_{\nu\lambda} = \partial_\mu g_{\nu\lambda} + \Gamma^\rho_{\mu\nu} g_{\rho\lambda} + \Gamma^\rho_{\mu\lambda} g_{\nu\rho} = 0. \tag{7.3}$$

Furthermore, we have seen (see the theorem following (4.31)) that given a symmetric connection, we can always make a coordinate change to bring the connection to zero at a particular point. Thus we see that in addition to making the metric flat at some point we can also make the connection vanish at that point through a coordinate transformation. Mathematically this is equivalent to saying that at $x = x_{\mathrm{p}}$ we can have

$$g_{\mu\nu}(x_{\mathrm{p}}) = \eta_{\mu\nu}, \qquad \partial_\rho g_{\mu\nu}|_{x=x_{\mathrm{p}}} = 0. \tag{7.4}$$

A frame where (7.4) holds is known as an inertial frame. However, we note that this does not say anything about the higher order derivatives of $g_{\mu\nu}$ at $x = x_{\mathrm{p}}$. In particular,

$$\partial_\sigma \partial_\rho g_{\mu\nu}|_{x=x_{\mathrm{p}}} \neq 0, \tag{7.5}$$

if the manifold is curved. On the other hand, if we go back to the definition of the curvature tensor, we note that

$$R^\mu_{\ \nu\lambda\rho} \sim -g^{\mu\sigma} \partial_\rho \partial_\nu g_{\lambda\sigma} + \cdots, \tag{7.6}$$

so that with a coordinate change we cannot rotate away the curvature. However, over infinitesimal distances the effect of curvature will be negligible.

7.3 Geodesic deviation

We have already seen that a free particle follows the geodesic path. Let us now consider a region of space containing a number of geodesics. Let us assume that the family of geodesics is parameterized by the proper time τ as well as by a second parameter h which distinguishes between the different geodesics. The geodesics, of course, obey the equation (see (5.83))

$$\frac{\partial^2 x^\mu(\tau, h)}{\partial \tau^2} - \Gamma^\mu_{\nu\lambda} \frac{\partial x^\nu}{\partial \tau} \frac{\partial x^\lambda}{\partial \tau} = 0. \tag{7.7}$$

If we define the tangent vector as

$$u^\mu(\tau, h) = \frac{\partial x^\mu(\tau, h)}{\partial \tau}, \tag{7.8}$$

then the geodesic equation can also be written as (see (5.86))

$$\frac{Du^\mu}{D\tau} = \frac{\partial u^\mu}{\partial \tau} - \Gamma^\mu_{\nu\lambda} u^\nu u^\lambda = 0. \tag{7.9}$$

Let us now define a second vector associated with the trajectories as

$$v^\mu(\tau, h) = \frac{\partial x^\mu(\tau, h)}{\partial h}. \tag{7.10}$$

This describes the difference in the coordinates of two infinitesimally close geodesics (for the same value of τ). Clearly from the commutativity of the ordinary derivatives we have

$$\frac{\partial v^\mu}{\partial \tau} = \frac{\partial^2 x^\mu}{\partial \tau \partial h} = \frac{\partial u^\mu}{\partial h}. \tag{7.11}$$

Therefore, we can write

$$\begin{aligned}
\frac{Dv^\mu}{D\tau} &= \frac{\partial v^\mu}{\partial \tau} - \Gamma^\mu_{\nu\lambda} \frac{\partial x^\nu}{\partial \tau} v^\lambda \\
&= \frac{\partial u^\mu}{\partial h} - \Gamma^\mu_{\nu\lambda} u^\nu v^\lambda,
\end{aligned} \tag{7.12}$$

which leads to

$$
\begin{aligned}
\frac{\mathrm{D}^2 v^\mu}{\mathrm{D}\tau^2} &= \frac{\partial}{\partial \tau}\frac{\mathrm{D}v^\mu}{\mathrm{D}\tau} - \Gamma^\mu_{\rho\sigma}\frac{\partial x^\rho}{\partial \tau}\frac{\mathrm{D}v^\sigma}{\mathrm{D}\tau} \\
&= \frac{\partial}{\partial \tau}\left(\frac{\partial u^\mu}{\partial h} - \Gamma^\mu_{\nu\lambda}u^\nu v^\lambda\right) \\
&\quad - \Gamma^\mu_{\rho\sigma}\, u^\rho \left(\frac{\partial u^\sigma}{\partial h} - \Gamma^\sigma_{\nu\lambda}u^\nu v^\lambda\right) \\
&= \frac{\partial}{\partial h}\frac{\partial u^\mu}{\partial \tau} - \partial_\sigma \Gamma^\mu_{\nu\lambda}u^\sigma u^\nu v^\lambda - \Gamma^\mu_{\nu\lambda}\frac{\partial u^\nu}{\partial \tau}\,v^\lambda \\
&\quad - \Gamma^\mu_{\nu\lambda}u^\nu \frac{\partial v^\lambda}{\partial \tau} - \Gamma^\mu_{\rho\sigma}u^\rho \frac{\partial u^\sigma}{\partial h} + \Gamma^\mu_{\rho\sigma}\Gamma^\sigma_{\nu\lambda}u^\rho u^\nu v^\lambda. \quad (7.13)
\end{aligned}
$$

Using the geodesic equation (7.9) we can write

$$
\begin{aligned}
\frac{\mathrm{D}^2 v^\mu}{\mathrm{D}\tau^2} &= \frac{\partial}{\partial h}\left(\Gamma^\mu_{\nu\lambda}u^\nu u^\lambda\right) - \partial_\sigma \Gamma^\mu_{\nu\lambda}u^\sigma u^\nu v^\lambda \\
&\quad - \Gamma^\mu_{\nu\lambda}\left(\Gamma^\nu_{\rho\sigma}u^\rho u^\sigma\right)v^\lambda - 2\Gamma^\mu_{\nu\lambda}u^\nu \frac{\partial u^\lambda}{\partial h} \\
&\quad + \Gamma^\mu_{\rho\sigma}\Gamma^\sigma_{\nu\lambda}u^\rho u^\nu v^\lambda \\
&= \partial_\sigma \Gamma^\mu_{\nu\lambda}v^\sigma u^\nu u^\lambda - \partial_\sigma \Gamma^\mu_{\nu\lambda}u^\sigma u^\nu v^\lambda \\
&\quad - \Gamma^\mu_{\nu\lambda}\Gamma^\nu_{\rho\sigma}u^\rho u^\sigma v^\lambda + \Gamma^\mu_{\rho\sigma}\Gamma^\sigma_{\nu\lambda}u^\rho u^\nu v^\lambda \\
&= \left(\partial_\lambda \Gamma^\mu_{\nu\rho} - \partial_\rho \Gamma^\mu_{\nu\lambda} - \Gamma^\mu_{\sigma\lambda}\Gamma^\sigma_{\rho\nu} + \Gamma^\mu_{\rho\sigma}\Gamma^\sigma_{\nu\lambda}\right)u^\rho u^\nu v^\lambda
\end{aligned}
$$

or,
$$
\frac{\mathrm{D}^2 v^\mu}{\mathrm{D}\tau^2} = R^\mu{}_{\nu\rho\lambda}u^\nu u^\rho v^\lambda. \quad (7.14)
$$

Thus we see that the second derivative of v^μ is proportional to the Riemann curvature tensor. Therefore, the separation between the geodesics changes (as τ changes) if the curvature does not vanish.

This shows that in the presence of a gravitational field a group of geodesics do not remain parallel. This is known as the geodesic deviation and we immediately see that by studying the deviation of the geodesics we can detect the presence of a gravitational field. This, of course, does not contradict the principle of equivalence because if we restrict ourselves to an infinitesimal volume in the manifold, then the term on the right hand side of (7.14) will be negligible and hence we cannot detect the deviation in the geodesics in such an infinitesimal region (locally).

7.4 Properties of the curvature tensor

The Riemann curvature tensor is a rank four tensor and hence has $4^4 = 256$ components. However, not all of them are independent or nontrivial. This is because the curvature tensor has many symmetry properties. For example, from the defining relation (see (5.17))

$$[D_\mu, D_\nu]A^\rho = R^\rho{}_{\sigma\mu\nu}A^\sigma, \tag{7.15}$$

we note that

$$R^\rho{}_{\sigma\mu\nu} = -R^\rho{}_{\sigma\nu\mu}. \tag{7.16}$$

That is, the curvature tensor is anti-symmetric in the last two indices. Furthermore, let us consider the scalar length of a vector ξ^μ defined by

$$\phi = \xi_\mu \xi^\mu = g_{\mu\nu}\xi^\mu \xi^\nu. \tag{7.17}$$

This leads to

$$\begin{aligned} D_\lambda D_\rho \phi &= D_\lambda \left(2g_{\mu\nu} \left(D_\rho \xi^\mu \right) \xi^\nu \right) \\ &= 2g_{\mu\nu} \left(\left(D_\lambda D_\rho \xi^\mu \right) \xi^\nu + \left(D_\rho \xi^\mu \right) \left(D_\lambda \xi^\nu \right) \right), \end{aligned} \tag{7.18}$$

where we have used the metric compatibility condition (7.3).

It now follows from (7.18) that

$$\begin{aligned} [D_\lambda, D_\rho]\phi &= 2g_{\mu\nu}\xi^\nu [D_\lambda, D_\rho]\xi^\mu \\ &= 2g_{\mu\nu}\xi^\nu R^\mu{}_{\sigma\lambda\rho}\xi^\sigma \\ &= 2\xi^\nu \xi^\sigma R_{\nu\sigma\lambda\rho}. \end{aligned} \tag{7.19}$$

On the other hand, we know that since ϕ is a scalar function (see (4.179))

$$[D_\lambda, D_\rho]\phi = [\partial_\lambda, \partial_\rho]\phi = 0, \tag{7.20}$$

so that, comparing with (7.19), we have

$$2\xi^\nu \xi^\sigma R_{\nu\sigma\lambda\rho} = 0. \tag{7.21}$$

Since this is true for any arbitrary vector ξ^μ in an arbitrary curved manifold, this implies

$$R_{\nu\sigma\lambda\rho} = -R_{\sigma\nu\lambda\rho}, \tag{7.22}$$

and we conclude that the Riemann-Christoffel curvature tensor is anti-symmetric in the first two indices as well.

Furthermore, let us take an arbitrary vector ξ_μ and construct from it the completely anti-symmetric third rank tensor

$$\begin{aligned}
\xi_{[\mu;\nu;\lambda]} &= (\xi_{\mu;\nu;\lambda} - \xi_{\mu;\lambda;\nu}) + \text{cyclic permutation of } \mu\nu\lambda \\
&= [D_\lambda, D_\nu]\xi_\mu + \text{cyclic permutation of } \mu\nu\lambda \\
&= -R^\sigma{}_{\mu\lambda\nu}\xi_\sigma + \text{cyclic permutation of } \mu\nu\lambda, \tag{7.23}
\end{aligned}$$

where we have used (5.19). On the other hand, we also know that (see (4.181))

$$\xi_{[\mu;\nu;\lambda]} = 0. \tag{7.24}$$

Therefore, since ξ_μ is an arbitrary vector, it follows from (7.23) that (it can be checked easily that the cyclic permutation of $\mu\nu\lambda$ is the same as the cyclic permutation of $\mu\lambda\nu$)

$$R^\sigma{}_{\mu\lambda\nu}\xi_\sigma + \text{cyclic permutation of } \mu\lambda\nu = 0,$$

$$\text{or,} \quad R^\sigma{}_{\mu\lambda\nu} + \text{cyclic permutation of } \mu\lambda\nu = 0,$$

$$\text{or,} \quad R_{\sigma\mu\lambda\nu} + \text{cyclic permutation of } \mu\lambda\nu = 0, \tag{7.25}$$

which can be written explicitly as

$$R_{\sigma\mu\lambda\nu} + R_{\sigma\lambda\nu\mu} + R_{\sigma\nu\mu\lambda} = 0. \tag{7.26}$$

Alternatively, using the anti-symmetry property in the first two indices (7.22), we can also write this relation as

$$R_{\mu\sigma\lambda\nu} + R_{\lambda\sigma\nu\mu} + R_{\nu\sigma\mu\lambda} = 0. \tag{7.27}$$

With these relations, let us now look at the combination of the curvature tensors

$$R_{\mu\nu\lambda\rho} - R_{\lambda\rho\mu\nu}. \tag{7.28}$$

Using the cyclic identities (7.26) and (7.27), we determine

$$
\begin{aligned}
R_{\mu\nu\lambda\rho} - R_{\lambda\rho\mu\nu} &= -(R_{\mu\lambda\rho\nu} + R_{\mu\rho\nu\lambda}) + (R_{\mu\rho\nu\lambda} + R_{\nu\rho\lambda\mu}) \\
&= -R_{\mu\lambda\rho\nu} + R_{\nu\rho\lambda\mu} \\
&= (R_{\rho\lambda\nu\mu} + R_{\nu\lambda\mu\rho}) - (R_{\nu\lambda\mu\rho} + R_{\nu\mu\rho\lambda}) \\
&= -(R_{\mu\nu\lambda\rho} - R_{\lambda\rho\mu\nu}) = 0. \tag{7.29}
\end{aligned}
$$

Since this is a tensor equation it follows that

$$R_{\mu\nu\lambda\rho} - R_{\lambda\rho\mu\nu} = 0,$$

or, $$R_{\mu\nu\lambda\rho} = R_{\lambda\rho\mu\nu}. \tag{7.30}$$

Parenthetically we note that this relation can also be derived simply from

$$
\begin{aligned}
R_{\mu\nu\lambda\rho} &= -R_{\mu\lambda\rho\nu} - R_{\mu\rho\nu\lambda} \\
&= -R_{\mu\lambda\rho\nu} + (R_{\nu\rho\lambda\mu} + R_{\lambda\rho\mu\nu}), \\
R_{\mu\nu\lambda\rho} &= -R_{\lambda\nu\rho\mu} - R_{\rho\nu\mu\lambda} \\
&= (R_{\lambda\rho\mu\nu} + R_{\lambda\mu\nu\rho}) - R_{\rho\nu\mu\lambda}. \tag{7.31}
\end{aligned}
$$

This leads to

$$R_{\mu\lambda\rho\nu} - R_{\rho\nu\mu\lambda} = 0, \tag{7.32}$$

which coincides with (7.30).

We can now collect all the algebraic properties of the curvature tensor as (see (7.16), (7.22), (7.26) and (7.30))

$$R_{\mu\nu\lambda\rho} = -R_{\mu\nu\rho\lambda} = -R_{\nu\mu\lambda\rho} = R_{\lambda\rho\mu\nu},$$

$$R_{\mu\nu\lambda\rho} + R_{\mu\lambda\rho\nu} + R_{\mu\rho\nu\lambda} = 0. \tag{7.33}$$

This reduces the number of independent components of the curvature tensor to 20 ($36 - 12 - 4 = 20$). This is still a large number of components since we know that they have to be constructed out of ten components of the metric tensor. In fact, these 20 components depend on the choice of the coordinate system used.

Let us recall that the curvature tensor is defined as (see (5.18))

$$R^{\mu}_{\ \nu\lambda\rho} = \partial_{\rho}\Gamma^{\mu}_{\nu\lambda} - \partial_{\lambda}\Gamma^{\mu}_{\nu\rho} + \Gamma^{\mu}_{\lambda\sigma}\Gamma^{\sigma}_{\rho\nu} - \Gamma^{\mu}_{\rho\sigma}\Gamma^{\sigma}_{\lambda\nu}, \tag{7.34}$$

with the connection given by the Christoffel symbol (4.49)

$$\Gamma^{\mu}_{\nu\lambda} = -\frac{1}{2}\,g^{\mu\sigma}(\partial_{\nu}g_{\lambda\sigma} + \partial_{\lambda}g_{\sigma\nu} - \partial_{\sigma}g_{\nu\lambda}). \tag{7.35}$$

In a locally flat Cartesian coordinate system we have (see (7.4))

$$g_{\mu\nu} = \eta_{\mu\nu}, \quad \partial_{\sigma}g_{\mu\nu} = 0. \tag{7.36}$$

Therefore, in this coordinate system, we can write (the connection vanishes in this coordinate system)

$$R^{\mu}{}_{\nu\lambda\rho} = -\frac{1}{2}\,\eta^{\mu\sigma}(\partial_\rho\partial_\nu g_{\lambda\sigma} + \partial_\rho\partial_\lambda g_{\sigma\nu} - \partial_\rho\partial_\sigma g_{\nu\lambda}$$

$$-\,\partial_\lambda\partial_\nu g_{\rho\sigma} - \partial_\lambda\partial_\rho g_{\sigma\nu} + \partial_\lambda\partial_\sigma g_{\nu\rho}),$$

$$\text{or,}\quad R_{\mu\nu\lambda\rho} = -\frac{1}{2}\,(\partial_\rho\partial_\nu g_{\lambda\mu} - \partial_\rho\partial_\mu g_{\nu\lambda} + \partial_\lambda\partial_\mu g_{\nu\rho} - \partial_\lambda\partial_\nu g_{\rho\mu})$$

$$= -\frac{1}{2}\,(\partial_\rho\partial_\nu g_{\lambda\mu} - \partial_\rho\partial_\mu g_{\lambda\nu} + \partial_\lambda\partial_\mu g_{\nu\rho} - \partial_\lambda\partial_\nu g_{\mu\rho}). \tag{7.37}$$

Written in this form the symmetry properties of the curvature tensor (7.33) are manifest and they continue to hold in any coordinate frame since the symmetry properties are respected by a general coordinate transformation.

Let us also note that we can obtain other symmetry relations of the curvature tensor which are not algebraic in nature. These are known as the Bianchi identities. (See (4.190) for the Bianchi identity in the electromagnetic theory.) To derive these, let us consider a general vector ξ_μ. From (5.19) we know that

$$[D_\lambda, D_\rho]\xi_\mu = \xi_{\mu;[\rho;\lambda]} = R_{\mu\nu\lambda\rho}\xi^\nu. \tag{7.38}$$

Let us further differentiate this covariantly which leads to

$$\xi_{\mu;[\rho;\lambda];\sigma} = R_{\mu\nu\lambda\rho;\sigma}\xi^\nu + R_{\mu\nu\lambda\rho}\xi^\nu{}_{;\sigma}. \tag{7.39}$$

On the other hand, we note that (see (5.21) and note that the metric tensor is covariantly flat)

$$\begin{aligned}\xi_{\mu;\sigma;[\rho,\lambda]} &= [D_\lambda, D_\rho]\xi_{\mu;\sigma} \\ &= R_{\mu\nu\lambda\rho}\,\xi^\nu{}_{;\sigma} + R_{\sigma\nu\lambda\rho}\,\xi_\mu{}^{;\nu}. \end{aligned} \tag{7.40}$$

If we anti-symmetrize both sides of (7.39) with respect to $\rho\lambda\sigma$ we obtain

$$\xi_{\mu;[\rho;\lambda;\sigma]} = \xi_{\mu;[\rho;\lambda];\sigma} + \text{cyclic permutation of } \rho\lambda\sigma$$

$$= \left[R_{\mu\nu\lambda\rho;\sigma}\xi^\nu + R_{\mu\nu\lambda\rho}\xi^\nu{}_{;\sigma}\right] + \text{cyclic permutation of } \rho\lambda\sigma, \tag{7.41}$$

while the anti-symmetrization of (7.40) leads to

$$\xi_{\mu;[\rho;\lambda;\sigma]} = \xi_{\mu;\sigma;[\rho;\lambda]} + \text{cyclic permutation of } \rho\lambda\sigma$$

$$= \left[R_{\mu\nu\lambda\rho}\xi^{\nu}{}_{;\sigma} + R_{\sigma\nu\lambda\rho}\xi_{\mu;}{}^{\nu} \right] + \text{cyclic permutation of } \rho\lambda\sigma.$$

$$(7.42)$$

The second term on the right hand side of (7.42) vanishes because of the cyclic identity (7.27). As a result, from (7.41) and (7.42) we obtain

$$\xi_{\mu;[\rho;\lambda;\sigma]} = R_{\mu\nu\lambda\rho;\sigma}\xi^{\nu} + R_{\mu\nu\lambda\rho}\xi^{\nu}{}_{;\sigma} + \text{cyclic permutation of } \rho\lambda\sigma$$

$$= R_{\mu\nu\lambda\rho}\xi^{\nu}{}_{;\sigma} + \text{cyclic permutation of } \rho\lambda\sigma, \qquad (7.43)$$

which yields

$$R_{\mu\nu\lambda\rho;\sigma}\xi^{\nu} + \text{cyclic permutation of } \rho\lambda\sigma = 0. \qquad (7.44)$$

Since this must be true for any arbitrary vector ξ^{ν}, we conclude that

$$R_{\mu\nu\lambda\rho;\sigma} + \text{cyclic permutation of } \rho\lambda\sigma = 0. \qquad (7.45)$$

This is known as the Bianchi identity (in gravitation) and is quite useful in the study of gravitation as we will see.

Let us note here that we can contract various tensor indices of the Riemann-Christoffel curvature tensor to obtain other curvature tensors of lower rank. However, because of the anti-symmetry of the curvature tensor in various indices, the number of contractions we can make is limited. Let us define

$$R_{\mu\lambda} = g^{\nu\rho}R_{\mu\nu\lambda\rho} = R_{\mu\nu\lambda}{}^{\nu} = R_{\mu}{}^{\nu}{}_{\lambda\nu}. \qquad (7.46)$$

It is clear from the definition in (7.46) that

$$R_{\mu\lambda} = R^{\nu}{}_{\mu\nu\lambda} = -R^{\nu}{}_{\mu\lambda\nu} = -R_{\mu}{}^{\nu}{}_{\nu\lambda}, \qquad (7.47)$$

where we have used the symmetry properties in (7.33). The second rank tensor $R_{\mu\nu}$ is known as the contracted Riemann curvature or the Ricci tensor. It is symmetric in the two indices as can be seen from

$$R_{\lambda\mu} = g^{\nu\rho} R_{\lambda\nu\mu\rho} = g^{\nu\rho} R_{\mu\rho\lambda\nu} = R_{\mu\lambda}, \qquad (7.48)$$

where we have used (7.30). Therefore, the Ricci tensor is a second rank symmetric tensor with only ten independent components (exactly like the metric tensor which is also a second rank symmetric tensor). Furthermore, if we contract the indices of the Ricci tensor, we obtain the Ricci scalar curvature, namely

$$R = g^{\mu\lambda} R_{\mu\lambda}. \qquad (7.49)$$

Note that this is the only nontrivial scalar curvature that we can construct from the Riemann tensor. The only other possibility which can be constructed using the Levi-Civita tensor vanishes,

$$\epsilon^{\mu\nu\lambda\rho} R_{\mu\nu\lambda\rho} = 0, \qquad (7.50)$$

because of the cyclicity identity (7.26) satisfied by the Riemann tensor.

Let us now examine the consequences of the Bianchi identity (7.45) for the Ricci tensor. To see this, we recall that

$$R_{\mu\nu\lambda\rho;\sigma} + \text{cyclic permutation of } \rho\lambda\sigma = 0. \qquad (7.51)$$

Since the metric is flat with respect to covariant differentiation (4.103), we now obtain

$$g^{\nu\rho} \left(R_{\mu\nu\lambda\rho;\sigma} + \text{cyclic permutation of } \rho\lambda\sigma \right) = 0,$$

or, $\quad g^{\nu\rho}(R_{\mu\nu\lambda\rho;\sigma} + R_{\mu\nu\rho\sigma;\lambda} + R_{\mu\nu\sigma\lambda;\rho}) = 0,$

or, $\quad R_{\mu\lambda;\sigma} - R_{\mu\sigma;\lambda} - R^{\rho}{}_{\mu\sigma\lambda;\rho} = 0. \qquad (7.52)$

Contracting further the indices μ and λ this leads to

$$g^{\mu\lambda}\left(R_{\mu\lambda;\sigma} - R_{\mu\sigma;\lambda} - R^{\rho}_{\ \mu\sigma\lambda;\rho}\right) = 0,$$

or, $$R_{;\sigma} - R^{\mu}_{\ \sigma;\mu} - R^{\rho}_{\ \sigma;\rho} = 0,$$

or, $$R^{\mu}_{\ \nu;\mu} - \frac{1}{2}\,R_{;\nu} = 0,$$

or, $$R^{\mu\nu}_{\ \ ;\mu} - \frac{1}{2}\,R_{;}^{\ \nu} = 0. \tag{7.53}$$

If we define a second rank symmetric tensor from the Ricci tensor
(7.46) and the Ricci scalar (7.49) as

$$G^{\mu\nu} = R^{\mu\nu} - \frac{1}{2}\,g^{\mu\nu}R, \tag{7.54}$$

then, it is clear from (7.53) that we can write

$$D_{\mu}G^{\mu\nu} = R^{\mu\nu}_{\ \ ;\mu} - \frac{1}{2}\,R_{;}^{\ \nu} = 0, \tag{7.55}$$

which shows that this tensor is conserved in a curved manifold. $G^{\mu\nu}$
is known as the Einstein curvature tensor and with these preparations
we are now ready to discuss Einstein's equations.

▶ **Example (Symmetry in inertial frame).** We recall from (7.37) that in an inertial frame defined by (7.4)

$$g_{\mu\nu} = \eta_{\mu\nu}, \quad \partial_{\sigma}g_{\mu\nu} = 0, \tag{7.56}$$

so that in this frame we have

$$\Gamma^{\sigma}_{\mu\nu} = 0, \tag{7.57}$$

we can write

$$R_{\mu\nu\lambda\rho} = -\frac{1}{2}\left(\partial_{\rho}\partial_{\nu}g_{\lambda\mu} - \partial_{\rho}\partial_{\mu}g_{\lambda\nu} + \partial_{\lambda}\partial_{\mu}g_{\nu\rho} - \partial_{\lambda}\partial_{\nu}g_{\mu\rho}\right). \tag{7.58}$$

Since the metric tensor is symmetric and the derivatives commute, namely,

$$g_{\mu\nu} = g_{\nu\mu}, \quad \partial_\mu \partial_\nu = \partial_\nu \partial_\mu, \tag{7.59}$$

it follows from the definition in (7.58) that (see (7.33))

$$R_{\mu\nu\lambda\rho} = -R_{\nu\mu\lambda\rho} = -R_{\mu\nu\rho\lambda} = R_{\lambda\rho\mu\nu}. \tag{7.60}$$

Similarly, from (7.58) we also obtain

$$R_{\mu\nu\lambda\rho} + R_{\mu\lambda\rho\nu} + R_{\mu\rho\nu\lambda}$$

$$= -\frac{1}{2}\Big(\partial_\rho\partial_\nu g_{\lambda\mu} - \partial_\rho\partial_\mu g_{\lambda\nu} + \partial_\lambda\partial_\mu g_{\nu\rho} - \partial_\lambda\partial_\nu g_{\mu\rho}$$

$$+ \partial_\nu\partial_\lambda g_{\rho\mu} - \partial_\nu\partial_\mu g_{\rho\lambda} + \partial_\rho\partial_\mu g_{\lambda\nu} - \partial_\rho\partial_\lambda g_{\mu\nu}$$

$$+ \partial_\lambda\partial_\rho g_{\nu\mu} - \partial_\lambda\partial_\mu g_{\nu\rho} + \partial_\nu\partial_\mu g_{\rho\lambda} - \partial_\nu\partial_\rho g_{\mu\lambda}\Big)$$

$$= 0. \tag{7.61}$$

Finally, we note that since in the inertial frame the connection vanishes (see (7.57)) we have

$$R_{\mu\nu\lambda\rho;\sigma} = \partial_\sigma R_{\mu\nu\lambda\rho}, \tag{7.62}$$

which leads to

$$R_{\mu\nu\lambda\rho;\sigma} + R_{\mu\nu\rho\sigma;\lambda} + R_{\mu\nu\sigma\lambda;\rho}$$

$$= -\frac{1}{2}\Big(\partial_\sigma(\partial_\rho\partial_\nu g_{\lambda\mu} - \partial_\rho\partial_\mu g_{\lambda\nu} + \partial_\lambda\partial_\mu g_{\nu\rho} - \partial_\lambda\partial_\nu g_{\mu\rho})$$

$$+ \partial_\lambda(\partial_\sigma\partial_\nu g_{\rho\mu} - \partial_\sigma\partial_\mu g_{\rho\nu} + \partial_\rho\partial_\mu g_{\nu\sigma} - \partial_\rho\partial_\nu g_{\mu\sigma})$$

$$+ \partial_\rho(\partial_\lambda\partial_\nu g_{\sigma\mu} - \partial_\lambda\partial_\mu g_{\sigma\nu} + \partial_\sigma\partial_\mu g_{\nu\lambda} - \partial_\sigma\partial_\nu g_{\mu\lambda})\Big)$$

$$= 0, \tag{7.63}$$

which coincides with (7.45). Thus, the symmetry properties of the curvature tensor are easily checked in a locally inertial frame and being tensor symmetries, they hold in any frame.

◀

▶ **Example (Curvature in two dimensional space).** Let us consider the two dimensional Euclidean space. In two dimensions $R_{\mu\nu\lambda\rho}$ must have the symmetries already mentioned in (7.60) and (7.61) which implies (in two dimensions)

$$R_{1111} = R_{1122} = R_{1112} = R_{2211} = R_{2212} = R_{2222} = 0, \qquad (7.64)$$

by antisymmetry (for example, $R_{1112} = -R_{1112} = 0$). Therefore, the only nontrivial component of the curvature in two dimensions, from symmetry considerations, follows to be

$$R_{1212} = R_{2121} = -R_{2112} = -R_{1221}. \qquad (7.65)$$

This is in agreement with the fact that there is only one linearly independent term in $R_{\mu\nu\lambda\rho}$ in two dimensions. We can relate this curvature to the Gaussian curvature discussed in chapter 1 (see (1.19)) in the following way.

From the definition of the Christoffel symbol (4.49) we can calculate and obtain

$$\Gamma^1_{11} = -\frac{1}{2}\, g^{11} \partial_1 g_{11} - \frac{1}{2}\, g^{12} \left(2\partial_1 g_{12} - \partial_2 g_{11} \right),$$

$$\Gamma^1_{12} = \Gamma^1_{21} = -\frac{1}{2}\, g^{11} \partial_2 g_{11} - \frac{1}{2}\, g^{12} \partial_1 g_{22},$$

$$\Gamma^1_{22} = -\frac{1}{2}\, g^{11} \left(2\partial_2 g_{12} - \partial_1 g_{22} \right) - \frac{1}{2}\, g^{12} \partial_2 g_{22},$$

$$\Gamma^2_{11} = -\frac{1}{2}\, g^{21} \partial_1 g_{11} - \frac{1}{2}\, g^{22} \left(2\partial_1 g_{12} - \partial_2 g_{11} \right),$$

$$\Gamma^2_{12} = \Gamma^2_{21} = -\frac{1}{2}\, g^{21} \partial_2 g_{11} - \frac{1}{2}\, g^{22} \partial_1 g_{22},$$

$$\Gamma^2_{22} = -\frac{1}{2}\, g^{21} \left(2\partial_2 g_{21} - \partial_1 g_{22} \right) - \frac{1}{2}\, g^{22} \partial_2 g_{22}. \qquad (7.66)$$

The Riemann-Christoffel curvature tensor can be calculated from the definition in (5.18) as well as

$$R_{\mu\nu\lambda\rho} = g_{\mu\sigma} R^{\sigma}{}_{\nu\lambda\rho}. \qquad (7.67)$$

However, we note that in two dimensions, it is much easier to calculate

$$g R_{\mu\nu\lambda\rho}, \qquad (7.68)$$

where g denotes the determinant of the metric tensor.

In two dimensions, the determinant of the metric tensor has the simple form

$$g = g_{11} g_{22} - g_{12} g_{21}, \qquad (7.69)$$

which leads to

$$gg^{11} = g_{22}, \quad gg^{22} = g_{11}, \quad gg^{12} = gg^{21} = -g_{12}. \tag{7.70}$$

In deriving (7.70) we have used the two dimensional identities $(g^{\mu\lambda}g_{\lambda\nu} = \delta^\mu_\nu)$

$$g^{11}g_{11} + g^{12}g_{21} = 1,$$
$$g^{22}g_{22} + g^{21}g_{12} = 1,$$
$$g^{11}g_{12} + g^{12}g_{22} = 0,$$
$$g^{21}g_{11} + g^{22}g_{21} = 0. \tag{7.71}$$

With this, we can now calculate (7.68) which leads to

$$
\begin{aligned}
gR_{1212} &= \frac{g}{2}\left(2\partial_1\partial_2 g_{12} - \partial_2^2 g_{11} - \partial_1^2 g_{22}\right) \\
&\quad - \frac{g_{22}}{4}\left[\partial_1 g_{11}\left(2\partial_2 g_{12} - \partial_1 g_{22}\right) - \left(\partial_2 g_{11}\right)^2\right] \\
&\quad + \frac{g_{12}}{4}\left[\partial_1 g_{11}\partial_2 g_{22} - 2\partial_2 g_{11}\partial_1 g_{22} + \left(2\partial_1 g_{12} - \partial_2 g_{11}\right)\left(2\partial_2 g_{12} - \partial_1 g_{22}\right)\right] \\
&\quad - \frac{g_{11}}{4}\left[\partial_2 g_{22}\left(2\partial_1 g_{12} - \partial_2 g_{11}\right) - \left(\partial_1 g_{22}\right)^2\right] \\
&\equiv g^2 K, \tag{7.72}
\end{aligned}
$$

where K is the Gaussian curvature defined in (1.19). Therefore, we conclude that, in two dimensions,

$$K = \frac{1}{g} R_{1212}, \tag{7.73}$$

is the unique curvature tensor. Alternatively, we can write

$$R_{1212} = gK, \tag{7.74}$$

and keeping in mind (7.69) we recognize that in two dimensions (7.74) allows us to write the Riemann-Christoffel tensor as

$$R_{\mu\nu\lambda\rho} = K\left(g_{\mu\lambda}g_{\nu\rho} - g_{\mu\rho}g_{\nu\lambda}\right), \tag{7.75}$$

which has all the symmetries (7.64) and (7.65) manifest. Furthermore, from (7.75), the Ricci tensor follows to be (in two dimensions)

$$R_{\mu\nu} = g^{\sigma\rho}R_{\mu\sigma\nu\rho} = Kg_{\mu\nu}, \tag{7.76}$$

leading to the Ricci scalar

$$R = g^{\mu\nu} R_{\mu\nu} = 2K. \tag{7.77}$$

◄

7.5 Einstein's equation

In the last chapter we extended equations for various particles and fields to a curved manifold (in the presence of gravitation). This involved simply taking the relativistic dynamical equations in a flat manifold and covariantizing them. A natural question that comes to mind is, of course, what is the equation for the gravitational potential itself. This is particularly interesting because we know the equation for the gravitational potential only in the non-relativistic limit. Classically we know that the static gravitational potential satisfies the Poisson equation. Namely, if $\phi(\mathbf{x})$ denotes the gravitational potential, then it satisfies the Laplace equation

$$\boldsymbol{\nabla}^2 \phi(\mathbf{x}) = 0, \tag{7.78}$$

in regions free of gravitational sources (masses). However, if there is a distribution of (non-relativistic) matter in a given region, then the gravitational potential satisfies the Poisson equation

$$\boldsymbol{\nabla}^2 \phi(\mathbf{x}) = 4\pi G_{\mathrm{N}} \rho(\mathbf{x}), \tag{7.79}$$

where G_{N} is Newton's constant and $\rho(\mathbf{x})$ describes the matter density in the given region. In CGS units, Newton's constant has the value

$$G_{\mathrm{N}} \approx 6.67 \times 10^{-8} \mathrm{cm}^3 \mathrm{g}^{-1} \mathrm{sec}^{-2}. \tag{7.80}$$

Equation (7.79) can be seen easily to hold by noting that the gravitational potential for a continuous distribution of matter can be written as

$$\phi(\mathbf{x}) = -G_{\mathrm{N}} \int \mathrm{d}^3 x' \, \frac{\rho(\mathbf{x}')}{|\mathbf{x} - \mathbf{x}'|}, \tag{7.81}$$

so that

$$
\begin{aligned}
\mathbf{\nabla}^2 \phi(\mathbf{x}) &= -G_N \int \mathrm{d}^3 x' \, \rho(\mathbf{x}') \, \mathbf{\nabla}^2 \left(\frac{1}{|\mathbf{x} - \mathbf{x}'|} \right) \\
&= -G_N \int \mathrm{d}^3 x' \, \rho(\mathbf{x}') \left(- 4\pi \delta^3(\mathbf{x} - \mathbf{x}') \right) \\
&= 4\pi G_N \rho(\mathbf{x}).
\end{aligned}
\tag{7.82}
$$

We have also seen in (6.21) that in the weak field limit, we can identify

$$
g_{00}(\mathbf{x}) = 1 + 2\phi(\mathbf{x}), \qquad g_{ij} = \eta_{ij}.
\tag{7.83}
$$

Thus if we assume that the metric tensor $g_{\mu\nu}$ represents the gravitational potential, then the equations for $g_{\mu\nu}$ must involve second order derivatives of the metric and must reduce to the respective equations (7.78) and (7.79) in the appropriate limit for the two cases (matter free and in the presence of a distribution of matter). Furthermore, we are looking for at least ten equations since $g_{\mu\nu}$ has ten components. These must be tensor equations since $g_{\mu\nu}$ is a tensor and the equations are expected to be covariant under a general coordinate transformation.

Let us now look at the geodesic deviation equation which we have derived in (7.14)

$$
\frac{\mathrm{D}^2 v^\mu}{\mathrm{D}\tau^2} = R^\mu{}_{\nu\rho\lambda} u^\nu u^\rho v^\lambda,
\tag{7.84}
$$

where

$$
u^\mu = \frac{\partial x^\mu}{\partial \tau}, \quad v^\mu = \frac{\partial x^\mu}{\partial h}.
\tag{7.85}
$$

Infinitesimally, therefore, we can write down the geodesic deviation as satisfying the equation

$$
\frac{\mathrm{D}^2 \delta x^\mu}{\mathrm{D}\tau^2} = R^\mu{}_{\nu\rho\lambda} \frac{\mathrm{d}x^\nu}{\mathrm{d}\tau} \frac{\mathrm{d}x^\rho}{\mathrm{d}\tau} \delta x^\lambda = K^\mu{}_\lambda \, \delta x^\lambda,
\tag{7.86}
$$

where we have identified

$$K^{\mu}{}_{\lambda} = R^{\mu}{}_{\nu\rho\lambda} \frac{dx^{\nu}}{d\tau} \frac{dx^{\rho}}{d\tau}. \tag{7.87}$$

Let us now consider the equivalent problem in classical Newtonian physics. Let us assume that a particle of unit mass is subjected to a static, space dependent gravitational force. In this case, from Newton's equation (see (6.6)) we have

$$\frac{d^2 x^i}{dt^2} = -\partial^i \phi\big|_x. \tag{7.88}$$

Similarly, a nearby geodesic (trajectory) satisfies

$$\frac{d^2 \tilde{x}^i}{dt^2} = -\partial^i \phi\big|_{\tilde{x}}. \tag{7.89}$$

If these trajectories are infinitesimally apart then we can write

$$\delta x^i = \tilde{x}^i - x^i, \tag{7.90}$$

and

$$\partial^i \phi(\mathbf{x})\big|_{\tilde{x}} = \partial^i \phi(\mathbf{x})\big|_x + \delta x^j \partial^i \partial_j \phi(\mathbf{x})\big|_x + O(\delta x^2), \tag{7.91}$$

which leads to

$$\frac{d^2 \tilde{x}^i}{dt^2} - \frac{d^2 x^i}{dt^2} = -\partial^i \phi(\mathbf{x})\big|_{\tilde{x}} + \partial^i \phi(\mathbf{x})\big|_x,$$

$$\text{or,} \quad \frac{d^2 \delta x^i}{dt^2} \simeq -\delta x^j \partial^i \partial_j \phi(\mathbf{x})\big|_x. \tag{7.92}$$

Comparing this with the geodesic deviation equation (7.86) in the curved space we can identify

$$K^i{}_j = -\partial^i \partial_j \phi. \tag{7.93}$$

We know that in empty space we can write (see (7.78))

$$\nabla^2 \phi = -\partial^i \partial_i \phi = K^i{}_i = 0. \tag{7.94}$$

Therefore, we feel comfortable in generalizing this equation to curved space, free of matter, as

$$K^\mu{}_\mu = R^\mu{}_{\nu\rho\mu} \frac{dx^\nu}{d\tau} \frac{dx^\rho}{d\tau} = 0,$$

or, $$-R_{\nu\rho} \frac{dx^\nu}{d\tau} \frac{dx^\rho}{d\tau} = 0. \tag{7.95}$$

Since this must be true for any arbitrary vector $\frac{dx^\mu}{d\tau}$, we have

$$R_{\nu\rho} = 0, \tag{7.96}$$

in a region free of matter (gravitational source).

In other words in empty space we can write the gravitational field equations as

$$R_{\mu\nu} = 0. \tag{7.97}$$

This is a tensor equation and since $R_{\mu\nu}$ has ten components, there are ten equations. Furthermore, the curvature involves second derivatives of the metric tensor and hence it is plausible that this is the correct equation in the case of empty space. We also note here that the more stringent equation

$$R^\mu{}_{\nu\lambda\rho} = 0, \tag{7.98}$$

involves more than ten equations and leads to the trivial flat spacetime solution for the metric and, therefore, has to be rejected.

To derive the equation in the presence of matter, let us note that the classical non-relativistic (static) equation (7.79) can be written in the weak field limit in the form

$$\nabla^2 \phi = 4\pi G_N \rho,$$

$$\text{or,} \quad \nabla^2 g_{00} = 8\pi G_N T_{00}. \tag{7.99}$$

Therefore, we recognize that generalization to curved space-time would involve the stress tensor on the right hand side of the equation (as source for the gravitational field). As we have seen in (2.113), $T^{\mu\nu}$ is a symmetric second rank tensor and hence it is tempting to generalize (7.97) to the case of non-empty space as

$$R^{\mu\nu} = \alpha T^{\mu\nu}, \tag{7.100}$$

where $T^{\mu\nu}$ is the stress tensor (associated with matter including radiation and other non-gravitational fields) and α is a normalization constant to be determined. This is a genuine tensor equation which reduces to the free space equation (7.97) in the source free case when

$$T^{\mu\nu} = 0. \tag{7.101}$$

However, this equation suffers from the defect that it is not a consistent equation. That is, whereas the stress tensor is conserved (see (6.135))

$$D_\mu T^{\mu\nu} = 0, \tag{7.102}$$

the Ricci tensor is not in general

$$D_\mu R^{\mu\nu} \neq 0. \tag{7.103}$$

Historically, equation (7.100) is what Einstein had suggested originally to describe the dynamics of the gravitational field. However, very quickly he as well as Poincaré, independently, realized the inconsistency in the equation and generalized the equation to

$$G^{\mu\nu} = \alpha T^{\mu\nu}, \tag{7.104}$$

where $G^{\mu\nu}$ is the Einstein tensor defined in (7.54). As we have seen in (7.55)

$$D_\mu G^{\mu\nu} = D_\mu \left(R^{\mu\nu} - \frac{1}{2} g^{\mu\nu} R \right) = 0, \tag{7.105}$$

so that the equation (7.104) is at least consistent. Furthermore, we note that in empty space (7.104) reduces to

$$G^{\mu\nu} = R^{\mu\nu} - \frac{1}{2} g^{\mu\nu} R = 0. \tag{7.106}$$

This seems to be different from (7.97), but we note from (7.106) that

$$g_{\mu\nu} G^{\mu\nu} = R - \frac{4}{2} R = 0,$$

$$\text{or,} \quad R = 0. \tag{7.107}$$

Hence using this in (7.106), we obtain

$$G^{\mu\nu} = R^{\mu\nu} = 0, \tag{7.108}$$

in the case of empty space. Therefore, in the source free limit, (7.104) indeed reduces to the free space equation (7.97).

The constant α in (7.104) is a normalization constant to be determined shortly. We note that the field equations also lead to the relation that

$$g_{\mu\nu} G^{\mu\nu} = g_{\mu\nu} \left(R^{\mu\nu} - \frac{1}{2} g^{\mu\nu} R \right) = \alpha g_{\mu\nu} T^{\mu\nu},$$

$$\text{or,} \quad R - \frac{4}{2} R = \alpha T,$$

$$\text{or,} \quad R = -\alpha T, \tag{7.109}$$

so that we can write the Einstein equation (7.104) also as

$$R^{\mu\nu} = \alpha \left(T^{\mu\nu} - \frac{1}{2} g^{\mu\nu} T \right). \tag{7.110}$$

Here $T = T^\mu_\mu$ denotes the trace of the stress tensor. Furthermore, the stress tensor is that associated with matter particles and fields (not including gravitation).

Let us now see whether the Einstein equation (7.104) (or (7.110)) reduces to Poisson equation (7.79) and if so what is the value of the normalization constant α for which this would be true. To simplify the derivation, we assume that matter consists of static dust particles (without any radiation field) for which we can write the stress tensor as

$$T^{\mu\nu} = \rho u^\mu u^\nu = \rho \, \frac{\mathrm{d}x^\mu}{\mathrm{d}\tau} \frac{\mathrm{d}x^\nu}{\mathrm{d}\tau}. \tag{7.111}$$

Since the dust particles are static $(t = \tau)$, then it follows that

$$
\begin{aligned}
T^{00} &= \rho, \\
T^{0i} &= T^{ij} = 0, \\
T &= g_{\mu\nu}T^{\mu\nu} = g_{00}T^{00} = g_{00}\rho \simeq \rho.
\end{aligned}
\tag{7.112}
$$

Here we have assumed that the metric changes only slightly from the Minkowski metric and correspondingly have kept only the dominant term in the trace in the last relation in (7.112).

Assuming that the gravitational potential is weak, we can approximate the Riemann curvature (5.18) as

$$R^\mu{}_{\nu\lambda\rho} \simeq \partial_\rho \Gamma^\mu_{\nu\lambda} - \partial_\lambda \Gamma^\mu_{\nu\rho}, \tag{7.113}$$

so that the Ricci tensor (7.46) takes the form

$$R_{\nu\rho} = \partial_\rho \Gamma^\mu_{\nu\mu} - \partial_\mu \Gamma^\mu_{\nu\rho}. \tag{7.114}$$

Furthermore, since only T_{00} is nonzero (see (7.112)), it is sufficient to look at the R_{00} equation in (7.110) to determine the constant α. From (7.114) we have

$$R_{00} = \partial_0 \Gamma^\mu_{0\mu} - \partial_\mu \Gamma^\mu_{00}. \tag{7.115}$$

However, since we are assuming the gravitational potential to be static, the metric tensor would not involve time and we can write

$$
\begin{aligned}
R_{00} &= -\partial_\mu \Gamma^\mu_{00} = -\partial_i \Gamma^i_{00} \\
&= -\partial_i \left(-\frac{1}{2} \, g^{ij} (\partial_0 g_{0j} + \partial_0 g_{j0} - \partial_j g_{00}) \right) \\
&= -\frac{1}{2} \, \partial_i \partial^i g_{00} = \frac{1}{2} \, \boldsymbol{\nabla}^2 g_{00}.
\end{aligned}
\tag{7.116}
$$

Furthermore, we recall that for a weak field, we can write

$$
g_{00} = 1 + 2\phi,
\tag{7.117}
$$

where ϕ is the gravitational potential so that the field equation in (7.110) leads to

$$
R^{00} \simeq R_{00} = \alpha \left(T^{00} - \frac{1}{2} \, g^{00} T \right),
$$

or, $\quad \dfrac{1}{2} \, \boldsymbol{\nabla}^2 (2\phi) = \alpha \left(\rho - \dfrac{1}{2} \, \rho \right) = \dfrac{\alpha}{2} \, \rho,$

or, $\quad \boldsymbol{\nabla}^2 \phi = \dfrac{\alpha}{2} \, \rho.$
$$\tag{7.118}$$

Comparing this with the Poisson equation (7.79)

$$
\boldsymbol{\nabla}^2 \phi = 4\pi G_{\mathrm{N}} \rho,
\tag{7.119}
$$

we determine

$$
\alpha = 8\pi G_{\mathrm{N}},
\tag{7.120}
$$

so that the field equations (7.104) and (7.110) in the presence of matter can be written as (G_{N} is Newton's constant)

$$G^{\mu\nu} \;=\; \left(R^{\mu\nu} - \frac{1}{2}\, g^{\mu\nu} R\right) = 8\pi G_{\mathrm{N}} T^{\mu\nu},$$

$$\text{or,} \quad R^{\mu\nu} \;=\; 8\pi G_{\mathrm{N}} \left(T^{\mu\nu} - \frac{1}{2}\, g^{\mu\nu} T\right). \tag{7.121}$$

Both forms of these equations are known as Einstein's equations and it is clear that they relate the distribution of matter and energy to the curvature associated with the space-time manifold.

Let us note here that different authors use different conventions for the metric as well as for the definitions of the Christoffel symbol, Riemann tensor, Ricci tensor etc. Consequently, the sign of the term on the right hand side of the Einstein equation (7.121) depends on the convention used to define various quantities. However, consistency requires that in the weak field limit they reduce to the Poisson equation (7.79).

7.6 Cosmological constant

Let us note here that the Einstein tensor $G^{\mu\nu}$ is not the only tensor that can be constructed from the metric tensor and its first and second derivatives which is conserved. In fact, it is clear that one can write a modified equation

$$R^{\mu\nu} - \frac{1}{2}\, g^{\mu\nu} R - \Lambda g^{\mu\nu} = 8\pi G_{\mathrm{N}} T^{\mu\nu}, \tag{7.122}$$

where Λ is a constant known as the cosmological constant and since $g^{\mu\nu}$ is constant under covariant differentiation (see (4.103)), this term does not violate any consistency condition. However, physically one can think of this term as a constant gravitational force acting on all particles. Namely, we can take it to the right hand side of the equation and absorb it into the definition of $T^{\mu\nu}$. Then in the flat space-time limit, it would correspond to a universal constant force acting on all particles. This term is known as the cosmological constant term and historically Einstein had introduced this term into the field equations for gravity in order to obtain a static solution of the field equations. However, we note here that if this modified equation were the true equation, then it has to reduce to the right Newtonian limit. Since Newton's equation works so well in the classical domain, this puts a limit on the magnitude of the cosmological constant Λ. In fact, just looking at the equation, we note that since the metric is dimensionless, the canonical dimension of Λ must be (recall that the curvature involves second order derivatives of the metric)

$$[\Lambda] = [L]^{-2}. \tag{7.123}$$

Therefore, its value must be related to a characteristic length in the problem. The only meaningful characteristic length in the study of the universe is its radius. The observational measurements give a value for this radius to be

$$
\begin{aligned}
\text{Radius} \quad &> \quad 10^{10} \text{ light-yrs} \\
&\approx \quad 10^{10} \times (3 \times 10^{10} \text{ cms/sec}) \times (3 \times 10^{7}\text{sec}) \\
&= \quad 9 \times 10^{27} \text{ cms} \simeq 10^{28} \text{ cms}. \tag{7.124}
\end{aligned}
$$

(The present observed value for the age of the universe is 13.7 billion years leading to a radius of 78 billion light-years.) The theoretical limit on the cosmological constant can then be written as and compared with the observed value

$$
\begin{aligned}
\Lambda_{\text{th}} &\leq \frac{1}{(\text{Radius})^2} = 10^{-56} \text{ cm}^{-2} \approx 10^{-66} \text{ (eV)}^2, \\
\Lambda_{\text{obs}} &\simeq 10^{-67} \text{ (eV)}^2. \tag{7.125}
\end{aligned}
$$

This simple argument gives the right order of magnitude for the cosmological constant. One of the burning questions in physics today is why the cosmological constant is so small and how to derive this within the context of physical theories.

7.7 Initial value problem

Let us now look at the structure of Einstein's equations. Although what we say is true in the presence of matter, for simplicity we will restrict only to the case of the gravitational field equations in empty space (7.97)

$$G_{\mu\nu}(g) = R_{\mu\nu}(g) = 0. \tag{7.126}$$

This is a tensor equation and there are ten equations involving the ten components of the metric tensor. Furthermore, this involves the first and the second order derivatives of the metric tensor as well as the metric tensor itself. However, unlike other simple physical equations which we come across in physics, here the equation is highly nonlinear. That is although the second order derivative terms are linear, the first order derivative terms are nonlinear. Therefore, the superposition principle which is so dear to us becomes inapplicable here.

The Cauchy initial value problem also suffers from difficulties. Since this is a second order differential equation, if we give the value of the metric and its first derivative at an initial time x^0, namely, if

$$g_{\mu\nu}(\mathbf{x}, x^0) = f_{\mu\nu}(\mathbf{x}), \qquad \left.\frac{\partial g_{\mu\nu}}{\partial t}\right|_{x^0} = h_{\mu\nu}(\mathbf{x}), \tag{7.127}$$

are known, then we would expect that we can determine the metric components at all later times uniquely. However, in the case of Einstein's equation this is not possible. This can be seen simply from the fact that although the equation

$$G^{\mu\nu} = 0, \tag{7.128}$$

seems like ten equations, the Einstein tensor satisfies the divergence free condition (7.55)

$$D_\mu G^{\mu\nu} = 0. \tag{7.129}$$

These are four constraint equations and hence the Einstein equation represents truly six independent equations. The situation here is analogous to Maxwell's equations

$$\partial_\mu F^{\mu\nu} = J^\nu, \tag{7.130}$$

which define a set of four equations and involve second order derivatives of the vector potential. Thus we expect that if the vector potential $A_\mu(x)$ and its time derivative $\dot{A}_\mu(x)$ are given at some initial time x^0, then we can determine the value of the vector potential uniquely at any later time. However, as we know this is not true. This is because, there exist identities of the form

$$\partial_\nu \partial_\mu F^{\mu\nu} = \partial_\nu J^\nu = 0. \tag{7.131}$$

Hence Maxwell's equations truly represent three independent equations. The reason for the existence of constraints in the case of Maxwell's equations is understood as the consequence of gauge invariance in this theory. Namely, under

$$A_\mu \to A_\mu + \partial_\mu \alpha(x), \tag{7.132}$$

Maxwell's equations remain invariant. This also explains why the Cauchy initial value problem cannot be solved uniquely in the case of Maxwell's theory because, if we determine one solution which satisfies the initial conditions, we can always make a gauge transformation (with the parameter of the transformation vanishing at the initial time) to obtain a different solution so that the solution cannot be unique. In this case, we know that the initial value problem can be solved uniquely only after choosing a gauge condition. A popular gauge condition which is commonly used in Maxwell's theory (respecting relativistic invariance) is the Landau gauge condition, namely,

$$\partial_\mu A^\mu = 0. \tag{7.133}$$

That we can always bring the vector potential to satisfy this condition can be seen as follows. Let us assume that

$$\partial_\mu A^\mu \neq 0. \tag{7.134}$$

In this case, we can make a gauge transformation

$$A_\mu \to A'_\mu = A_\mu + \partial_\mu \alpha, \tag{7.135}$$

such that

$$\partial_\mu A'^\mu = \partial_\mu \left(A^\mu + \partial^\mu \alpha(x)\right) = 0,$$

or, $\Box \alpha(x) = -\partial_\mu A^\mu(x),$

or, $\alpha(x) = \displaystyle\int \mathrm{d}^4 x' \; G(x, x') \partial'_\mu A^\mu(x'), \tag{7.136}$

where the Green's function $G(x - x')$ satisfies

$$\Box_x G(x - x') = -\delta^4(x - x'). \tag{7.137}$$

Therefore, we see that there always exists a gauge transformation which would bring the vector field to satisfy the Landau gauge condition.

The situation is quite similar in the case of Einstein's equation. The presence of the constraints can be traced to the fact that the equations are covariant under a general coordinate transformation,

$$x^\mu \to x'^\mu(x). \tag{7.138}$$

This again shows the difficulty in solving the Cauchy initial value problem uniquely. Namely, if we determine a metric tensor $g_{\mu\nu}(x)$ which satisfies the initial conditions and solves the equation, we can always make a coordinate change to obtain another solution (still maintaining the initial conditions). Therefore, as in Maxwell's theory we have to choose a coordinate condition in order to solve the Cauchy initial value problem uniquely.

A particularly convenient choice of the coordinate system is obtained by imposing

$$\Gamma^\lambda = g^{\mu\nu} \Gamma^\lambda_{\mu\nu} = 0. \tag{7.139}$$

We note that if $\Gamma^\lambda \neq 0$, then under a coordinate transformation

$$x^\mu \to x'^\mu(x), \tag{7.140}$$

we have (see (4.22))

$$\Gamma'^{\ \lambda}_{\mu\nu} = \frac{\partial x'^\lambda}{\partial x^\rho} \frac{\partial x^\sigma}{\partial x'^\mu} \frac{\partial x^\tau}{\partial x'^\nu} \Gamma^\rho_{\sigma\tau} + \frac{\partial^2 x'^\lambda}{\partial x^\sigma \partial x^\tau} \frac{\partial x^\sigma}{\partial x'^\mu} \frac{\partial x^\tau}{\partial x'^\nu}. \tag{7.141}$$

Requiring that

$$\Gamma'^\lambda = g'^{\mu\nu}(x')\Gamma'^{\ \lambda}_{\mu\nu} = 0, \tag{7.142}$$

we obtain

$$g'^{\mu\nu}(x') \left[\frac{\partial x'^\lambda}{\partial x^\rho} \frac{\partial x^\sigma}{\partial x'^\mu} \frac{\partial x^\tau}{\partial x'^\nu} \Gamma^\rho_{\sigma\tau} + \frac{\partial^2 x'^\lambda}{\partial x^\sigma \partial x^\tau} \frac{\partial x^\sigma}{\partial x'^\mu} \frac{\partial x^\tau}{\partial x'^\nu} \right] = 0,$$

$$\text{or,} \quad \frac{\partial x'^\lambda}{\partial x^\rho} g^{\sigma\tau}(x)\Gamma^\rho_{\sigma\tau}(x) + \frac{\partial^2 x'^\lambda}{\partial x^\sigma \partial x^\tau} g^{\sigma\tau}(x) = 0,$$

$$\text{or,} \quad \frac{\partial x'^\lambda}{\partial x^\rho} \Gamma^\rho(x) + g^{\sigma\tau}(x)\frac{\partial^2 x'^\lambda}{\partial x^\sigma \partial x^\tau} = 0,$$

$$\text{or,} \quad \frac{\partial x'^\lambda}{\partial x^\rho} \Gamma^\rho(x) = -g^{\sigma\tau}(x)\frac{\partial^2 x'^\lambda}{\partial x^\sigma \partial x^\tau}. \tag{7.143}$$

This shows that if $\Gamma^\rho \neq 0$, then we can always find a coordinate transformation which will bring it to satisfy this condition.

To see what this condition means, let us note that

$$\begin{aligned}
\Gamma^\lambda &= g^{\mu\nu}\Gamma^\lambda_{\mu\nu} \\
&= g^{\mu\nu}\left(-\frac{1}{2} g^{\lambda\rho}(\partial_\mu g_{\nu\rho} + \partial_\nu g_{\rho\mu} - \partial_\rho g_{\mu\nu}) \right) \\
&= -\frac{1}{2} g^{\mu\nu} g^{\lambda\rho}(\partial_\mu g_{\nu\rho} + \partial_\nu g_{\rho\mu} - \partial_\rho g_{\mu\nu}) \\
&= -g^{\lambda\rho} g^{\mu\nu} \partial_\mu g_{\nu\rho} + \frac{1}{2} g^{\lambda\rho} g^{\mu\nu} \partial_\rho g_{\mu\nu}
\end{aligned}$$

$$= -g^{\lambda\rho}\left(-(\partial_\mu g^{\mu\nu})g_{\nu\rho}\right) + \frac{1}{2}\,g^{\lambda\rho}g^{\mu\nu}\partial_\rho g_{\mu\nu}$$

$$= \partial_\mu g^{\mu\lambda} + \frac{1}{\sqrt{-g}}\,\partial_\rho(\sqrt{-g})g^{\lambda\rho}$$

$$= \frac{1}{\sqrt{-g}}\,\partial_\mu(\sqrt{-g}\,g^{\mu\lambda}), \tag{7.144}$$

where we have used (4.148). Therefore, the condition (7.139) can be written as

$$\Gamma^\lambda = \partial_\mu(\sqrt{-g}\,g^{\mu\lambda}) = 0, \tag{7.145}$$

which is quite similar to the Landau gauge. This is known as the harmonic coordinate condition (harmonic gauge). The reason for this nomenclature is that if this is true, then we have

$$\begin{aligned}
\Box\phi(x) &= D_\lambda D^\lambda\phi(x) = D_\lambda(\partial^\lambda\phi(x)) \\
&= \frac{1}{\sqrt{-g}}\,\partial_\lambda(\sqrt{-g}\,\partial^\lambda\phi(x)) \\
&= \frac{1}{\sqrt{-g}}\,\partial_\lambda\left(\sqrt{-g}\,g^{\lambda\rho}\partial_\rho\phi(x)\right) \\
&= \frac{1}{\sqrt{-g}}\left[\partial_\lambda(\sqrt{-g}\,g^{\lambda\rho})\partial_\rho\phi(x) + \sqrt{-g}\,g^{\lambda\rho}\partial_\lambda\partial_\rho\phi(x)\right] \\
&= \partial^\lambda\partial_\lambda\phi(x). \tag{7.146}
\end{aligned}$$

Therefore, if a scalar function satisfies

$$\partial^\lambda\partial_\lambda\phi(x) = 0, \tag{7.147}$$

then with this gauge condition it also satisfies

$$\Box\phi(x) = D_\lambda D^\lambda\phi(x) = 0. \tag{7.148}$$

Functions satisfying this condition are known as harmonic functions. In particular it is easy to see that with this condition, the coordinates themselves become harmonic functions. To see this note that although the coordinates carry a vector index, they are really not vectors under a general coordinate transformation. In fact

$$x^\mu \to x'^\mu = x'^\mu(x), \tag{7.149}$$

can be thought of as defining four functional scalar relations. Thus under a general coordinate transformation, the coordinates can be thought of as scalars. In the harmonic gauge, this leads to

$$\Box x^\mu = D_\lambda D^\lambda x^\mu = \partial^\lambda \partial_\lambda x^\mu = 0. \tag{7.150}$$

Therefore, with this gauge condition the coordinates, themselves, become harmonic functions and hence the condition

$$\Gamma^\lambda = 0, \tag{7.151}$$

is referred to as the harmonic coordinate condition.

7.8 Einstein's equation from an action

Let us now ask whether we can obtain Einstein's equation (7.121) from a variational principle. To do this, of course, we have to define an action for the theory. And to simplify, we will restrict ourselves only to the matter free case first. The method generalizes easily to incorporate matter which we also discuss afterwards.

Let us note that the action has to be a scalar under a coordinate transformation. We have already seen that (see discussion around (3.35))

$$\int d^4x \ \sqrt{-g}, \tag{7.152}$$

represents an invariant volume element in a curved manifold. Therefore, we simply have to look for a scalar Lagrangian density (of weight zero) formed out of gravitational field (metric). Furthermore, if we restrict ourselves only to quantities involving at most second order derivatives of the dynamical variable (as is conventional in dynamical theories), then the choice is unique (this excludes terms of the form, say $R_{\mu\nu}R^{\mu\nu}, R^2$ etc.)

$$\mathcal{L}_E = \beta R = \beta g^{\mu\nu} R_{\mu\nu}, \tag{7.153}$$

where β is a normalization constant to be determined. Therefore, we can write the invariant action in the form

$$S_E = \int d^4x \sqrt{-g} \, \mathcal{L}_E = \beta \int d^4x \sqrt{-g} \, g^{\mu\nu} R_{\mu\nu}. \tag{7.154}$$

Let us now define the problem of variation in the case of a gravitational theory. The metric tensor in this case is the fundamental dynamical variable and plays the role analogous to the trajectory of a particle in classical mechanics. The question, therefore, is what is the dynamical equation satisfied by the metric tensor which would leave the Einstein action (7.154) unchanged if we change the metric infinitesimally. Namely, under the infinitesimal variation

$$g_{\mu\nu} \to g_{\mu\nu} + \delta g_{\mu\nu},$$

$$\lim_{x \to \infty} g_{\mu\nu} \to \eta_{\mu\nu},$$

$$\lim_{x \to \infty} \delta g_{\mu\nu} \to 0, \tag{7.155}$$

we would like to know the condition (equation) under which the Einstein action would be stationary. To determine the variation of the action (7.154) we need various relations which we collect below.

First, let us note that

$$g^{\mu\nu} g_{\nu\lambda} = \delta^\mu_\lambda,$$

$$\text{or,} \quad \delta g^{\mu\nu} g_{\nu\lambda} + g^{\mu\nu} \delta g_{\nu\lambda} = 0,$$

$$\text{or,} \quad \delta g^{\mu\nu} = -g^{\mu\sigma} \delta g_{\sigma\lambda} g^{\lambda\nu}. \tag{7.156}$$

Furthermore (see also (4.148)),

$$\delta(\sqrt{-g}) = \frac{1}{2\sqrt{-g}} \delta(-g)$$

$$= -\frac{1}{2\sqrt{-g}} \delta \left(e^{\text{Tr} \ln g_{\mu\nu}} \right)$$

$$= \frac{(-g)}{2\sqrt{-g}} g^{\mu\nu} \delta g_{\nu\mu}$$

$$= \frac{\sqrt{-g}}{2} g^{\mu\nu} \delta g_{\mu\nu}. \tag{7.157}$$

Let us also note from (4.49) that

$$\Gamma^\rho_{\mu\nu} = -\frac{1}{2} g^{\rho\sigma} (\partial_\mu g_{\nu\sigma} + \partial_\nu g_{\sigma\mu} - \partial_\sigma g_{\mu\nu}), \tag{7.158}$$

which leads to

$$\delta\Gamma^\rho_{\mu\nu} = -\frac{1}{2} \delta g^{\rho\sigma} (\partial_\mu g_{\nu\sigma} + \partial_\nu g_{\sigma\mu} - \partial_\sigma g_{\mu\nu})$$

$$-\frac{1}{2} g^{\rho\sigma} (\partial_\mu \delta g_{\nu\sigma} + \partial_\nu \delta g_{\sigma\mu} - \partial_\sigma \delta g_{\mu\nu}), \tag{7.159}$$

and we will analyze the two terms on the right hand side separately. Using (7.156), the first term leads to

$$\frac{1}{2} g^{\rho\alpha} \delta g_{\alpha\beta} g^{\beta\sigma} (\partial_\mu g_{\nu\sigma} + \partial_\nu g_{\sigma\mu} - \partial_\sigma g_{\mu\nu}) = -g^{\rho\alpha} \delta g_{\alpha\beta} \Gamma^\beta_{\mu\nu}. \tag{7.160}$$

To simplify the second term in (7.159), let us use the identities

$$D_\mu \delta g_{\nu\sigma} = \partial_\mu \delta g_{\nu\sigma} + \Gamma^\beta_{\mu\nu} \delta g_{\beta\sigma} + \Gamma^\beta_{\mu\sigma} \delta g_{\nu\beta},$$

$$D_\nu \delta g_{\sigma\mu} = \partial_\nu \delta g_{\sigma\mu} + \Gamma^\beta_{\nu\sigma} \delta g_{\beta\mu} + \Gamma^\beta_{\nu\mu} \delta g_{\sigma\beta},$$

$$D_\sigma \delta g_{\mu\nu} = \partial_\sigma \delta g_{\mu\nu} + \Gamma^\beta_{\sigma\mu} \delta g_{\beta\nu} + \Gamma^\beta_{\sigma\nu} \delta g_{\mu\beta}, \tag{7.161}$$

which lead to

$$(D_\mu \delta g_{\nu\sigma} + D_\nu \delta g_{\sigma\mu} - D_\sigma \delta g_{\mu\nu})$$

$$= (\partial_\mu \delta g_{\nu\sigma} + \partial_\nu \delta g_{\sigma\mu} - \partial_\sigma \delta g_{\mu\nu}) + 2\Gamma^\beta_{\mu\nu} \delta g_{\beta\sigma}. \qquad (7.162)$$

It follows from this that

$$(\partial_\mu \delta g_{\nu\sigma} + \partial_\nu \delta g_{\sigma\mu} - \partial_\sigma \delta g_{\mu\nu})$$

$$= (D_\mu \delta g_{\nu\sigma} + D_\nu \delta g_{\sigma\mu} - D_\sigma \delta g_{\mu\nu}) - 2\Gamma^\beta_{\mu\nu} \delta g_{\beta\sigma}. \qquad (7.163)$$

Consequently, the second term in (7.159) gives

$$-\frac{1}{2} g^{\rho\sigma} \left(D_\mu \delta g_{\nu\sigma} + D_\nu \delta g_{\sigma\mu} - D_\sigma \delta g_{\mu\nu} - 2\Gamma^\beta_{\mu\nu} \delta g_{\beta\sigma} \right)$$

$$= -\frac{1}{2} g^{\rho\sigma} (D_\mu \delta g_{\nu\sigma} + D_\nu \delta g_{\sigma\mu} - D_\sigma \delta g_{\mu\nu}) + g^{\rho\alpha} \delta g_{\alpha\beta} \Gamma^\beta_{\mu\nu}. \quad (7.164)$$

Adding (7.160) and (7.164) we obtain

$$\delta\Gamma^\rho_{\mu\nu} = -\frac{1}{2} g^{\rho\sigma} (D_\mu \delta g_{\nu\sigma} + D_\nu \delta g_{\sigma\mu} - D_\sigma \delta g_{\mu\nu}), \qquad (7.165)$$

which shows that the variation of the connection (Christoffel symbol) is a tensor as we have noted earlier.

We are now ready to calculate the change in the Ricci tensor. We know from (5.18) that the Riemann curvature tensor is defined as

$$R^\mu_{\ \nu\lambda\rho} = \partial_\rho \Gamma^\mu_{\nu\lambda} - \partial_\lambda \Gamma^\mu_{\nu\rho} + \Gamma^\mu_{\lambda\sigma} \Gamma^\sigma_{\nu\rho} - \Gamma^\mu_{\rho\sigma} \Gamma^\sigma_{\nu\lambda}, \qquad (7.166)$$

so that

$$\begin{aligned} R_{\nu\rho} &= \delta^\lambda_\mu R^\mu_{\ \nu\lambda\rho} \\ &= \partial_\rho \Gamma^\mu_{\nu\mu} - \partial_\mu \Gamma^\mu_{\nu\rho} + \Gamma^\mu_{\mu\sigma} \Gamma^\sigma_{\nu\rho} - \Gamma^\mu_{\rho\sigma} \Gamma^\sigma_{\nu\mu}. \end{aligned} \qquad (7.167)$$

Therefore, we have

$$\delta R_{\nu\rho} = \partial_\rho \delta\Gamma^\mu_{\nu\mu} - \partial_\mu \delta\Gamma^\mu_{\nu\rho} + \delta\Gamma^\mu_{\mu\sigma}\Gamma^\sigma_{\nu\rho} + \Gamma^\mu_{\mu\sigma}\delta\Gamma^\sigma_{\nu\rho}$$
$$-\delta\Gamma^\mu_{\rho\sigma}\Gamma^\sigma_{\nu\mu} - \Gamma^\mu_{\rho\sigma}\delta\Gamma^\sigma_{\nu\mu}$$
$$= \left(\partial_\rho \delta\Gamma^\mu_{\nu\mu} - \Gamma^\mu_{\rho\sigma}\delta\Gamma^\sigma_{\nu\mu} + \Gamma^\sigma_{\rho\nu}\delta\Gamma^\mu_{\sigma\mu} + \Gamma^\sigma_{\rho\mu}\delta\Gamma^\mu_{\nu\sigma}\right)$$
$$- \left(\partial_\mu \delta\Gamma^\mu_{\nu\rho} - \Gamma^\mu_{\mu\sigma}\delta\Gamma^\sigma_{\nu\rho} + \Gamma^\sigma_{\mu\nu}\delta\Gamma^\mu_{\sigma\rho} + \Gamma^\sigma_{\mu\rho}\delta\Gamma^\mu_{\nu\sigma}\right)$$
$$= D_\rho\left(\delta\Gamma^\mu_{\nu\mu}\right) - D_\mu\left(\delta\Gamma^\mu_{\nu\rho}\right)$$

$$\text{or,} \quad \delta R_{\nu\rho} = D_\rho\left(\delta\Gamma^\lambda_{\nu\lambda}\right) - D_\lambda\left(\delta\Gamma^\lambda_{\nu\rho}\right). \tag{7.168}$$

This is known as the Palatini identity. Let us note further that the Ricci scalar is given by (7.49)

$$R = g^{\mu\nu} R_{\mu\nu}, \tag{7.169}$$

which leads to

$$\delta R = \delta g^{\mu\nu} R_{\mu\nu} + g^{\mu\nu} \delta R_{\mu\nu}$$
$$= -g^{\mu\lambda}g^{\rho\nu}\delta g_{\lambda\rho} R_{\mu\nu} + g^{\mu\nu}\left(D_\nu\left(\delta\Gamma^\lambda_{\mu\lambda}\right) - D_\lambda\left(\delta\Gamma^\lambda_{\mu\nu}\right)\right)$$
$$= -\delta g_{\mu\nu} R^{\mu\nu} + D_\nu\left(g^{\mu\nu}\delta\Gamma^\lambda_{\mu\lambda}\right) - D_\lambda\left(g^{\mu\nu}\delta\Gamma^\lambda_{\mu\nu}\right)$$
$$= -\delta g_{\mu\nu} R^{\mu\nu} + \frac{1}{\sqrt{-g}}\,\partial_\nu\left(\sqrt{-g}\,g^{\mu\nu}\delta\Gamma^\lambda_{\mu\lambda}\right)$$
$$- \frac{1}{\sqrt{-g}}\,\partial_\lambda\left(\sqrt{-g}\,g^{\mu\nu}\delta\Gamma^\lambda_{\mu\nu}\right), \tag{7.170}$$

where we have used (7.156), (7.165) and (7.168).

Therefore, we finally obtain the variation of the action (7.154) to be

$$
\begin{aligned}
\delta S_E \;=\;& \beta \int \mathrm{d}^4 x \left(\delta(\sqrt{-g}) R + \sqrt{-g}\, \delta R \right) \\[2mm]
=\;& \beta \int \mathrm{d}^4 x \left(\frac{1}{2}\, \sqrt{-g}\, g^{\mu\nu} \delta g_{\mu\nu} R \right. \\[2mm]
& + \sqrt{-g} \left\{ -\delta g_{\mu\nu} R^{\mu\nu} + \frac{1}{\sqrt{-g}}\, \partial_\nu \left(\sqrt{-g}\, g^{\mu\nu} \delta \Gamma^{\lambda}_{\mu\lambda} \right) \right. \\[2mm]
& \left. \left. -\frac{1}{\sqrt{-g}}\, \partial_\lambda \left(\sqrt{-g}\, g^{\mu\nu} \delta \Gamma^{\lambda}_{\mu\nu} \right) \right\} \right).
\end{aligned}
\tag{7.171}
$$

We note that the last two terms in (7.171) are total divergences and if we assume the connections to vanish at infinity, then we can ignore the last two terms and rearrange the other terms to write the change in the action as

$$
\delta S_E = \beta \int \mathrm{d}^4 x\, \sqrt{-g} \left(\frac{1}{2}\, g^{\mu\nu} R - R^{\mu\nu} \right) \delta g_{\mu\nu}.
\tag{7.172}
$$

Since we require that the action be stationary for arbitrary infinitesimal variations of the metric

$$
\delta S_E = 0,
\tag{7.173}
$$

this leads to

$$
\frac{1}{2}\, g^{\mu\nu} R - R^{\mu\nu} = 0,
$$

$$
\text{or,} \quad R^{\mu\nu} - \frac{1}{2}\, g^{\mu\nu} R = G^{\mu\nu} = 0.
\tag{7.174}
$$

This is the correct equation for empty space (7.97) and we have derived it from an action through a variational approach. Note that the normalization constant β remains undetermined at this point and gets determined only when matter is included.

In the presence of matter, we can generalize the action (7.154) to

$$S = S_E + S_{\text{matter}}$$

$$= \beta \int d^4x \sqrt{-g}\, R + \int d^4x \sqrt{-g}\, \mathcal{L}_{\text{matter}}. \qquad (7.175)$$

If we now look for a variation of the complete action in (7.175), we can write

$$\delta S = \beta \int d^4x \sqrt{-g}\, \delta g_{\mu\nu} \left(\frac{1}{2}\, g^{\mu\nu} R - R^{\mu\nu} \right)$$

$$+ \int d^4x \sqrt{-g}\, \left(-\frac{1}{2} \right) \delta g_{\mu\nu} T_{\text{matter}}^{\mu\nu}$$

$$= \int d^4x \sqrt{-g}\, \delta g_{\mu\nu} \left[\beta \left(\frac{1}{2} g^{\mu\nu} R - R^{\mu\nu} \right) - \frac{1}{2} T_{\text{matter}}^{\mu\nu} \right]. \, (7.176)$$

Here, although we have not yet derived, we are using the fact that we can write the change in the action for matter as

$$\delta S_{\text{matter}} = -\frac{1}{2} \int d^4x \sqrt{-g}\, \delta g_{\mu\nu} T^{\mu\nu}. \qquad (7.177)$$

In fact, this is quite plausible recalling the fact that, in the case of charged particles coupled to the Maxwell field, the variation of the matter action (under an infinitesimal change of the gauge field) can be written as

$$\delta S_{\text{matter}} = \int d^4x\, \delta A_\mu\, J^\mu, \qquad (7.178)$$

where J^μ represents the source for the Maxwell theory. Clearly if the action (7.175) is to be stationary for an arbitrary infinitesimal variations of the metric, namely,

$$\delta S = 0, \qquad (7.179)$$

then it follows from (7.176) that

$$\beta \left(\frac{1}{2} g^{\mu\nu} R - R^{\mu\nu} \right) - \frac{1}{2} T^{\mu\nu}_{\text{matter}} = 0,$$

$$\text{or,} \quad R^{\mu\nu} - \frac{1}{2} g^{\mu\nu} R = G^{\mu\nu} = -\frac{1}{2\beta} T^{\mu\nu}_{\text{matter}}. \tag{7.180}$$

Comparing this with Einstein's equation (7.121)

$$G^{\mu\nu} = R^{\mu\nu} - \frac{1}{2} g^{\mu\nu} R = 8\pi G_{\text{N}} T^{\mu\nu}_{\text{matter}}, \tag{7.181}$$

we determine

$$\beta = -\frac{1}{16\pi G_{\text{N}}}. \tag{7.182}$$

This determines the normalization constant uniquely and thus we can write the Einstein Lagrangian density (7.153) for the matter free case as

$$\mathcal{L}_E = -\frac{1}{16\pi G_{\text{N}}} R. \tag{7.183}$$

Written as an action the analogy with the Maxwell theory is obvious. The analogous gauge invariance in this case is the invariance under a general (local) coordinate transformation. Let us also note that a cosmological constant can also be incorporated into the Einstein action (Lagrangian density) simply as

$$\mathcal{L}_E = -\frac{1}{16\pi G_{\text{N}}} (R + \Lambda). \tag{7.184}$$

CHAPTER 8

Schwarzschild solution

As a simple application of Einstein's equations, let us determine the gravitational field (metric) of a static, spherically symmetric star. Many stars conform to this condition. There are also many others that behave differently. For example, a star may have asymmetries associated with it, it may be rotating or it may be pulsating. However, the static, spherically symmetric star is a simple example for which the metric can be solved exactly. Therefore, it leads to theoretical predictions which can be verified as tests of general relativity.

8.1 Line element

Although Einstein's equations are highly nonlinear, the reason why we can solve them for a static, spherically symmetric star is that the symmetry present in the problem restricts the form of the solution greatly. For example, since the gravitating mass (source) is static, the metric components would be independent of time. Furthermore, the spherical symmetry of the problem requires that the components of the metric can depend only on the radial coordinate r. Let us recall that in spherical coordinates, the flat space-time can be characterized by the line element

$$\mathrm{d}\tau^2 = \mathrm{d}t^2 - (\mathrm{d}r^2 + r^2(\mathrm{d}\theta^2 + \sin^2\theta \mathrm{d}\phi^2)). \tag{8.1}$$

We can generalize this line element to a static, isotropic curved space as

$$\mathrm{d}\tau^2 = A(r)\mathrm{d}t^2 - \left(B(r)\mathrm{d}r^2 + C(r)r^2\mathrm{d}\theta^2 + D(r)r^2\sin^2\theta\mathrm{d}\phi^2\right). \tag{8.2}$$

The following assumptions have gone into writing the line element in this form. First of all since the metric components are independent of time, the line element should be invariant if we let $dt \to -dt$. This implies that linear terms in dt cannot occur. Isotropy similarly tells that if we let $d\theta \to -d\theta$ or $d\phi \to -d\phi$, the line element should be invariant. Thus terms of the form $drd\theta$, $drd\phi$ or $d\theta d\phi$ cannot occur either. This restricts the form of the metric to be diagonal.

Let us now look at the line element (8.2) at a fixed time and radius. At the north pole ($d\phi = 0$) with $\epsilon = rd\theta$, we have

$$d\tau^2 = -C(r)\epsilon^2. \tag{8.3}$$

On the other hand, if we look at the line element in the same slice of space-time but at the equator ($\theta = \frac{\pi}{2}$) with $\epsilon = rd\phi$, then

$$d\tau^2 = -D(r)\epsilon^2. \tag{8.4}$$

However, if the space is isotropic then these two lengths must be equal which requires

$$C(r) = D(r). \tag{8.5}$$

Thus we can write the line element (8.2) as

$$d\tau^2 = A(r)dt^2 - B(r)dr^2 - C(r)r^2(d\theta^2 + \sin^2\theta d\phi^2). \tag{8.6}$$

There is a simpler way of understanding this result. The spherically symmetric line element (8.1)

$$d\tau^2 = dt^2 - dr^2 - r^2 d\Omega^2, \tag{8.7}$$

with the angular element

$$d\Omega^2 = (d\theta^2 + \sin^2\theta d\phi^2), \tag{8.8}$$

generalizes to a static, spherically symmetric curved space as

$$d\tau^2 = A(r)dt^2 - B(r)dr^2 - C(r)r^2 d\Omega^2. \tag{8.9}$$

We note here that the function $C(r)$ in (8.6) is redundant in the sense that it can be scaled away. Namely, if we let

$$r \to \tilde{r} = [C(r)]^{\frac{1}{2}} r, \tag{8.10}$$

then

$$\mathrm{d}\tilde{r} = \mathrm{d}r \left[(C(r))^{\frac{1}{2}} + \frac{1}{2} \frac{rC'(r)}{(C(r))^{\frac{1}{2}}} \right],$$

$$\text{or,} \quad \mathrm{d}r = f(\tilde{r})\mathrm{d}\tilde{r}, \tag{8.11}$$

where we have identified (prime denotes a derivative with respect to r)

$$f(\tilde{r}) = \frac{2 (C(r))^{\frac{1}{2}}}{2C(r) + rC'(r)}. \tag{8.12}$$

Let us also define

$$A(r) = \tilde{A}(\tilde{r}), \quad B(r)\mathrm{d}r^2 = \tilde{B}(\tilde{r})\mathrm{d}\tilde{r}^2. \tag{8.13}$$

Thus with this scaling the form of the line element (8.6) becomes

$$\mathrm{d}\tau^2 = \tilde{A}(\tilde{r})\mathrm{d}t^2 - \tilde{B}(\tilde{r})\mathrm{d}\tilde{r}^2 - \tilde{r}^2(\mathrm{d}\theta^2 + \sin^2\theta\mathrm{d}\phi^2). \tag{8.14}$$

This shows that with a proper choice of the coordinate system the line element for a static, spherically symmetric gravitational field can be written as

$$\mathrm{d}\tau^2 = A(r)\mathrm{d}t^2 - B(r)\mathrm{d}r^2 - r^2(\mathrm{d}\theta^2 + \sin^2\theta\mathrm{d}\phi^2), \tag{8.15}$$

which is known as the general Schwarzschild line element.

Birkhoff has demonstrated that the requirement of a static gravitational potential is superfluous for the Schwarzschild line element. In fact, he has shown that even when a spherically symmetric star or a gravitational mass is undergoing a radial motion, the line element

can be written in the Schwarzschild form (8.15). This is known as
Birkhoff's theorem and is important in the sense that it rules out
the possibility of gravitational waves being emitted by radial pulsars since the metric in this case can be equivalently thought of as
static. Therefore, to be able to emit gravitational waves, if such a
phenomenon exists, a star must undergo different deformations.

8.2 Connection

As we see now, the Schwarzschild line element (8.15) is given in terms
of two unknown functions $A(r)$ and $B(r)$. The metric components
can be read off from the line element (8.15) to be

$$
\begin{aligned}
g_{00} &= g_{tt} = A(r), \\
g_{11} &= g_{rr} = -B(r), \\
g_{22} &= g_{\theta\theta} = -r^2, \\
g_{33} &= g_{\phi\phi} = -r^2 \sin^2 \theta.
\end{aligned}
\tag{8.16}
$$

This is a diagonal metric and hence the nontrivial components of the
inverse metric can also be easily written down as

$$
\begin{aligned}
g^{00} &= g^{tt} = \frac{1}{A(r)}, \\
g^{11} &= g^{rr} = -\frac{1}{B(r)}, \\
g^{22} &= g^{\theta\theta} = -\frac{1}{r^2}, \\
g^{33} &= g^{\phi\phi} = -\frac{1}{r^2 \sin^2 \theta}.
\end{aligned}
\tag{8.17}
$$

We can solve Einstein's equations far away from the star to
determine the forms of the functions $A(r)$, $B(r)$. That is, outside
the star we can solve the empty space equation (7.97)

$$
R_{\mu\nu} = 0,
\tag{8.18}
$$

subject to the boundary condition that infinitely far away from the star, the metric reduces to the Minkowski form (8.1). To solve Einstein's equations we must, of course, calculate the connections and the curvature tensor. For example, from the definition of the Christoffel symbol (4.49) we have

$$\Gamma^{\mu}_{\nu\lambda} = -\frac{1}{2}\, g^{\mu\rho}(\partial_{\nu}g_{\lambda\rho} + \partial_{\lambda}g_{\rho\nu} - \partial_{\rho}g_{\nu\lambda}), \tag{8.19}$$

and since we know the metric components, these can be calculated. But this method is tedious (as we have pointed out earlier) and hence let us try to determine the components of the connection from the geodesic equation which has the form (see (5.130))

$$g_{\mu\nu}\,\frac{\mathrm{d}^2 x^{\nu}}{\mathrm{d}\tau^2} + \frac{1}{2}\,(\partial_{\lambda}g_{\rho\mu} + \partial_{\rho}g_{\mu\lambda} - \partial_{\mu}g_{\lambda\rho})\,\frac{\mathrm{d}x^{\lambda}}{\mathrm{d}\tau}\,\frac{\mathrm{d}x^{\rho}}{\mathrm{d}\tau} = 0. \tag{8.20}$$

For $\mu = 0$ (t-equation), this leads to

$$g_{00}\,\frac{\mathrm{d}^2 x^0}{\mathrm{d}\tau^2} + \frac{1}{2}\,(\partial_{\lambda}g_{\rho 0} + \partial_{\rho}g_{0\lambda} - \partial_0 g_{\lambda\rho})\,\frac{\mathrm{d}x^{\lambda}}{\mathrm{d}\tau}\,\frac{\mathrm{d}x^{\rho}}{\mathrm{d}\tau} = 0,$$

or, $$A(r)\,\frac{\mathrm{d}^2 x^0}{\mathrm{d}\tau^2} + \partial_1 g_{00}\,\frac{\mathrm{d}x^0}{\mathrm{d}\tau}\,\frac{\mathrm{d}x^1}{\mathrm{d}\tau} = 0,$$

or, $$\frac{\mathrm{d}^2 x^0}{\mathrm{d}\tau^2} + \frac{A'(r)}{A(r)}\,\frac{\mathrm{d}x^0}{\mathrm{d}\tau}\,\frac{\mathrm{d}x^1}{\mathrm{d}\tau} = 0. \tag{8.21}$$

Comparing this with the $\mu = 0$ equation of (see (5.83))

$$\frac{\mathrm{d}^2 x^{\mu}}{\mathrm{d}\tau^2} - \Gamma^{\mu}_{\lambda\rho}\,\frac{\mathrm{d}x^{\lambda}}{\mathrm{d}\tau}\,\frac{\mathrm{d}x^{\rho}}{\mathrm{d}\tau} = 0, \tag{8.22}$$

we determine

$$\Gamma^0_{01} = \Gamma^0_{10} = -\frac{1}{2}\,\frac{A'(r)}{A(r)}. \tag{8.23}$$

Here prime denotes differentiation with respect to r.

For $\mu = 1$ (r-equation), from (8.20) we obtain

$$g_{11} \frac{d^2x^1}{d\tau^2} + \frac{1}{2} \left(\partial_\lambda g_{\rho 1} + \partial_\rho g_{\lambda 1} - \partial_1 g_{\lambda \rho} \right) \frac{dx^\lambda}{d\tau} \frac{dx^\rho}{d\tau} = 0,$$

or, $$-B \frac{d^2x^1}{d\tau^2} - \frac{1}{2} \partial_1 g_{00} \left(\frac{dx^0}{d\tau} \right)^2 + \frac{1}{2} \partial_1 g_{11} \left(\frac{dx^1}{d\tau} \right)^2$$

$$- \frac{1}{2} \partial_1 g_{22} \left(\frac{dx^2}{d\tau} \right)^2 - \frac{1}{2} \partial_1 g_{33} \left(\frac{dx^3}{d\tau} \right)^2 = 0,$$

or, $$-B \frac{d^2x^1}{d\tau^2} - \frac{1}{2} A' \left(\frac{dx^0}{d\tau} \right)^2 - \frac{1}{2} B' \left(\frac{dx^1}{d\tau} \right)^2$$

$$+ r \left(\frac{dx^2}{d\tau} \right)^2 + r \sin^2 \theta \left(\frac{dx^3}{d\tau} \right)^2 = 0,$$

or, $$\frac{d^2x^1}{d\tau^2} + \frac{1}{2} \frac{A'}{B} \left(\frac{dx^0}{d\tau} \right)^2 + \frac{1}{2} \frac{B'}{B} \left(\frac{dx^1}{d\tau} \right)^2$$

$$- \frac{r}{B} \left(\frac{dx^2}{d\tau} \right)^2 - \frac{r}{B} \sin^2 \theta \left(\frac{dx^3}{d\tau} \right)^2 = 0. \qquad (8.24)$$

Thus, comparing with the $\mu = 1$ equation of (8.22), we obtain

$$\begin{aligned}
\Gamma^1_{00} &= -\tfrac{1}{2} \tfrac{A'}{B}, & \Gamma^1_{11} &= -\tfrac{1}{2} \tfrac{B'}{B}, \\
\Gamma^1_{22} &= \tfrac{r}{B}, & \Gamma^1_{33} &= \tfrac{r \sin^2 \theta}{B}.
\end{aligned} \qquad (8.25)$$

For $\mu = 2$ (θ-equation), equation (8.20) leads to

$$g_{22} \frac{d^2x^2}{d\tau^2} + \frac{1}{2} \left(\partial_\lambda g_{\rho 2} + \partial_\rho g_{2\lambda} - \partial_2 g_{\lambda \rho} \right) \frac{dx^\lambda}{d\tau} \frac{dx^\rho}{d\tau} = 0,$$

or, $$-r^2 \frac{d^2x^2}{d\tau^2} + \partial_1 g_{22} \frac{dx^1}{d\tau} \frac{dx^2}{d\tau} - \frac{1}{2} \partial_2 g_{33} \left(\frac{dx^3}{d\tau} \right)^2 = 0,$$

or, $$-r^2 \frac{d^2x^2}{d\tau^2} - 2r \frac{dx^1}{d\tau} \frac{dx^2}{d\tau} + r^2 \sin\theta \cos\theta \left(\frac{dx^3}{d\tau} \right)^2 = 0,$$

or, $$\frac{d^2x^2}{d\tau^2} + \frac{2}{r} \frac{dx^1}{d\tau} \frac{dx^2}{d\tau} - \sin\theta \cos\theta \left(\frac{dx^3}{d\tau} \right)^2 = 0. \qquad (8.26)$$

This determines

$$\Gamma^2_{12} = \Gamma^2_{21} = -\frac{1}{r}, \qquad \Gamma^2_{33} = \sin\theta\cos\theta. \tag{8.27}$$

Similarly for $\mu = 3$ (ϕ-equation), we obtain from (8.20)

$$g_{33}\frac{\mathrm{d}^2x^3}{\mathrm{d}\tau^2} + \frac{1}{2}\left(\partial_\lambda g_{\rho 3} + \partial_\rho g_{3\lambda} - \partial_3 g_{\lambda\rho}\right)\frac{\mathrm{d}x^\lambda}{\mathrm{d}\tau}\frac{\mathrm{d}x^\rho}{\mathrm{d}\tau} = 0,$$

or, $\quad -r^2\sin^2\theta\,\frac{\mathrm{d}^2x^3}{\mathrm{d}\tau^2} + \partial_1 g_{33}\frac{\mathrm{d}x^1}{\mathrm{d}\tau}\frac{\mathrm{d}x^3}{\mathrm{d}\tau} + \partial_2 g_{33}\frac{\mathrm{d}x^2}{\mathrm{d}\tau}\frac{\mathrm{d}x^3}{\mathrm{d}\tau} = 0,$

or, $\quad -r^2\sin^2\theta\,\frac{\mathrm{d}^2x^3}{\mathrm{d}\tau^2} - 2r\sin^2\theta\,\frac{\mathrm{d}x^1}{\mathrm{d}\tau}\frac{\mathrm{d}x^3}{\mathrm{d}\tau}$

$$-2r^2\sin\theta\cos\theta\,\frac{\mathrm{d}x^2}{\mathrm{d}\tau}\frac{\mathrm{d}x^3}{\mathrm{d}\tau} = 0,$$

or, $\quad \dfrac{\mathrm{d}^2x^3}{\mathrm{d}\tau^2} + \dfrac{2}{r}\dfrac{\mathrm{d}x^1}{\mathrm{d}\tau}\dfrac{\mathrm{d}x^3}{\mathrm{d}\tau} + 2\cot\theta\,\dfrac{\mathrm{d}x^2}{\mathrm{d}\tau}\dfrac{\mathrm{d}x^3}{\mathrm{d}\tau} = 0, \tag{8.28}$

which yields

$$\Gamma^3_{13} = \Gamma^3_{31} = -\frac{1}{r}, \quad \Gamma^3_{23} = \Gamma^3_{32} = -\cot\theta. \tag{8.29}$$

Let us collect all the nontrivial components of the connection which are given by

$$
\begin{aligned}
\Gamma^0_{01} &= \Gamma^0_{10} = -\tfrac{1}{2}\tfrac{A'}{A}, & & \\
\Gamma^1_{00} &= -\tfrac{1}{2}\tfrac{A'}{B}, & \Gamma^1_{11} &= -\tfrac{1}{2}\tfrac{B'}{B}, \\
\Gamma^1_{22} &= \tfrac{r}{B}, & \Gamma^1_{33} &= \tfrac{r\sin^2\theta}{B}, \\
\Gamma^2_{12} &= \Gamma^2_{21} = -\tfrac{1}{r}, & \Gamma^2_{33} &= \sin\theta\cos\theta, \\
\Gamma^3_{13} &= \Gamma^3_{31} = -\tfrac{1}{r}, & \Gamma^3_{23} &= \Gamma^3_{32} = -\cot\theta.
\end{aligned}
\tag{8.30}
$$

8.3 Solution of the Einstein equation

It is now easy to calculate the Ricci tensor from the connections (8.30). We recall that the Riemann-Christoffel curvature tensor (5.18) is defined to be

$$R^\rho{}_{\mu\lambda\nu} = \partial_\nu \Gamma^\rho_{\mu\lambda} - \partial_\lambda \Gamma^\rho_{\mu\nu} + \Gamma^\rho_{\lambda\sigma}\Gamma^\sigma_{\mu\nu} - \Gamma^\rho_{\nu\sigma}\Gamma^\sigma_{\mu\lambda}, \tag{8.31}$$

so that

$$\begin{aligned}
R_{\mu\nu} &= \delta^\lambda_\rho R^\rho{}_{\mu\lambda\nu} \\
&= \partial_\nu \Gamma^\lambda_{\mu\lambda} - \partial_\lambda \Gamma^\lambda_{\mu\nu} + \Gamma^\lambda_{\lambda\sigma}\Gamma^\sigma_{\mu\nu} - \Gamma^\lambda_{\nu\sigma}\Gamma^\sigma_{\mu\lambda}.
\end{aligned} \tag{8.32}$$

It follows now that

$$\begin{aligned}
R_{00} &= \partial_0 \Gamma^\lambda_{0\lambda} - \partial_\lambda \Gamma^\lambda_{00} + \Gamma^\lambda_{\lambda\sigma}\Gamma^\sigma_{00} - \Gamma^\lambda_{0\sigma}\Gamma^\sigma_{0\lambda} \\
&= -\partial_i \Gamma^i_{00} + \Gamma^\lambda_{\lambda 1}\Gamma^1_{00} - \Gamma^0_{0\sigma}\Gamma^\sigma_{00} - \Gamma^1_{0\sigma}\Gamma^\sigma_{01} \\
&= -\partial_1 \Gamma^1_{00} + \left(\Gamma^0_{01} + \Gamma^1_{11} + \Gamma^2_{21} + \Gamma^3_{31}\right)\Gamma^1_{00} - \Gamma^0_{01}\Gamma^1_{00} - \Gamma^1_{00}\Gamma^0_{01} \\
&= -\partial_1 \Gamma^1_{00} + \left(-\Gamma^0_{01} + \Gamma^1_{11} + \Gamma^2_{21} + \Gamma^3_{31}\right)\Gamma^1_{00} \\
&= \frac{1}{2}\frac{\partial}{\partial r}\left(\frac{A'}{B}\right) + \left(\frac{1}{2}\frac{A'}{A} - \frac{1}{2}\frac{B'}{B} - \frac{1}{r} - \frac{1}{r}\right)\left(-\frac{1}{2}\frac{A'}{B}\right) \\
&= \frac{1}{2}\frac{\partial}{\partial r}\left(\frac{A'}{B}\right) - \frac{1}{2}\frac{A'}{B}\left(\frac{1}{2}\frac{A'}{A} - \frac{1}{2}\frac{B'}{B} - \frac{2}{r}\right), \tag{8.33}
\end{aligned}$$

so that the 00-component of (8.18) leads to

$$R_{00} = \frac{1}{2}\frac{A''}{B} - \frac{1}{2}\frac{A'B'}{B^2} - \frac{1}{2}\frac{A'}{B}\left(\frac{1}{2}\frac{A'}{A} - \frac{1}{2}\frac{B'}{B} - \frac{2}{r}\right) = 0,$$

or, $$\frac{A''}{2B} - \frac{1}{4}\frac{A'}{B}\left(\frac{A'}{A} + \frac{B'}{B}\right) + \frac{A'}{rB} = 0. \tag{8.34}$$

Similarly, we have

$$
\begin{aligned}
R_{11} &= \partial_1 \Gamma_{1\lambda}^\lambda - \partial_\lambda \Gamma_{11}^\lambda + \Gamma_{\lambda\sigma}^\lambda \Gamma_{11}^\sigma - \Gamma_{1\sigma}^\lambda \Gamma_{1\lambda}^\sigma \\
&= \partial_1 \left(\Gamma_{10}^0 + \Gamma_{11}^1 + \Gamma_{12}^2 + \Gamma_{13}^3 \right) - \partial_1 \Gamma_{11}^1 \\
&\quad + \Gamma_{11}^1 \left(\Gamma_{01}^0 + \Gamma_{11}^1 + \Gamma_{21}^2 + \Gamma_{31}^3 \right) \\
&\quad - \Gamma_{10}^0 \Gamma_{10}^0 - \Gamma_{11}^1 \Gamma_{11}^1 - \Gamma_{12}^2 \Gamma_{12}^2 - \Gamma_{13}^3 \Gamma_{13}^3 \\
&= \frac{\partial}{\partial r} \left(-\frac{1}{2} \frac{A'}{A} - \frac{2}{r} \right) + \left(-\frac{1}{2} \frac{B'}{B} \right) \left(-\frac{1}{2} \frac{A'}{A} - \frac{2}{r} \right) \\
&\quad - \frac{1}{4} \left(\frac{A'}{A} \right)^2 - \frac{1}{r^2} - \frac{1}{r^2} \\
&= -\frac{A''}{2A} + \frac{1}{2} \left(\frac{A'}{A} \right)^2 + \frac{2}{r^2} + \frac{1}{4} \left(\frac{A'}{A} \right) \left(\frac{B'}{B} \right) + \frac{B'}{rB} \\
&\quad - \frac{1}{4} \left(\frac{A'}{A} \right)^2 - \frac{2}{r^2} \\
&= -\frac{A''}{2A} + \frac{1}{4} \left(\frac{A'}{A} \right) \left(\frac{A'}{A} + \frac{B'}{B} \right) + \frac{B'}{rB}. \tag{8.35}
\end{aligned}
$$

Thus, the 11-component of (8.18) leads to

$$
R_{11} = 0,
$$

or, $$
\frac{A''}{2A} - \frac{1}{4} \left(\frac{A'}{A} \right) \left(\frac{A'}{A} + \frac{B'}{B} \right) - \frac{B'}{rB} = 0. \tag{8.36}
$$

We also have

$$
\begin{aligned}
R_{22} &= \partial_2 \Gamma_{2\lambda}^\lambda - \partial_\lambda \Gamma_{22}^\lambda + \Gamma_{\lambda\sigma}^\lambda \Gamma_{22}^\sigma - \Gamma_{2\sigma}^\lambda \Gamma_{2\lambda}^\sigma \\
&= \partial_2 \Gamma_{23}^3 - \partial_1 \Gamma_{22}^1 + \Gamma_{22}^1 \left(\Gamma_{01}^0 + \Gamma_{11}^1 + \Gamma_{21}^2 + \Gamma_{31}^3 \right) \\
&\quad - \Gamma_{22}^1 \Gamma_{21}^2 - \Gamma_{21}^2 \Gamma_{22}^1 - \Gamma_{23}^3 \Gamma_{23}^3
\end{aligned}
$$

$$= \mathrm{cosec}^2\theta - \frac{\partial}{\partial r}\left(\frac{r}{B}\right) + \left(\frac{r}{B}\right)\left(-\frac{1}{2}\frac{A'}{A} - \frac{1}{2}\frac{B'}{B} + \frac{1}{r} - \frac{1}{r}\right)$$

$$- \cot^2\theta$$

$$= 1 - \frac{1}{B} + \frac{rB'}{B^2} - \frac{1}{2}\frac{r}{B}\left(\frac{A'}{A} + \frac{B'}{B}\right)$$

$$= 1 - \frac{1}{B} - \frac{r}{2B}\left(\frac{A'}{A} - \frac{B'}{B}\right). \tag{8.37}$$

Thus, the 22-component of (8.18) yields

$$R_{22} = 0,$$

or, $$\frac{1}{B} + \frac{r}{2B}\left(\frac{A'}{A} - \frac{B'}{B}\right) - 1 = 0. \tag{8.38}$$

Multiplying (8.34) by $\frac{B}{A}$ and subtracting from (8.36) we have

$$-\frac{B'}{rB} - \frac{A'}{rA} = 0,$$

or, $$\frac{A'}{A} = -\frac{B'}{B},$$

or, $$AB = \text{constant } = k. \tag{8.39}$$

If we now impose the boundary condition that at large distances the metric becomes Minkowskian (see (8.1)), we have

$$\lim_{r\to\infty} A(r) \;\to\; 1,$$

$$\lim_{r\to\infty} B(r) \;\to\; 1. \tag{8.40}$$

This therefore, determines the constant of integration in (8.39) to be

$$k = 1, \tag{8.41}$$

and we have

$$A(r)B(r) = 1,$$

or, $B(r) = \dfrac{1}{A(r)}.$ (8.42)

If we now substitute this relation into (8.38), we obtain

$$A(r) + \frac{rA}{2}\left(\frac{A'}{A} + \frac{A'}{A}\right) - 1 = 0,$$

or, $A(r) + rA'(r) = 1,$

or, $\dfrac{\mathrm{d}(rA(r))}{\mathrm{d}r} = 1,$

or, $rA(r) = r + \text{constant} = r + m,$ (8.43)

so that

$$\begin{aligned} A(r) &= 1 + \frac{m}{r}, \\ B(r) &= \frac{1}{A(r)} = \left(1 + \frac{m}{r}\right)^{-1}. \end{aligned}$$ (8.44)

Here m is a constant of integration to be determined.

We can now write down the Schwarzschild line element (8.15) in the form

$$\mathrm{d}\tau^2 = \left(1 + \frac{m}{r}\right)\mathrm{d}t^2 - \left(1 + \frac{m}{r}\right)^{-1}\mathrm{d}r^2 - r^2(\mathrm{d}\theta^2 + \sin^2\theta\,\mathrm{d}\phi^2).$$ (8.45)

Let us emphasize here that there are ten equations of Einstein

$$R_{\mu\nu} = 0,$$ (8.46)

and we have used only three of them to determine the form of the Schwarzschild line element. Therefore, it remains to be shown that the other seven equations are consistent with the solution in (8.45). (Namely, since there are ten equations involving two unknown functions, it is an overdetermined system and we should check for the compatibility of the solution.) In fact it can be easily shown that

$$R_{\mu\nu} \equiv 0, \quad \text{for} \quad \mu \neq \nu,$$

$$R_{33} = \sin^2\theta \, R_{22} = 0, \tag{8.47}$$

so that all the ten equations are consistent with the line element (8.45).

To determine the constant of integration m, let us note that very far away from a star of mass M we have seen that the metric has the form (see (6.21))

$$g_{00} = 1 + 2\phi(r) = 1 - \frac{2G_\text{N}M}{r}, \tag{8.48}$$

where M denotes the mass of the star. Comparing this with the solution in (8.45) we determine the constant of integration to be

$$m = -2G_\text{N}M, \tag{8.49}$$

so that the Schwarzschild line element (8.45) takes the final form

$$d\tau^2 = \left(1 - \frac{2G_\text{N}M}{r}\right)dt^2 - \left(1 - \frac{2G_\text{N}M}{r}\right)^{-1}dr^2 - r^2(d\theta^2 + \sin^2\theta d\phi^2). \tag{8.50}$$

This determines the form of the line element and, therefore, the metric uniquely.

One striking feature of the Schwarzschild metric in (8.50) is that at $r = 2G_\text{N}M$,

$$g_{00} = 0, \quad g_{rr} \to \infty. \tag{8.51}$$

That is, the Schwarzschild metric is singular at the Schwarzschild radius defined by

$$r_S = 2G_\text{N}M. \tag{8.52}$$

For most objects, this radius lies inside the object. For example, since (in order to have the appropriate dimension, the actual constant appearing in (8.50) (or (8.52)) is the Newton's constant G_N defined in (7.80) divided by c^2 which we have set to unity in our convention, but which can now be restored for calculations)

$$G_N \simeq 7 \times 10^{-29} \text{ cm gm}^{-1},$$

$$M(\text{earth}) \simeq 6 \times 10^{24} \text{ kg} = 6 \times 10^{27} \text{ gm}, \tag{8.53}$$

the Schwarzschild radius (8.52) for earth has the value

$$
\begin{aligned}
r_S(\text{earth}) &= 2G_N M(\text{earth}) \simeq 2 \times 7 \times 10^{-29} \times 6 \times 10^{27} \text{ cm} \\
&= .84 \text{ cm}, \tag{8.54}
\end{aligned}
$$

which is well inside the earth $(r(\text{earth}) \simeq 6 \times 10^6 \text{m} \simeq 6 \times 10^3 \text{ km})$. For the sun, we have

$$M(\text{sun}) \simeq 2 \times 10^{30} \text{ kg} = 2 \times 10^{33} \text{ gm}, \tag{8.55}$$

so that

$$
\begin{aligned}
r_S(\text{sun}) &= 2G_N M(\text{sun}) \simeq 2 \times 7 \times 10^{-29} \times 2 \times 10^{33} \text{ cm} \\
&= 2.8 \times 10^5 \text{cm} \simeq 3 \text{ km}. \tag{8.56}
\end{aligned}
$$

We can compare this with the radius of the sun which is given by

$$r_{\text{sun}} \simeq 7 \times 10^5 \text{ km}. \tag{8.57}$$

Let us, for the sake of curiosity, calculate the Schwarzschild radius for the proton. We know that

$$m(\text{proton}) \simeq 1.6 \times 10^{-24} \text{ gm}, \tag{8.58}$$

and, therefore,

$$
\begin{aligned}
r_S(\text{proton}) &\simeq 2 \times 7 \times 10^{-29} \times 1.6 \times 10^{-24} \text{ cm} \\
&\simeq 2.2 \times 10^{-52} \text{ cm}. \tag{8.59}
\end{aligned}
$$

We can compare this with the radius of proton which has the value

$$r(\text{proton}) \simeq 1.2 \times 10^{-13} \text{ cm.} \tag{8.60}$$

Thus we see that for most objects the Schwarzschild radius indeed lies inside the object which is, of course, beyond the region of validity of the solution (which holds true only in the exterior of the star). Therefore, we can ignore this singularity in the metric in such cases since the solution in the interior of the star is given by a different set of equations.

For some other astronomical objects like the neutron stars, however, the Schwarzschild radius is quite close to the physical radius of the object. In this case, we cannot neglect the singularity in the metric. However, the curvature tensor in this case can be shown to be singularity free. Thus the apparent singularity in the metric is an artifact of our choice of coordinates and can be rotated away by a suitable choice of coordinates. We will see this later. However, we note that if we compress matter into a star hard enough so that its physical radius is smaller than its Schwarzschild radius, then we obtain a black hole which we will also discuss later. The Schwarzschild radius is often referred to as the event horizon for reasons that will become clear shortly. We note that if $r < r_S = 2G_N M$, i.e., when the Schwarzschild radius is larger than the size of the object and we go inside the Schwarzschild radius, then

$$g_{00} < 0, \quad g_{11} > 0. \tag{8.61}$$

Namely, the time and the radial coordinates seem to exchange their roles. This leads to a lot of interesting phenomena. Let us emphasize here that the Schwarzschild solution does have a genuine singularity at the origin $r = 0$ when the star is a point mass object. This is not surprising since such a singularity occurs even in classical gravity.

▶ Example (Schwarzschild solution in the presence of a cosmological constant).
As we have seen in (8.50), the Schwarzschild line element has the form

$$\mathrm{d}\tau^2 = A(r)\mathrm{d}t^2 - B(\tau)\mathrm{d}r^2 - r^2\mathrm{d}\Omega^2$$

$$\equiv \left(1 - \frac{2G_N M}{r}\right)dt^2 - \left(1 - \frac{2G_N M}{r}\right)^{-1} dr^2 - r^2 d\Omega^2. \tag{8.62}$$

We obtained the line element by solving the Einstein's equations away from the star where

$$R_{\mu\nu} = 0, \quad \mu, \nu = 0, 1, 2, 3, \tag{8.63}$$

for only the components R_{00}, R_{11}, and R_{22}. Since there are 10 independent components of $R_{\mu\nu}$, we still have to show that the line element (8.62) also solves the remaining seven equations

$$R_{33} = R_{01} = R_{02} = R_{03} = R_{12} = R_{13} = R_{23} = 0. \tag{8.64}$$

To proceed, we recall that the components of the connection have been determined in (8.30) to have the form (see also the determination of A, B in (8.44) and (8.49))

$$\Gamma^0_{01} = \Gamma^0_{10} = -\frac{A'}{2A} = -\frac{G_N M}{r^2 - 2G_N M r},$$

$$\Gamma^1_{00} = -\frac{A'}{2B} = -\frac{G_N M}{r}\left(1 - \frac{2G_N M}{r}\right),$$

$$\Gamma^1_{11} = -\frac{B'}{2B} = -\frac{G_N M}{r^2 - 2G_N M r},$$

$$\Gamma^1_{22} = \frac{r}{B} = r - 2G_N M,$$

$$\Gamma^2_{12} = \Gamma^2_{21} = -\frac{1}{r},$$

$$\Gamma^3_{13} = \Gamma^3_{31} = -\frac{1}{r},$$

$$\Gamma^1_{33} = (r - 2G_N M)\sin^2\theta,$$

$$\Gamma^2_{33} = \sin\theta\cos\theta,$$

$$\Gamma^3_{23} = \Gamma^3_{32} = -\cot\theta. \tag{8.65}$$

Furthermore, from the definition of the Ricci curvature tensor in (8.32)

$$R_{\mu\nu} = \partial_\nu\Gamma^\lambda_{\mu\lambda} - \partial_\lambda\Gamma^\lambda_{\mu\nu} + \Gamma^\lambda_{\lambda\sigma}\Gamma^\sigma_{\mu\nu} - \Gamma^\lambda_{\nu\sigma}\Gamma^\sigma_{\mu\lambda}, \tag{8.66}$$

we obtain

$$R_{33} = \partial_3 \Gamma^\lambda_{3\lambda} - \partial_\lambda \Gamma^\lambda_{33} + \Gamma^\lambda_{\lambda\sigma} \Gamma^\sigma_{33} - \Gamma^\lambda_{3\sigma} \Gamma^\sigma_{3\lambda}$$

$$= -A \sin^2 \theta - r A' \sin^2 \theta - \cos 2\theta + \frac{1}{2} A' r \sin^2 \theta - A \sin^2 \theta$$

$$\quad + \frac{1}{2} \sin 2\theta \cot \theta + A \sin^2 \theta - \frac{1}{2} \sin^2 \theta r A'$$

$$= -A \sin^2 \theta - (1 - A) \sin^2 \theta - 1 + 2 \sin^2 \theta + \cos^2 \theta$$

$$= -\sin^2 \theta - 1 + \sin^2 \theta + \sin^2 \theta + \cos^2 \theta$$

$$= 0,$$

$$R_{01} = \partial_0 \Gamma^\lambda_{1\lambda} - \partial_\lambda \Gamma^\lambda_{10} + \Gamma^\lambda_{\lambda\sigma} \Gamma^\sigma_{10} - \Gamma^\lambda_{0\sigma} \Gamma^\sigma_{1\lambda}$$

$$= \Gamma^\lambda_{\lambda 0} \Gamma^0_{01} - \Gamma^0_{10} \Gamma^0_{00} - \Gamma^1_{11} \Gamma^1_{01}$$

$$= 0,$$

$$R_{02} = \partial_2 \Gamma^\lambda_{0\lambda} - \partial_\lambda \Gamma^\lambda_{02} + \Gamma^\lambda_{\lambda\sigma} \Gamma^\sigma_{02} - \Gamma^\lambda_{2\sigma} \Gamma^\sigma_{0\lambda}$$

$$= 0,$$

$$R_{03} = \partial_3 \Gamma^\lambda_{0\lambda} - \partial_\lambda \Gamma^\lambda_{03} + \Gamma^\lambda_{\lambda\sigma} \Gamma^\sigma_{03} - \Gamma^\lambda_{3\sigma} \Gamma^\sigma_{0\lambda}$$

$$= 0,$$

$$R_{12} = \partial_2 \Gamma^\lambda_{1\lambda} - \partial_\lambda \Gamma^\lambda_{12} + \Gamma^\lambda_{\lambda\sigma} \Gamma^\sigma_{12} - \Gamma^\lambda_{2\sigma} \Gamma^\sigma_{1\lambda}$$

$$= \Gamma^3_{32} \Gamma^2_{12} - \Gamma^1_{21} \Gamma^1_{11} - \Gamma^2_{22} \Gamma^2_{12} - \Gamma^3_{23} \Gamma^3_{13} - \Gamma^0_{20} \Gamma^0_{10}$$

$$= 0,$$

$$R_{13} = \partial_3 \Gamma^\lambda_{1\lambda} - \partial_\lambda \Gamma^\lambda_{13} + \Gamma^\lambda_{\lambda\sigma} \Gamma^\sigma_{13} - \Gamma^\lambda_{3\sigma} \Gamma^\sigma_{1\lambda}$$

$$= 0,$$

$$R_{23} = \partial_3 \Gamma^\lambda_{2\lambda} - \partial_\lambda \Gamma^\lambda_{23} + \Gamma^\lambda_{\lambda\sigma} \Gamma^\sigma_{23} - \Gamma^\lambda_{3\sigma} \Gamma^\sigma_{2\lambda}$$

$$= 0. \tag{8.67}$$

Here we have used the fact that the connections are independent of the azimuthal angle ϕ and that only some of the connections depend on θ. This shows that the line element (8.62) also solves the remaining seven Einstein's equations.

In the presence of a cosmological constant, Einstein's equations (7.122) (see also (7.121)) become

$$R_{\mu\nu} = 8\pi G_{\rm N} (T_{\mu\nu} - \frac{1}{2} g_{\mu\nu} T) + \Lambda g_{\mu\nu}, \tag{8.68}$$

where Λ denotes the cosmological constant. Far away from the star, we have $T^{\mu\nu} = T = 0$, and (8.68) takes the form

$$R_{\mu\nu} = \Lambda g_{\mu\nu}. \tag{8.69}$$

For the spherically symmetric line element (8.15), we have already determined the Ricci curvature tensors and the relevant equations in this case, take the forms (compare with (8.34), (8.36) and (8.38) respectively)

$$R_{00} = \frac{A''}{2B} - \frac{A'}{4B}\left(\frac{A'}{A} + \frac{B'}{B}\right) + \frac{A'}{rB} = \Lambda A,$$

$$R_{11} = -\frac{A''}{2A} - \frac{A'}{4A}\left(\frac{A'}{A} + \frac{B'}{B}\right) + \frac{B'}{rB} = -\Lambda B,$$

$$R_{22} = 1 - \frac{1}{B} - \frac{r}{2B}\left(\frac{A'}{A} - \frac{B'}{B}\right) = -\Lambda r^2. \tag{8.70}$$

It follows from (8.70) that

$$\frac{B}{A}R_{00} + R_{11} = \frac{1}{r}\left(\frac{A'}{A} + \frac{B'}{B}\right) = 0,$$

$$\text{or,}\quad \frac{d}{dr}\ln(AB) = 0, \tag{8.71}$$

which determines

$$A(r)B(r) = \text{constant} = 1, \tag{8.72}$$

as before. Using this in the R_{22} equation, we obtain

$$1 - A(r) - rA(r) \times \frac{A'(r)}{A(r)} = -\Lambda r^2,$$

$$\text{or,}\quad \frac{d}{dr}(rA) = 1 + \Lambda r^2,$$

$$\text{or,}\quad A(r) = 1 + \frac{1}{3}\Lambda r^3 + \frac{m}{r}, \tag{8.73}$$

where m is an integration constant. It follows now from (8.72) that

$$B(r) = \frac{1}{A(r)} = \left(1 + \frac{1}{3}\Lambda r^3 + \frac{m}{r}\right)^{-1}. \tag{8.74}$$

As in the case of the absence of the cosmological constant, we can determine $m = -2G_N M$, so that far away from the star we get the familiar solution $g_{00} = 1 - \frac{2G_N M}{r}$ (see (8.49)), if Λ is extremely small (as is normally the case).
    The Schwarzschild line element then becomes

$$\mathrm{d}\tau^2 = \left(1 + \frac{1}{3}\Lambda r^3 - \frac{2G_{\mathrm{N}}M}{r}\right)\mathrm{d}t^2 - \left(1 + \frac{1}{3}\Lambda r^3 - \frac{2G_{\mathrm{N}}M}{r}\right)^{-1}\mathrm{d}r^2 - r^2\mathrm{d}\Omega^2.$$

$$(8.75)$$

◀

▶ **Example (3-dimensional space of constant curvature).** A space is said to be of constant curvature if the Riemann-Christoffel curvature can be written in the form

$$R_{\mu\nu\lambda\rho} = K\left(g_{\mu\lambda}g_{\nu\rho} - g_{\mu\rho}g_{\nu\lambda}\right),\qquad(8.76)$$

where K denotes the constant curvature of the space (see also (7.75)).

Let us consider an isotropic 3-dimensional space of constant curvature K. For an isotropic space we can write the line element as

$$\mathrm{d}\tau^2 = A(r)\mathrm{d}r^2 + r^2\mathrm{d}\Omega^2 = A(r)\mathrm{d}r^2 + r^2(\mathrm{d}\theta^2 + \sin^2\theta\,\mathrm{d}\phi^2).\qquad(8.77)$$

The nontrivial components of the metric tensor for this space are given by

$$g_{11} = A(r),\quad g_{22} = r^2,\quad g_{33} = r^2\sin^2\theta,\qquad(8.78)$$

and the Lagrangian for the geodesic, in this case, is given by (a dot denotes a derivative with respect to τ)

$$L = \frac{1}{2}(A\dot{r}^2 + r^2\dot{\theta}^2 + r^2\sin^2\theta\dot{\phi}^2).\qquad(8.79)$$

From the Euler-Lagrange equations we can determine the affine connections as in section **8.2** and the nontrivial components have the forms

$$\Gamma^1_{11} = -\frac{A'}{2A},$$

$$\Gamma^1_{22} = \frac{r}{A},$$

$$\Gamma^1_{33} = \frac{r}{A}\sin^2\theta,$$

$$\Gamma^2_{12} = \Gamma^2_{21} = -\frac{1}{r},$$

$$\Gamma^3_{13} = \Gamma^3_{31} = -\frac{1}{r},$$

$$\Gamma^2_{33} = \sin\theta\cos\theta,\qquad(8.80)$$

$$\Gamma^3_{23} = \Gamma^3_{32} = -\cot\theta.\qquad(8.81)$$

We can now compute the components of the Riemann-Christoffel curvature tensor

$$R_{\mu\nu\lambda\rho} = g_{\mu\sigma}(\partial_\rho\Gamma^\sigma_{\nu\lambda} - \partial_\lambda\Gamma^\sigma_{\nu\rho} + \Gamma^\beta_{\nu\rho}\Gamma^\sigma_{\beta\lambda} - \Gamma^\beta_{\nu\lambda}\Gamma^\sigma_{\beta\rho}). \tag{8.82}$$

For example, for $\mu = \lambda = 1, \nu = \rho = 2$ we have

$$R_{1212} = g_{1\sigma}(\partial_2\Gamma^\sigma_{21} - \partial_1\Gamma^\sigma_{22} + \Gamma^\beta_{22}\Gamma^\sigma_{\beta1} - \Gamma^\beta_{21}\Gamma^\sigma_{\beta2})$$

$$= -A\partial_r\frac{r}{A} - A\frac{r}{A}\frac{A'}{2A} + A\frac{1}{r}\frac{r}{A}$$

$$= \frac{rA'}{2A}. \tag{8.83}$$

On the other hand, since the space is of constant curvature, we conclude from (8.76) that

$$R_{1212} = K(g_{11}g_{22} - g_{12}g_{21}) = Kg_{11}g_{22} = KAr^2. \tag{8.84}$$

Comparing (8.83) and (8.84) we obtain

$$\frac{rA'}{2A} = KAr^2,$$

$$\text{or,} \quad \frac{A'}{A^2} = 2Kr,$$

$$\text{or,} \quad A(r) = \frac{1}{c - Kr^2}, \tag{8.85}$$

where c is a constant of integration. Requiring that the space corresponds to flat space (in spherical coordinates) when the curvature vanishes, namely,

$$\lim_{K\to0} A(r) \to 1, \tag{8.86}$$

we determine the constant of integration to be $c = 1$ and

$$A(r) = \frac{1}{1 - Kr^2}, \tag{8.87}$$

so that the line element (8.77) takes the form

$$d\tau^2 = \frac{dr^2}{1 - Kr^2} + r^2\,d\Omega^2. \tag{8.88}$$

There are three cases to consider now. For $K = 0$, we get the familiar "flat-space" geometry described in spherical coordinates,

$$\mathrm{d}\tau^2 = \mathrm{d}r^2 + r^2\,\mathrm{d}\Omega^2. \tag{8.89}$$

For $K > 0$, and $Kr^2 < 1$ (equivalently, $r < K^{-1/2}$, which is the radius of the universe) we can define

$$r = \frac{\sin(\sqrt{K}\rho)}{\sqrt{K}}, \quad \mathrm{d}\rho^2 = \frac{\mathrm{d}r^2}{1 - Kr^2}, \tag{8.90}$$

and the line element (8.88) becomes

$$\mathrm{d}\tau^2 = \mathrm{d}\rho^2 + \frac{1}{K}\sin^2(\sqrt{K}\rho)\,\mathrm{d}\Omega^2, \tag{8.91}$$

which describes a spherical space with an oscillating radius $\frac{1}{K}\sin^2(\sqrt{K}\rho)$. (Note that for small $\sqrt{K}\rho$, the line element becomes $\mathrm{d}\tau^2 \approx \mathrm{d}\rho^2 + \rho^2\,\mathrm{d}\Omega^2$.) Finally, for $K < 0$, with $K = -|K|$, we can define

$$r = \frac{\sinh(\sqrt{|K|}\rho)}{\sqrt{|K|}}, \quad \mathrm{d}\rho^2 = \frac{\mathrm{d}r^2}{1 + |K|r^2}, \tag{8.92}$$

and the line element takes the form

$$\mathrm{d}\tau^2 = \mathrm{d}\rho^2 + \frac{1}{|K|}\sinh^2(\sqrt{|K|}\rho)\,\mathrm{d}\Omega^2. \tag{8.93}$$

This describes a pseudosphere of expanding radius.

◀

▶ **Example (Angular momentum conservation).** Given the Schwarzschild line element (8.50), the Lagrangian for the geodesic in such a space is determined to be (see (6.50) and a dot denotes a derivative with respect to τ)

$$L = \frac{1}{2}(A\dot{t}^2 - B\dot{r}^2 + r^2(\dot{\theta}^2 + \sin^2\theta\dot{\phi}^2)), \tag{8.94}$$

where, as we have seen in (8.44) and (8.49)

$$A(r) = 1 - \frac{2G_N M}{r} = B^{-1}(r). \tag{8.95}$$

The generalized conjugate momenta follow from (8.94) to be,

$$p_t = \frac{\partial L}{\partial \dot{t}} = A(r)\dot{t}, \qquad\qquad p_r = \frac{\partial L}{\partial \dot{r}} = -B(r)\dot{r},$$

$$p_\theta = \frac{\partial L}{\partial \dot{\theta}} = r^2\dot{\theta}, \qquad\qquad p_\phi = \frac{\partial L}{\partial \dot{\phi}} = r^2\sin^2\theta\dot{\phi}. \tag{8.96}$$

The geodesic equations for the θ, ϕ components can now be written as (see (6.51))

$$\frac{\mathrm{d}p_\theta}{\mathrm{d}\tau} = r^2 \sin\theta \cos\theta \dot\phi^2,$$

$$\frac{\mathrm{d}p_\phi}{\mathrm{d}\tau} = 0. \tag{8.97}$$

Let us consider the total (orbital) angular momentum of a particle, moving along the geodesic, defined by

$$L_{\mathrm{orb}}^2 = p_\theta^2 + \frac{p_\phi^2}{\sin^2\theta}. \tag{8.98}$$

Using (8.97) as well as the definitions of the conjugate momenta in (8.96), we obtain

$$\begin{aligned}
\frac{\mathrm{d}L_{\mathrm{orb}}^2}{\mathrm{d}\tau} &= \frac{\mathrm{d}}{\mathrm{d}\tau}\left(p_\theta^2 + \frac{p_\phi^2}{\sin^2\theta}\right) \\
&= 2p_\theta \frac{\mathrm{d}p_\theta}{\mathrm{d}\tau} + \frac{2p_\phi}{\sin^2\theta}\frac{\mathrm{d}p_\phi}{\mathrm{d}\tau} - \frac{2p_\phi^2}{\sin^3\theta}\cos\theta\,\dot\theta \\
&= 2p_\theta(r^2\sin\theta\cos\theta\dot\phi^2) - \frac{2p_\phi^2\cos\theta\,\dot\theta}{\sin^3\theta} \\
&= 2(r^2\dot\theta)(r^2\sin\theta\cos\theta\dot\phi^2) - \frac{2(r^2\sin^2\theta\dot\phi)^2\cos\theta\,\dot\theta}{\sin^3\theta} \\
&= 0, \tag{8.99}
\end{aligned}$$

proving that $L_{\mathrm{orb}}^2 = $ constant. Namely, the orbital angular momentum of a particle moving along a Schwarzschild geodesic is conserved.

◄

8.4 Properties of the Schwarzschild solution

Let us try to understand the features of the Schwarzschild solution (8.50). It is an exact solution and the line element has the form

$$\mathrm{d}\tau^2 = \left(1 - \frac{2G_{\mathrm{N}}M}{r}\right)\mathrm{d}t^2 - \left(1 - \frac{2G_{\mathrm{N}}M}{r}\right)^{-1}\mathrm{d}r^2 - r^2(\mathrm{d}\theta^2 + \sin^2\theta\mathrm{d}\phi^2). \tag{8.100}$$

First of all, let us note that if the mass of the star is zero ($M = 0$), then the line element simply reduces to that of flat space-time

of relativity (8.1) as it should. (This is also the asymptotic form
(space) of the Schwarzschild solution.) Consequently we can identify
the time and spatial coordinates to be those as measured in that
coordinate frame. In the presence of a star we can continue to identify
these coordinates as before (namely, as observed by an asymptotic
observer), the effect of curvature manifests mainly in the term $\frac{2G_{\rm N}M}{r}$.
Let us also note that the Schwarzschild solution mainly investigates
the effects of a single star in the external region without considering
the effects of other masses in the universe. This is a reasonable
thing to do in the vicinity of a star where the effect of other stars is
negligible (if the other stars are far away).

Furthermore, it is a property of static spaces that the space-time
splits into a time direction and spatial hypersurfaces. The angular
coordinates θ, ϕ, of course, take values $0 \leq \theta \leq \pi$ and $0 \leq \phi \leq 2\pi$
as in flat space. The radial distance, however, is restricted to lie
between $r_A \leq r \leq \infty$ where r_A is the larger of r_S and $r_{\rm star}$. Since
the metric is static we can assume $-\infty < t < \infty$. Let us also note
that the static nature of the space allows us to identify the three
dimensional spaces as fixed in time. We can, in fact, study the three
dimensional surfaces (at any fixed time) given by

$$\mathrm{d}s^2 = \left(1 - \frac{2G_{\rm N}M}{r}\right)^{-1} \mathrm{d}r^2 + r^2(\mathrm{d}\theta^2 + \sin^2\theta \mathrm{d}\phi^2). \tag{8.101}$$

Furthermore, if we restrict ourselves to surfaces with constant r, then
we can write

$$\mathrm{d}\sigma^2 = r^2(\mathrm{d}\theta^2 + \sin^2\theta \mathrm{d}\phi^2). \tag{8.102}$$

This is, of course, the surface of a sphere embedded in a Euclidean
space. The circumference at the equator (i.e., $\theta = \frac{\pi}{2}$) is given by

$$\int \mathrm{d}\sigma = r \int_0^{2\pi} \mathrm{d}\phi = 2\pi r. \tag{8.103}$$

Thus by measuring the circumference of the great circles we can in
fact obtain the value of the radial coordinate uniquely. This is like
in flat space.

However, if we look at the radial increment of the line element (8.101) for fixed θ and ϕ, we obtain

$$\mathrm{d}R = \left(1 - \frac{2G_\mathrm{N}M}{r}\right)^{-\frac{1}{2}} \mathrm{d}r,$$

or, $\mathrm{d}R \geq \mathrm{d}r.$ \hfill (8.104)

This shows that r no longer correctly measures the radial distance. However, we note that as we go infinitely far away from the star we have

$$\mathrm{d}R \to \mathrm{d}r, \hspace{4cm} (8.105)$$

and r does measure the radial distance in this limit. This can be understood on physical grounds simply as due to the effects of curvature. That is, the space becomes curved near the star such that any radial cross-section projects onto flat space-time as a circle of radius r. However, because of the curvature, the radial distance between the projected circles which gives $\mathrm{d}r$ is smaller than the actual length between the two corresponding radial cross sections as shown in Fig. 8.1. Similarly we note from (8.50) that a clock at rest records the proper time to be

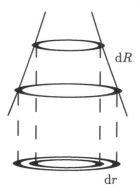

Figure 8.1: Radial increment on a curved space compared with its projection on a plane.

$$d\tau = \left(1 - \frac{2G_\mathrm{N}M}{r}\right)^{\frac{1}{2}} dt. \tag{8.106}$$

These results are quite similar in form to the relations in special relativity with which we are familiar, namely, $(c = 1)$

$$d\tau = \left(1 - \frac{2G_\mathrm{N}M}{r}\right)^{\frac{1}{2}} dt, \qquad d\tau_\mathrm{SR} = \left(1 - v^2\right)^{\frac{1}{2}} dt,$$

$$dR = \left(1 - \frac{2G_\mathrm{N}M}{r}\right)^{-\frac{1}{2}} dr, \qquad dL_\mathrm{SR} = \left(1 - v^2\right)^{-\frac{1}{2}} d\ell. \tag{8.107}$$

However, there are important differences. First of all we note that the Lorentz contraction or dilation in the case of special relativity is a constant factor whereas in the case of the Schwarzschild solution it is position dependent. Furthermore, the relations in the case of the Schwarzschild solution are coordinate dependent. If we use a different coordinate system, these relations would change.

8.5 Isotropic coordinates

In general most soluble problems in gravity have more than one simple coordinate system where exact results can be obtained. Another such coordinate system for the Schwarzschild solution is the isotropic coordinate system. Here we would like to write the line element in the isotropic form

$$d\tau^2 = A(r)dt^2 - B(r)ds^2, \tag{8.108}$$

where ds^2 denotes the line element for the flat three dimensional Euclidean space. The reason behind looking for an isotropic line element is that the spatial element ds^2 is invariant under changes in the spatial coordinate system. Namely, we can express the space coordinates by Cartesian coordinates x, y, z or by spherical coordinates or by cylindrical coordinates and the line element would still have the same form. Furthermore, if we look at the cosine of the angle between two vectors in this three dimensional space

$$\cos\theta = -\frac{\xi^i \eta_i}{|\xi||\eta|}, \tag{8.109}$$

it is the same as in the flat Euclidean space (the metric cancels out in the numerator and in the denominator). That is, this coordinate system is angle preserving and for this reason this line element is often also referred to as the conformal line element. The coordinate system in which the line element takes this particular form is known as the isotropic coordinate system.

Let us assume that there exists a coordinate system parameterized by $\rho(r)$, θ and ϕ such that (unlike the line element in (8.50), here $B(r)$ is not the inverse of $A(r)$)

$$
\begin{aligned}
\mathrm{d}\tau^2 &= A(r)\mathrm{d}t^2 - B(r)\mathrm{d}s^2 \\
&= \left(1 - \frac{2G_\mathrm{N}M}{r}\right)\mathrm{d}t^2 - B(r)(\mathrm{d}\rho^2 + \rho^2(\mathrm{d}\theta^2 + \sin^2\theta\mathrm{d}\phi^2)).
\end{aligned}
\tag{8.110}
$$

Comparing this with the Schwarzschild line element (8.50)

$$\mathrm{d}\tau^2 = \left(1 - \frac{2G_\mathrm{N}M}{r}\right)\mathrm{d}t^2 - \left(1 - \frac{2G_\mathrm{N}M}{r}\right)^{-1}\mathrm{d}r^2 - r^2(\mathrm{d}\theta^2 + \sin^2\theta\mathrm{d}\phi^2), \tag{8.111}$$

the angular components lead to

$$r^2 = B(r)\rho^2(r). \tag{8.112}$$

Similarly comparison of the radial component gives

$$B(r)\mathrm{d}\rho^2 = \left(1 - \frac{2G_\mathrm{N}M}{r}\right)^{-1}\mathrm{d}r^2,$$

$$\text{or,} \quad r^2\frac{\mathrm{d}\rho^2}{\rho^2} = \left(1 - \frac{2G_\mathrm{N}M}{r}\right)^{-1}\mathrm{d}r^2,$$

or,
$$\frac{d\rho}{\rho} = \frac{dr}{r\left(1 - \frac{2G_N M}{r}\right)^{\frac{1}{2}}} = \frac{dr}{(r^2 - 2G_N Mr)^{\frac{1}{2}}},$$

$$= \frac{dr}{\sqrt{(r - G_N M)^2 - (G_N M)^2}}. \tag{8.113}$$

This can be integrated to give

$$\ln \rho(r) = \ln \left| r - G_N M + \sqrt{r^2 - 2G_N Mr} \right| + C, \tag{8.114}$$

where C is a constant which can be determined from the boundary condition that as $r \to \infty$ the line element (8.110) should correspond to that of the flat space-time and hence (asymptotically $B(r) \to 1$)

$$\lim_{r \to \infty} \rho(r) \to r. \tag{8.115}$$

This leads to

$$\lim_{r \to \infty} \ln \rho(r) \to \ln(r + r) + C = \ln r + \ln 2 + C = \ln r,$$

or, $C = -\ln 2,$ \hfill (8.116)

and determines the form of the solution in (8.114) uniquely to be

$$\ln \rho(r) = \ln \left| r - G_N M + \sqrt{r^2 - 2G_N Mr} \right| - \ln 2,$$

or, $2\rho(r) = (r - G_N M) + \sqrt{r^2 - 2G_N Mr},$

or, $\rho(r) = \dfrac{1}{2} \left[(r - G_N M) + \sqrt{r^2 - 2G_N Mr} \right].$ \hfill (8.117)

Once we know the coordinate dependence of $\rho(r)$, we can express the metric coefficient $B(r)$ as a function of ρ. We note from (8.117) that

$$(2\rho(r) - (r - G_{\mathrm{N}}M))^2 = r^2 - 2G_{\mathrm{N}}Mr,$$

or, $\quad 4\rho^2(r) - 4\rho(r)(r - G_{\mathrm{N}}M) + (r - G_{\mathrm{N}}M)^2 = r^2 - 2G_{\mathrm{N}}Mr,$

or, $\quad 4\rho^2 - 4\rho(r - G_{\mathrm{N}}M) + (G_{\mathrm{N}}M)^2 = 0,$

or, $\quad (2\rho + G_{\mathrm{N}}M)^2 - 4\rho r = 0,$

or, $\quad \dfrac{r}{\rho} = \left(1 + \dfrac{G_{\mathrm{N}}M}{2\rho}\right)^2,$ $\hfill (8.118)$

which leads to (see (8.112))

$$B(r) = \frac{r^2}{\rho^2(r)} = \left(1 + \frac{G_{\mathrm{N}}M}{2\rho}\right)^4. \qquad (8.119)$$

Furthermore, using (8.118) we note that

$$
\begin{aligned}
A(r) &= 1 - \frac{2G_{\mathrm{N}}M}{r} = 1 - \frac{2G_{\mathrm{N}}M}{\rho\left(1 + \frac{G_{\mathrm{N}}M}{2\rho}\right)^2} \\
&= \frac{\rho\left(1 + \frac{G_{\mathrm{N}}M}{\rho} + \left(\frac{G_{\mathrm{N}}M}{2\rho}\right)^2\right) - 2G_{\mathrm{N}}M}{\rho\left(1 + \frac{G_{\mathrm{N}}M}{2\rho}\right)^2} \\
&= \frac{\left(1 - \frac{G_{\mathrm{N}}M}{2\rho}\right)^2}{\left(1 + \frac{G_{\mathrm{N}}M}{2\rho}\right)^2}.
\end{aligned}
\qquad (8.120)
$$

Therefore, the conformal line element can be written in the unique form

$$d\tau^2 = \frac{\left(1 - \frac{G_{\mathrm{N}}M}{2\rho}\right)^2}{\left(1 + \frac{G_{\mathrm{N}}M}{2\rho}\right)^2}\, dt^2 - \left(1 + \frac{G_{\mathrm{N}}M}{2\rho}\right)^4 ds^2. \qquad (8.121)$$

Tests of general relativity

There are several tests of general relativity that we can do with the Schwarzschild solution obtained in the last chapter. We have already talked about the gravitational red shift in section **6.3**. That essentially tests the principle of equivalence. In this chapter we describe some other experiments which directly test general relativity.

9.1 Radar echo experiment

sun

planet

earth

Figure 9.1: Radar echo experiment.

Let us consider a small planet such as Mercury or Venus between the earth and the sun and assume that there is an observer on earth in the gravitational field of the sun as shown in Fig. 9.1. Let the coordinates of the observer and the planet at a given time be (r_1, θ_1, ϕ_1) and (r_2, θ_2, ϕ_2) respectively $(r_1 > r_2)$ with the sun defining the origin of the coordinate system. Furthermore, let us assume that when they are aligned, the observer on earth sends a radar signal which is reflected by the planet and is received back on earth. We are interested in calculating the time interval between the transmission and the reception of the radar signal. In special relativity light signals travel with the speed of light $(c = 1)$ and hence the time interval is

265

the same as twice the radial separation between the two planets given by (see (8.104))

$$\Delta\tau_{\mathrm{sr}} = 2 \int_{r_2}^{r_1} \mathrm{d}R = 2 \int_{r_2}^{r_1} \mathrm{d}r \left(1 - \frac{2G_{\mathrm{N}}M}{r}\right)^{-\frac{1}{2}}. \tag{9.1}$$

Assuming the gravitational potential of the sun to be small we can approximate this as

$$\begin{aligned} \Delta\tau_{\mathrm{sr}} &\approx 2 \int_{r_2}^{r_1} \mathrm{d}r \left(1 + \frac{G_{\mathrm{N}}M}{r}\right) \\ &= 2 \left[(r_1 - r_2) + G_{\mathrm{N}}M \ln\frac{r_1}{r_2}\right]. \end{aligned} \tag{9.2}$$

On the other hand, looking at the Schwarzschild metric (8.50) we know that since for the photon $\mathrm{d}\tau = 0$, we have (motion is along the radial direction so that $\mathrm{d}\theta = 0 = \mathrm{d}\phi$)

$$\left(1 - \frac{2G_{\mathrm{N}}M}{r}\right)\mathrm{d}t^2 = \left(1 - \frac{2G_{\mathrm{N}}M}{r}\right)^{-1}\mathrm{d}r^2,$$

$$\text{or,} \quad \frac{\mathrm{d}r}{\mathrm{d}t} = \left(1 - \frac{2G_{\mathrm{N}}M}{r}\right). \tag{9.3}$$

This is the coordinate velocity of light. Therefore, the coordinate time interval taken by the signal for the same process is given by

$$\begin{aligned} \Delta t &= 2 \int_{r_2}^{r_1} \frac{\mathrm{d}r}{1 - \frac{2G_{\mathrm{N}}M}{r}} \approx 2 \int_{r_2}^{r_1} \mathrm{d}r \left(1 + \frac{2G_{\mathrm{N}}M}{r}\right) \\ &= 2 \left[(r_1 - r_2) + 2G_{\mathrm{N}}M \ln\frac{r_1}{r_2}\right]. \end{aligned} \tag{9.4}$$

The proper time interval recorded by the observer on earth is obtained from the coordinate time interval to be

$$
\begin{aligned}
\Delta\tau &= g_{00}^{\frac{1}{2}}(r_1)\Delta t \\
&= \left(1 - \frac{2G_{\mathrm{N}}M}{r_1}\right)^{\frac{1}{2}} \Delta t \\
&\approx \left(1 - \frac{G_{\mathrm{N}}M}{r_1}\right) \times 2\left[(r_1 - r_2) + 2G_{\mathrm{N}}M\ln\frac{r_1}{r_2}\right] \\
&= 2\left[(r_1 - r_2) + 2G_{\mathrm{N}}M\ln\frac{r_1}{r_2} - \frac{G_{\mathrm{N}}M(r_1 - r_2)}{r_1}\right].
\end{aligned}
\tag{9.5}
$$

Here we have restricted only to the dominant terms. However, this already shows that there is a difference in the time interval predicted by special relativity and general relativity. In fact

$$
\Delta\tau - \Delta\tau_{\mathrm{sr}} \simeq 2G_{\mathrm{N}}M\left[\ln\frac{r_1}{r_2} - \frac{r_1 - r_2}{r_1}\right].
\tag{9.6}
$$

This time interval has been measured experimentally and the result agrees with the theoretical prediction up to about 20 percent. However, one can criticize such experiments on many grounds. First of all the planet involved is not a point object. The earth is not stationary, but is moving in its orbit. Both the planet as well as the earth produce gravitational fields which have not been taken into consideration in the calculation. In recent years, however, such questions have been taken into account and experiments yield a 5 percent agreement with the theoretical result. These tests come from the tracking of the satellite systems where the satellites are relatively small and can be treated as massless objects. Thus the time delay experiment marks a good success of general relativity.

9.2 Motion of a particle in a Schwarzschild background

The Schwarzschild metric (8.50) is given by the line element

$$
\mathrm{d}\tau^2 = \left(1 - \frac{2G_{\mathrm{N}}M}{r}\right)\mathrm{d}t^2 - \left(1 - \frac{2G_{\mathrm{N}}M}{r}\right)^{-1}\mathrm{d}r^2 - r^2(\mathrm{d}\theta^2 + \sin^2\theta\,\mathrm{d}\phi^2),
\tag{9.7}
$$

where G_N is Newton's constant and M is the mass of the star producing the gravitational field. Let us now consider the motion of a particle of unit mass in the gravitational field of this star. Since the particle is massive, we can use its proper time to label the geodesics. We will discuss the motion of the photon separately in the next section. For the present problem, therefore, we can write a Lagrangian for the particle motion as (up to a factor $\frac{1}{2}$)

$$L = g_{\mu\nu}\dot{x}^\mu\dot{x}^\nu$$
$$= \left(1 - \frac{2G_N M}{r}\right)\dot{t}^2 - \left(1 - \frac{2G_N M}{r}\right)^{-1}\dot{r}^2 - r^2(\dot{\theta}^2 + \sin^2\theta\dot{\phi}^2), \tag{9.8}$$

where a dot denotes differentiation with respect to the proper time τ. Let us note that t and ϕ are cyclic variables and, consequently, we would have constants of motion. However, before we write down all four equations, let us look at the θ equation which leads to

$$\frac{\mathrm{d}}{\mathrm{d}\tau}\frac{\partial L}{\partial\dot{\theta}} - \frac{\partial L}{\partial\theta} = 0,$$

or, $$\frac{\mathrm{d}}{\mathrm{d}\tau}\left(-2r^2\dot{\theta}\right) + 2r^2\sin\theta\cos\theta\,\dot{\phi}^2 = 0,$$

or, $$-2\frac{\mathrm{d}}{\mathrm{d}\tau}\left(r^2\dot{\theta}\right) + r^2\sin 2\theta\,\dot{\phi}^2 = 0. \tag{9.9}$$

Therefore, we see that if the particle motion initially is in the equatorial plane, namely, if initially we have

$$\theta = \pi/2, \quad \dot{\theta} = 0, \tag{9.10}$$

then,

$$\frac{\mathrm{d}}{\mathrm{d}\tau}\left(r^2\dot{\theta}\right) = 0,$$

or, $$r^2\dot{\theta} = \text{constant} \ = 0, \tag{9.11}$$

where the vanishing of the constant follows from the fact that its value is zero initially. Therefore, the angle θ for the particle does not change with time and the motion of the particle always lies in the equatorial plane.

Consequently, to simplify the problem, we would consider the motion only in the equatorial plane ($\theta = \frac{\pi}{2}$). The Lagrangian (9.8), in this case, has the form

$$L = \left(1 - \frac{2G_{\mathrm{N}}M}{r}\right)\dot{t}^2 - \left(1 - \frac{2G_{\mathrm{N}}M}{r}\right)^{-1}\dot{r}^2 - r^2\dot{\phi}^2. \tag{9.12}$$

The ϕ equation following from this Lagrangian leads to the relation

$$\frac{\mathrm{d}}{\mathrm{d}\tau}\frac{\partial L}{\partial \dot{\phi}} - \frac{\partial L}{\partial \phi} = 0,$$

or, $$\frac{\mathrm{d}}{\mathrm{d}\tau}\left(-2r^2\dot{\phi}\right) = 0,$$

or, $$r^2\dot{\phi} = \text{constant} = \ell. \tag{9.13}$$

Similarly, the t equation yields

$$\frac{\mathrm{d}}{\mathrm{d}\tau}\frac{\partial L}{\partial \dot{t}} - \frac{\partial L}{\partial t} = 0,$$

or, $$\frac{\mathrm{d}}{\mathrm{d}\tau}\left(2\left(1 - \frac{2G_{\mathrm{N}}M}{r}\right)\dot{t}\right) = 0,$$

or, $$\left(1 - \frac{2G_{\mathrm{N}}M}{r}\right)\dot{t} = \text{constant} = k. \tag{9.14}$$

These relations lead to the two conserved quantities associated with the problem alluded to earlier. In particular, we note that the ϕ equation can be thought of as leading to the conservation of angular momentum.

Furthermore, the r equation has the form

$$\frac{d}{d\tau}\left(\frac{\partial L}{\partial \dot{r}}\right) - \frac{\partial L}{\partial r} = 0,$$

or, $$\frac{d}{d\tau}\left(-2\left(1 - \frac{2G_{\mathrm{N}}M}{r}\right)^{-1}\dot{r}\right) - \frac{2G_{\mathrm{N}}M}{r^2}\dot{t}^2$$

$$-\left(1 - \frac{2G_{\mathrm{N}}M}{r}\right)^{-2}\frac{2G_{\mathrm{N}}M}{r^2}\dot{r}^2 + 2r\dot{\phi}^2 = 0,$$

or, $$-\left(1 - \frac{2G_{\mathrm{N}}M}{r}\right)^{-1}\ddot{r} + \left(1 - \frac{2G_{\mathrm{N}}M}{r}\right)^{-2}\frac{2G_{\mathrm{N}}M}{r^2}\dot{r}^2$$

$$-\frac{G_{\mathrm{N}}M}{r^2}\dot{t}^2 - \left(1 - \frac{2G_{\mathrm{N}}M}{r}\right)^{-2}\frac{G_{\mathrm{N}}M}{r^2}\dot{r}^2 + r\dot{\phi}^2 = 0,$$

or, $$\left(1 - \frac{2G_{\mathrm{N}}M}{r}\right)^{-1}\ddot{r} - \left(1 - \frac{2G_{\mathrm{N}}M}{r}\right)^{-2}\frac{G_{\mathrm{N}}M}{r^2}\dot{r}^2$$

$$+\frac{G_{\mathrm{N}}M}{r^2}\dot{t}^2 - r\dot{\phi}^2 = 0. \tag{9.15}$$

Although we can try to solve the radial equation (9.15), it is often more convenient to solve for r from the equation we obtain from the line element (9.7) (with $\theta = \frac{\pi}{2}$), namely,

$$1 = \left(1 - \frac{2G_{\mathrm{N}}M}{r}\right)\dot{t}^2 - \left(1 - \frac{2G_{\mathrm{N}}M}{r}\right)^{-1}\dot{r}^2 - r^2\dot{\phi}^2. \tag{9.16}$$

Dividing (9.16) throughout by $\dot{\phi}^2$ we obtain

$$\left(1 - \frac{2G_{\mathrm{N}}M}{r}\right)\frac{\dot{t}^2}{\dot{\phi}^2} - \left(1 - \frac{2G_{\mathrm{N}}M}{r}\right)^{-1}\left(\frac{dr}{d\phi}\right)^2 - r^2 = \frac{1}{\dot{\phi}^2},$$

or, $$\left(\frac{dr}{d\phi}\right)^2 - \left(1 - \frac{2G_{\mathrm{N}}M}{r}\right)^2\frac{\dot{t}^2}{\dot{\phi}^2} + r^2\left(1 - \frac{2G_{\mathrm{N}}M}{r}\right)$$

$$= -\left(1 - \frac{2G_{\mathrm{N}}M}{r}\right)\frac{1}{\dot{\phi}^2},$$

or, $$\left(\frac{dr}{d\phi}\right)^2 - \frac{k^2 r^4}{\ell^2} + r^2\left(1 - \frac{2G_{\mathrm{N}}M}{r}\right) = -\left(1 - \frac{2G_{\mathrm{N}}M}{r}\right)\frac{r^4}{\ell^2},$$

$$\text{or,} \quad \left(\frac{dr}{d\phi}\right)^2 = -r^2\left(1 - \frac{2G_N M}{r}\right) + \frac{r^4}{\ell^2}\left((k^2 - 1) + \frac{2G_N M}{r}\right),$$

$$(9.17)$$

where we have used (9.13) and (9.14). If we now define a new variable

$$u = \frac{1}{r}, \qquad \frac{dr}{d\phi} = -\frac{1}{u^2}\frac{du}{d\phi}, \qquad (9.18)$$

then equation (9.17) simplifies to

$$\frac{1}{u^4}\left(\frac{du}{d\phi}\right)^2 = -\frac{1}{u^2}(1 - 2G_N M u) + \frac{1}{\ell^2 u^4}((k^2 - 1) + 2G_N M u),$$

$$\text{or,} \quad \left(\frac{du}{d\phi}\right)^2 = -u^2(1 - 2G_N M u) + \frac{1}{\ell^2}((k^2 - 1) + 2G_N M u)$$

$$= \frac{1}{\ell^2}\left((k^2 - 1) + 2G_N M u\right) - u^2(1 - 2G_N M u). \quad (9.19)$$

Upon integration, this leads to

$$\int d\phi = \int \frac{du\, \ell}{[((k^2 - 1) + 2G_N M u) - \ell^2 u^2(1 - 2G_N M u)]^{\frac{1}{2}}},$$

$$\text{or,} \quad \phi = \phi_0 + \int \frac{du\, \ell}{[((k^2 - 1) + 2G_N M u) - \ell^2 u^2(1 - 2G_N M u)]^{\frac{1}{2}}}.$$

$$(9.20)$$

As a result, we see that the value of ϕ can be determined as a function of r or conversely the value of r as a function of ϕ. This is an exact result. However, it is not easy to see from this relation the elliptic form of the orbital motion of the particle (planet) that we expect. To see this, we differentiate (9.19) with respect to ϕ which yields

$$2\frac{du}{d\phi}\frac{d^2u}{d\phi^2} = \frac{2G_N M}{\ell^2}\frac{du}{d\phi} - 2u\frac{du}{d\phi} + 6G_N M u^2 \frac{du}{d\phi},$$

$$\text{or,} \quad \frac{du}{d\phi}\left(\frac{d^2u}{d\phi^2} - \frac{G_N M}{\ell^2} + u - 3G_N M u^2\right) = 0. \quad (9.21)$$

One of the solutions of (9.21) is

$$\frac{\mathrm{d}u}{\mathrm{d}\phi} = 0,$$

or, $r(\phi) = \text{constant},$ (9.22)

which corresponds to circular motion. However, the more interesting solution of (9.21) comes from

$$\frac{\mathrm{d}^2 u}{\mathrm{d}\phi^2} = \frac{G_\mathrm{N} M}{\ell^2} - u + 3 G_\mathrm{N} M u^2.$$ (9.23)

To understand the meaning of this equation, let us look at the Newtonian problem of a point particle of unit mass moving in the equatorial plane in the gravitational field of a star for which the Lagrangian is given by (here a dot denotes a derivative with respect to t)

$$L = \frac{1}{2}\dot{r}^2 + \frac{1}{2} r^2 \dot{\phi}^2 + \frac{G_\mathrm{N} M}{r}.$$ (9.24)

The ϕ equation following from this gives (as also in (9.13))

$$r^2 \dot{\phi} = \text{constant} = \ell.$$ (9.25)

The r equation, on the other hand, has the form

$$\ddot{r} - r\dot{\phi}^2 + \frac{G_\mathrm{N} M}{r^2} = 0,$$

or, $\ddot{r} - \dfrac{\ell^2}{r^3} + \dfrac{G_\mathrm{N} M}{r^2} = 0.$ (9.26)

As before, let us change variable as

$$u = \frac{1}{r},$$ (9.27)

so that we have

$$\dot{r} = \frac{\mathrm{d}r}{\mathrm{d}t} = \frac{\mathrm{d}\phi}{\mathrm{d}t}\frac{\mathrm{d}r}{\mathrm{d}\phi} = \ell u^2 \left(-\frac{1}{u^2}\frac{\mathrm{d}u}{\mathrm{d}\phi} \right) = -\ell\frac{\mathrm{d}u}{\mathrm{d}\phi},$$

$$\ddot{r} = \frac{\mathrm{d}^2 r}{\mathrm{d}t^2} = \frac{\mathrm{d}\phi}{\mathrm{d}t}\frac{\mathrm{d}}{\mathrm{d}\phi}\left(\frac{\mathrm{d}r}{\mathrm{d}t} \right) = \ell u^2 \frac{\mathrm{d}}{\mathrm{d}\phi}\left(-\ell\frac{\mathrm{d}u}{\mathrm{d}\phi} \right)$$

$$= -\ell^2 u^2 \frac{\mathrm{d}^2 u}{\mathrm{d}\phi^2}. \tag{9.28}$$

Thus the classical Kepler problem (9.26) leads to

$$-\ell^2 u^2 \frac{\mathrm{d}^2 u}{\mathrm{d}\phi^2} - \ell^2 u^3 + G_{\mathrm{N}} M u^2 = 0,$$

or, $$\frac{\mathrm{d}^2 u}{\mathrm{d}\phi^2} = \frac{G_{\mathrm{N}} M}{\ell^2} - u. \tag{9.29}$$

This equation is known to lead to elliptic orbits and is similar to (9.23) which we have derived from the Schwarzschild metric. Comparing the two we conclude that the term $3G_{\mathrm{N}} M u^2$ in (9.23) must represent the relativistic correction to the particle motion. There are now two special cases which we consider below.

9.2.1 Vertical free fall. Let us consider the case where a massive particle ($m = 1$) falls radially in the gravitational field of a massive star. In this case

$$\dot{\phi} = 0, \tag{9.30}$$

which leads to (see (9.13))

$$\ell = 0, \tag{9.31}$$

and consequently we cannot divide by $\dot{\phi}^2$ as we had done earlier in (9.17). In this case the equation from the line element in (9.16) becomes ($\dot{\phi} = 0$)

$$\left(1 - \frac{2G_{\mathrm{N}}M}{r}\right)\dot{t}^2 - \left(1 - \frac{2G_{\mathrm{N}}M}{r}\right)^{-1}\dot{r}^2 = 1,$$

or, $$\frac{k^2}{\left(1 - \frac{2G_{\mathrm{N}}M}{r}\right)} - \frac{\dot{r}^2}{\left(1 - \frac{2G_{\mathrm{N}}M}{r}\right)} = 1,$$

or, $$\dot{r}^2 - k^2 + \left(1 - \frac{2G_{\mathrm{N}}M}{r}\right) = 0. \tag{9.32}$$

This clarifies the meaning of the constant k. For if the particle falls from rest at some value of $r = r_0$, namely, if

$$\dot{r}|_{r=r_0} = 0, \tag{9.33}$$

then (9.32) implies

$$k = \left(1 - \frac{2G_{\mathrm{N}}M}{r_0}\right)^{\frac{1}{2}}, \tag{9.34}$$

where the positive sign of the square root is taken to signify that coordinate time increases with τ (see (9.14)). Equation (9.34) shows that the constant k is not a universal constant, rather its value depends on the particular geodesic of the particle. For example, if the geodesic corresponds to the particle being at rest at a coordinate infinitely far away from the star then

$$k = 1. \tag{9.35}$$

Putting in this form of k into (9.32) we have

$$\dot{r}^2 - 2G_{\mathrm{N}}M\left(\frac{1}{r} - \frac{1}{r_0}\right) = 0,$$

or, $$\frac{1}{2}\dot{r}^2 = G_{\mathrm{N}}M\left(\frac{1}{r} - \frac{1}{r_0}\right). \tag{9.36}$$

Differentiating this equation with respect to τ, we obtain

$$\dot{r}\ddot{r} = -\frac{G_{\mathrm{N}}M}{r^2}\,\dot{r},$$

$$\text{or,}\quad \ddot{r} = -\frac{G_{\mathrm{N}}M}{r^2}. \tag{9.37}$$

Both these equations in (9.36) and (9.37) are similar to the classical Newtonian equation. The first one is analogous to the statement that a particle falling in a gravitational field gains an amount of kinetic energy equal to the change in its potential energy. The second form merely represents Newton's equation for the radial fall of a particle in the gravitational field of a massive object. However, we must remember the differences also. In particular, r does not measure the radial distance as we have seen and dots here refer to differentiation with respect to proper time.

Equation (9.36) also allows us to calculate the proper time experienced by a particle falling from rest at $r = r_0$. Thus if we assume $r = r_0$ at $\tau = 0$, then

$$
\begin{aligned}
\tau &= \int \mathrm{d}\tau = \int \frac{\mathrm{d}r'}{\left[2G_{\mathrm{N}}M\left(\frac{1}{r'} - \frac{1}{r_0}\right)\right]^{\frac{1}{2}}} \\
&= \frac{1}{\sqrt{2G_{\mathrm{N}}M}} \int_{r}^{r_0} \mathrm{d}r' \left(\frac{r_0 r'}{r_0 - r'}\right)^{\frac{1}{2}}.
\end{aligned} \tag{9.38}
$$

Here the limits of integration are chosen such that the proper time is positive (that is, the radial velocity actually decreases vectorially). The limit of integration can actually be extended up to the Schwarzschild radius $r = r_S = 2G_{\mathrm{N}}M$ unless the physical radius of the star is larger than the Schwarzchild radius. We also note that the proper time experienced by the particle in going up to the Schwarzchild radius is finite since there is no singularity in the integrand.

On the other hand, if we look at the coordinate time we note from (9.14) that

$$\dot{t} = \frac{\mathrm{d}t}{\mathrm{d}\tau} = \frac{k}{\left(1 - \frac{2G_{\mathrm{N}}M}{r}\right)} = \frac{\left(1 - \frac{2G_{\mathrm{N}}M}{r_0}\right)^{\frac{1}{2}}}{\left(1 - \frac{2G_{\mathrm{N}}M}{r}\right)}. \tag{9.39}$$

Integrating this we have

$$
\begin{aligned}
t &= \int d\tau \frac{\left(1 - \frac{2G_N M}{r_0}\right)^{\frac{1}{2}}}{\left(1 - \frac{2G_N M}{r}\right)} \\
&= \int_r^{r_0} \frac{dr'}{\sqrt{2G_N M}} \frac{\left(1 - \frac{2G_N M}{r_0}\right)^{\frac{1}{2}}}{\left(1 - \frac{2G_N M}{r'}\right)} \left(\frac{r_0 r'}{r_0 - r'}\right)^{\frac{1}{2}}.
\end{aligned}
\tag{9.40}
$$

This clearly shows that the coordinate time diverges as we approach the Schwarzschild radius. In fact, the integrand is singular at the Schwarzschild radius.

Let us next calculate the coordinate speed of the particle from the definition in (9.36)

$$
\frac{dr}{d\tau} = \dot{r} = \left[2G_N M \frac{(r_0 - r)}{r_0 r}\right]^{\frac{1}{2}}.
\tag{9.41}
$$

We can obtain the coordinate speed from this as

$$
\begin{aligned}
v(r) = \frac{dr}{dt} &= \frac{d\tau}{dt} \frac{dr}{d\tau} = \left(\dot{t}\right)^{-1} \dot{r} \\
&= \frac{\left(1 - \frac{2G_N M}{r}\right)}{k} \dot{r} \\
&= \frac{\left(1 - \frac{2G_N M}{r}\right)}{\left(1 - \frac{2G_N M}{r_0}\right)^{\frac{1}{2}}} \left[2G_N M \frac{(r_0 - r)}{r_0 r}\right]^{\frac{1}{2}},
\end{aligned}
\tag{9.42}
$$

where we have used (9.39).

This shows that although $\dot{r}$ remains finite as we approach the Schwarzschild radius, the coordinate speed goes to zero in that limit. As a result, the particle takes infinitely long to reach the Schwarzschild radius. Furthermore, setting

$$
\frac{dv}{dr} = 0,
\tag{9.43}
$$

from (9.42) we obtain

$$
\left(\frac{2G_{\mathrm{N}}M}{1 - \frac{2G_{\mathrm{N}}M}{r_0}} \right)^{\frac{1}{2}} \left[\frac{2G_{\mathrm{N}}M}{r^2} \left(\frac{r_0 - r}{r_0 r} \right)^{\frac{1}{2}} \right.
$$

$$
\left. + \left(1 - \frac{2G_{\mathrm{N}}M}{r} \right) \frac{1}{2} \left(\frac{r_0 - r}{r_0 r} \right)^{-\frac{1}{2}} \left(-\frac{1}{r^2} \right) \right] = 0,
$$

or, $2G_{\mathrm{N}}M - \dfrac{1}{2} \left(1 - \dfrac{2G_{\mathrm{N}}M}{r} \right) \dfrac{r_0 r}{r_0 - r} = 0,$

or, $2G_{\mathrm{N}}M(r_0 - r) - \dfrac{1}{2} \left(1 - \dfrac{2G_{\mathrm{N}}M}{r} \right) r_0 r = 0,$

or, $r = \dfrac{3G_{\mathrm{N}}M r_0}{2G_{\mathrm{N}}M + \frac{r_0}{2}}.$ (9.44)

This analysis shows that if the particle falls from rest at infinity $(r_0 \to \infty)$, its coordinate velocity increases until

$$
r = 6G_{\mathrm{N}}M, \tag{9.45}
$$

after which it decreases and goes to zero at the Schwarzschild radius. Let us compare this behavior with the classical Newtonian result which gives

$$
v(r) = \left(\frac{2G_{\mathrm{N}}M}{r} \right)^{\frac{1}{2}}, \tag{9.46}
$$

and hence leads to the conclusion that the speed of the particle keeps increasing and diverges at the origin.

9.2.2 Circular orbit. For a circular orbit we have $\dot{r} = 0 = \ddot{r}$, and, consequently, the r equation (9.15) takes the form

$$\frac{G_N M}{r^2} \, \dot{t}^2 = r \dot{\phi}^2,$$

or, $$\frac{\dot{\phi}^2}{\dot{t}^2} = \left(\frac{d\phi}{dt}\right)^2 = \frac{G_N M}{r^3},$$

or, $$\int dt = \int \left(\frac{r^3}{G_N M}\right)^{\frac{1}{2}} d\phi = \left(\frac{r^3}{G_N M}\right)^{\frac{1}{2}} \int d\phi. \qquad (9.47)$$

Therefore, the period of the orbit is given by

$$\Delta t = 2\pi \left(\frac{r^3}{G_N M}\right)^{\frac{1}{2}}. \qquad (9.48)$$

This is exactly like Kepler's law which says that the square of the period of an orbit is proportional to the cube of the radius. However, we must remember that here r does not denote the true radius of the orbit and Δt measures the coordinate orbit period.

If an observer is at rest at a distance r_0, then the proper time of the orbit that he would measure is (see (9.7))

$$\Delta \tau = \sqrt{g_{00}(r_0)} \, \Delta t = \left(1 - \frac{2 G_N M}{r_0}\right)^{\frac{1}{2}} \Delta t. \qquad (9.49)$$

We note here that as $r_0 \to \infty$, $\Delta t = \Delta \tau$. Hence by measuring the period of the orbit we can measure the radius of the orbit.

At this point, we can inquire about the period that an observer moving with the particle or the planet would measure. We note from (9.39) that

$$\Delta \tau = \left(1 - \frac{2 G_N M}{r}\right) k^{-1} \Delta t, \qquad (9.50)$$

so that the value of the period depends on the constant k. To determine the constant k we note that in this case $\dot{r} = 0$ and hence from (9.16) we obtain

$$1 = \left(1 - \frac{2 G_N M}{r}\right) \dot{t}^2 - r^2 \dot{\phi}^2. \qquad (9.51)$$

On the other hand, we know from (9.47) that for circular orbits

$$\frac{G_N M}{r^2} \dot{t}^2 = r\dot{\phi}^2. \tag{9.52}$$

Using this in (9.51) we obtain

$$\left(1 - \frac{2G_N M}{r}\right) \dot{t}^2 - \frac{G_N M}{r} \dot{t}^2 = 1,$$

or, $\quad \left(1 - \frac{3G_N M}{r}\right) \dot{t}^2 = 1,$

or, $\quad \left(1 - \frac{3G_N M}{r}\right) \frac{k^2}{\left(1 - \frac{2G_N M}{r}\right)^2} = 1,$

or, $\quad k^2 = \dfrac{\left(1 - \frac{2G_N M}{r}\right)^2}{\left(1 - \frac{3G_N M}{r}\right)}, \tag{9.53}$

so that (9.50) leads to

$$\begin{aligned}
\Delta\tau &= \left(1 - \frac{2G_N M}{r}\right) k^{-1} \Delta t \\
&= \left(1 - \frac{2G_N M}{r}\right) \frac{\left(1 - \frac{3G_N M}{r}\right)^{\frac{1}{2}}}{\left(1 - \frac{2G_N M}{r}\right)} \Delta t,
\end{aligned}$$

or, $\quad \Delta\tau = \left(1 - \frac{3G_N M}{r}\right)^{\frac{1}{2}} \Delta t. \tag{9.54}$

Note that since k^2 is positive, bound orbits are possible only if $r \geq 3G_N M$. Furthermore, when $r = 3GM$, $\Delta\tau = 0$. This shows that photons can be bound in circular orbits only if $r = 3G_N M$ (since the proper time vanishes for photons). We will come back to this point in the next section.

As a simple application of these results, let us consider two observers on a planet which is moving in a circular orbit of radius r

around a star. One of the observers takes off on his space craft and maintains his position at a fixed point near the orbit of the planet and then joins the planet at the end of a period. Thus the time of absence as measured by the observer on the space craft is given by (see (9.49))

$$\Delta\tau_{\text{sp}} = \left(1 - \frac{2G_{\text{N}}M}{r}\right)^{\frac{1}{2}} \Delta t, \qquad (9.55)$$

whereas the observer on the planet would measure the time of absence as (see (9.54))

$$\Delta\tau_{\text{pl}} = \left(1 - \frac{3G_{\text{N}}M}{r}\right)^{\frac{1}{2}} \Delta t, \qquad (9.56)$$

which leads to

$$\frac{\Delta\tau_{\text{sp}}}{\Delta\tau_{\text{pl}}} = \left(\frac{1 - \frac{2G_{\text{N}}M}{r}}{1 - \frac{3G_{\text{N}}M}{r}}\right)^{\frac{1}{2}} > 1. \qquad (9.57)$$

This shows that the observer moving in the space craft would find himself older (if they are twins) on return. This should be contrasted with the result from special relativity where the twin on the space craft would believe that he would be younger on return.

9.3 Motion of light rays in a Schwarzschild background

To consider the motion of light rays in a Schwarzschild background, we have to make two assumptions. First of all we assume that the light rays behave like any other particle and hence their motion in a gravitational field is given by the geodesic equation (see (5.83))

$$\frac{\mathrm{d}^2 x^\mu}{\mathrm{d}\tau^2} - \Gamma^\mu_{\nu\lambda} \frac{\mathrm{d}x^\nu}{\mathrm{d}\tau} \frac{\mathrm{d}x^\lambda}{\mathrm{d}\tau} = 0. \qquad (9.58)$$

Furthermore, we assume that the proper time interval for photons generalizes from Minkowski space-time to be

$$d\tau^2 = 0. \tag{9.59}$$

Clearly, therefore, we cannot use τ to label the geodesics. Instead we would use a different affine parameter, say λ to label the geodesics. Thus the geodesic equations for the photon are similar to other particles and in particular if we specialize to $\theta = \frac{\pi}{2}$, the equations are given by (see (9.13), (9.14) and (9.15))

$$r^2 \dot{\phi} = \ell = \text{constant},$$

$$\left(1 - \frac{2G_N M}{r}\right) \dot{t} = k = \text{constant},$$

$$\left(1 - \frac{2G_N M}{r}\right)^{-1} \ddot{r} - \left(1 - \frac{2G_N M}{r}\right)^{-2} \frac{G_N M \dot{r}^2}{r^2}$$

$$+ \frac{G_N M}{r^2} \dot{t}^2 - r \dot{\phi}^2 = 0, \tag{9.60}$$

where we have to remember that the dots now refer to differentiation with respect to the affine parameter λ. The only difference from the particle equation, in this case, lies in the constraint relation

$$d\tau^2 = 0, \tag{9.61}$$

which translates to

$$g_{\mu\nu} \frac{dx^\mu}{d\lambda} \frac{dx^\nu}{d\lambda} = 0,$$

or, $$\left(1 - \frac{2G_N M}{r}\right) \dot{t}^2 - \left(1 - \frac{2G_N M}{r}\right)^{-1} \dot{r}^2 - r^2 \dot{\phi}^2 = 0,$$

or, $$\left(1 - \frac{2G_N M}{r}\right) \frac{\dot{t}^2}{\dot{\phi}^2} - \left(1 - \frac{2G_N M}{r}\right)^{-1} \left(\frac{dr}{d\phi}\right)^2 - r^2 = 0,$$

or, $$\left(1 - \frac{2G_N M}{r}\right) \frac{k^2}{\left(1 - \frac{2G_N M}{r}\right)^2 \ell^2} r^4 - \left(1 - \frac{2G_N M}{r}\right)^{-1} \left(\frac{dr}{d\phi}\right)^2$$

$$- r^2 = 0,$$

or, $\left(\dfrac{\mathrm{d}r}{\mathrm{d}\phi}\right)^2 + \left(1 - \dfrac{2G_{\mathrm{N}}M}{r}\right) r^2 - \dfrac{k^2 r^4}{\ell^2} = 0.$ (9.62)

As before, we define $u = \frac{1}{r}$ so that equation (9.62) becomes

$$\left(\frac{\mathrm{d}u}{\mathrm{d}\phi}\right)^2 + (1 - 2G_{\mathrm{N}}Mu)u^2 - \frac{k^2}{\ell^2} = 0. \qquad (9.63)$$

Differentiating this equation with respect to ϕ we obtain

$$2\,\frac{\mathrm{d}u}{\mathrm{d}\phi}\frac{\mathrm{d}^2 u}{\mathrm{d}\phi^2} + 2u\,\frac{\mathrm{d}u}{\mathrm{d}\phi} - 6G_{\mathrm{N}}Mu^2\,\frac{\mathrm{d}u}{\mathrm{d}\phi} = 0,$$

or, $\dfrac{\mathrm{d}u}{\mathrm{d}\phi}\left(\dfrac{\mathrm{d}^2 u}{\mathrm{d}\phi^2} + u - 3G_{\mathrm{N}}Mu^2\right) = 0.$ (9.64)

The radial geodesic equations, therefore, lead to

$$\frac{\mathrm{d}u}{\mathrm{d}\phi} = 0, \quad \frac{\mathrm{d}^2 u}{\mathrm{d}\phi^2} + u - 3G_{\mathrm{N}}Mu^2 = 0. \qquad (9.65)$$

The first equation implies a circular orbit where the radius vector does not change. Setting $\dot{r} = \ddot{r} = 0$ in the radial equation in (9.60) we see that for this to be true

$$\frac{G_{\mathrm{N}}M}{r^2}\,\dot{t}^2 = r\dot{\phi}^2. \qquad (9.66)$$

On the other hand, the line element equation (9.62) for this orbit gives

$$\left(1 - \frac{2G_{\mathrm{N}}M}{r}\right)\,\dot{t}^2 = r^2\dot{\phi}^2. \qquad (9.67)$$

Comparing the two equations we determine

$$\left(1 - \frac{3G_{\mathrm{N}}M}{r}\right)\,\dot{t}^2 = 0,$$

or, $r = 3G_{\mathrm{N}}M.$ (9.68)

This shows that circular orbits for photons are possible only if the radius of the orbit satisfies the above condition. But we note here that this value of the orbit radius for most objects (stars) lies inside the physical radius of the object (see discussion after (8.52)) and hence such an orbit is not physically meaningful. Thus the radial equation yielding the true orbit that we can observe for photons is given by the second equation in (9.65)

$$\frac{d^2u}{d\phi^2} + u - 3G_N Mu^2 = 0. \tag{9.69}$$

The last term, of course, is a relativistic correction (as we have seen earlier at the end of section **9.1**) and is small compared to the second term as we can see from the fact that

$$\frac{3G_N Mu^2}{u} = 3G_N Mu = \frac{3}{2} \times \frac{2G_N M}{r} = \frac{3}{2}\frac{r_s}{r} \ll 1, \tag{9.70}$$

even if we consider the light ray near the physical surface of the planet or the star. Thus neglecting the last term we can ask what is the trajectory of the light ray to the lowest order satisfying

$$\frac{d^2u}{d\phi^2} + u = 0. \tag{9.71}$$

The solution of this (harmonic) equation is given by (with a particular choice of the phase)

$$u(\phi) = A \sin \phi,$$

or, $$\quad \frac{1}{r} = A \sin \phi,$$

or, $$\quad r \sin \phi = \frac{1}{A} = \text{constant}. \tag{9.72}$$

We recognize that $r \sin \phi$ represents the value of the y coordinate (of the photon). Thus the solution to the lowest order represents a straight line parallel to the x-axis and to lowest order light rays travel in a straight line. We see here that $\frac{1}{A}$ measures the distance of closest approach of the photon to the star. Denoting $\frac{1}{A}$ by r_0, the lowest order solution of (9.69) can be written as

$$u = \frac{1}{r_0} \sin \phi. \qquad (9.73)$$

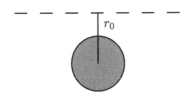

Figure 9.2: Photon trajectory near a star at the lowest order.

Equation (9.73) shows that light rays are unaffected by the gravitational field in the lowest order as shown in Fig. 9.2. To find the effects of the gravitational field we have to solve equation (9.69)

$$\frac{\mathrm{d}^2 u}{\mathrm{d}\phi^2} + u = 3G_{\mathrm{N}} M u^2 = \epsilon u^2, \qquad (9.74)$$

to the next order. Here $\epsilon = 3G_{\mathrm{N}} M$ is a small parameter which can be used as a parameter of perturbation in determining the solution. We expect the true solution to differ only slightly from the lowest order solution in (9.73). Thus perturbatively we can write (to the next order)

$$u \approx \frac{1}{r_0} \sin \phi + \epsilon u_1, \qquad (9.75)$$

and putting this into the equation and keeping terms only up to linear order in ϵ we have

$$-\frac{1}{r_0}\sin\phi + \epsilon\,\frac{d^2u_1}{d\phi^2} + \left(\frac{1}{r_0}\sin\phi + \epsilon u_1\right) = \epsilon\left(\frac{1}{r_0}\sin\phi + \epsilon u_1\right)^2,$$

or, $$\epsilon\left(\frac{d^2u_1}{d\phi^2} + u_1\right) = \epsilon\,\frac{1}{r_0^2}\sin^2\phi = \epsilon\,\frac{1}{2r_0^2}\,(1 - \cos 2\phi),$$

or, $$\frac{d^2u_1}{d\phi^2} + u_1 = \frac{1}{2r_0^2}\,(1 - \cos 2\phi). \tag{9.76}$$

This suggests that we can choose the solution to be of the form

$$u_1 = a + b\cos 2\phi, \tag{9.77}$$

which, when substituted into (9.76), leads to

$$-4b\cos 2\phi + a + b\cos 2\phi = \frac{1}{2r_0^2}\,(1 - \cos 2\phi),$$

or, $$a - 3b\cos 2\phi = \frac{1}{2r_0^2}\,(1 - \cos 2\phi). \tag{9.78}$$

This determines

$$a = \frac{1}{2r_0^2}, \quad b = \frac{1}{6r_0^2}, \tag{9.79}$$

so that we can write the solution (9.75) with the leading order correction as

$$u(\phi) \approx \frac{1}{r_0}\sin\phi + \epsilon\left(\frac{1}{2r_0^2} + \frac{1}{6r_0^2}\cos 2\phi\right),$$

or, $$\frac{1}{r} \approx \frac{1}{r_0}\sin\phi + \frac{3G_N M}{2r_0^2}\left(1 + \frac{1}{3}\cos 2\phi\right). \tag{9.80}$$

At infinite separations if we assume the light ray to come in at $\phi \simeq 0$ then it will go back to infinity at $\phi \simeq \pi$. If the relativistic corrections are not present, then of course, the angle of incidence is exactly $\phi = 0$ and the angle of exit is $\phi = \pi$. Let us assume that

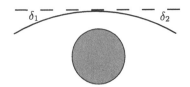

Figure 9.3: Bending of a light ray near a star at the leading order.

because of the gravitational effects, the light ray comes in at $\phi = -\delta_1$ and goes away at $\phi = \pi + \delta_2$ as shown in Fig. 9.3, where δ_1 and δ_2 are small angles. Clearly this implies that at incidence we have

$$\frac{1}{r_0}\sin(-\delta_1) + \frac{3G_{\mathrm{N}}M}{2r_0^2}\left(1 + \frac{1}{3}\cos(-2\delta_1)\right) = 0,$$

or, $$-\frac{\delta_1}{r_0} + \frac{3G_{\mathrm{N}}M}{2r_0^2}\left(1 + \frac{1}{3}\right) = 0,$$

or, $$\delta_1 = \frac{2G_{\mathrm{N}}M}{r_0}. \tag{9.81}$$

Similarly, at the other end we have

$$\frac{1}{r_0}\sin(\pi + \delta_2) + \frac{3G_{\mathrm{N}}M}{2r_0^2}\left(1 + \frac{1}{3}\cos(2\pi + 2\delta_2)\right) = 0,$$

or, $$-\frac{\sin\delta_2}{r_0} + \frac{3G_{\mathrm{N}}M}{2r_0^2}\left(1 + \frac{1}{3}\cos 2\delta_2\right) = 0,$$

or, $$-\frac{\delta_2}{r_0} + \frac{3G_{\mathrm{N}}M}{2r_0^2}\left(1 + \frac{1}{3}\right) = 0,$$

or, $$\delta_2 = \frac{2G_{\mathrm{N}}M}{r_0}. \tag{9.82}$$

This shows that the bending in the path of the light ray (from the leading order trajectory) as it comes in is the same as that when it goes out. This is, of course, expected from the spherical symmetry of the problem. The total amount of deflection in the path of the light ray, in the presence of the star, is given by

$$\delta_1 + \delta_2 = \frac{4G_\mathrm{N}M}{r_0}, \tag{9.83}$$

and is larger when the impact parameter is smaller. The bending of light, as it is commonly called, can be experimentally tested and is usually done by studying the light rays grazing the surface of the sun. In this case the theoretical prediction for the deflection is $1.75''$. Earlier measurements of this phenomenon involved photographing the stellar images during solar eclipse and then comparing them with those taken six months later when the sun is not in the field of view. This shows a deflection of $1.5'' \sim 3.0''$. We must remember that this is a very difficult experiment and various factors change during the six month interval. Consequently measurements cannot give very accurate results. In recent years one uses radio telescopes to measure the radio signals from point like sources such as quasars as they pass near the sun. Such measurements give a value of $1.57'' - 1.82''$ for the deflection of the signal with an error of $.2''$. This is a fairly good test of the theoretical prediction. However, errors need to be reduced still further.

9.4 Perihelion advance of Mercury

We have seen in (9.29) that the general equation for planetary motion in classical physics is given by (recall that the planet has been assumed to be of unit mass)

$$\frac{d^2u}{d\phi^2} = \frac{G_\mathrm{N}M}{\ell^2} - u. \tag{9.84}$$

Therefore, the general solution can be written as

$$u(\phi) = \frac{1}{r} = \frac{G_\mathrm{N}M}{\ell^2}\left(1 + e\cos(\phi + \delta)\right), \tag{9.85}$$

where e and δ are constants. By a suitable choice of the coordinate axis, we can choose the phase δ to be zero in which case the solution takes the form

$$u(\phi) = \frac{G_{\mathrm{N}}M}{\ell^2} (1 + e \cos \phi). \tag{9.86}$$

To determine the constant e we note that the total energy of the classical motion is given by (see (9.24))

$$
\begin{aligned}
E &= \frac{1}{2} \dot{r}^2 + \frac{1}{2} r^2 \dot{\phi}^2 - \frac{G_{\mathrm{N}}M}{r} \\
&= \frac{1}{2} \left(\frac{\mathrm{d}\phi}{\mathrm{d}t} \frac{\mathrm{d}r}{\mathrm{d}\phi} \right)^2 + \frac{1}{2} \frac{\ell^2}{r^2} - \frac{G_{\mathrm{N}}M}{r} \\
&= \frac{1}{2} \ell^2 \left(\frac{\mathrm{d}u}{\mathrm{d}\phi} \right)^2 + \frac{1}{2} \ell^2 u^2 - G_{\mathrm{N}}Mu,
\end{aligned}
\tag{9.87}
$$

where $u = \frac{1}{r}$ and we have used (9.25). We can rearrange this to write

$$\left(\frac{\mathrm{d}u}{\mathrm{d}\phi} \right)^2 = \frac{2E}{\ell^2} + \frac{2G_{\mathrm{N}}M}{\ell^2} u - u^2. \tag{9.88}$$

Substituting the actual form of the solution from (9.86) we obtain

$$
\left(-\frac{G_{\mathrm{N}}M}{\ell^2} e \sin \phi \right)^2 = \frac{2E}{\ell^2} + 2 \left(\frac{G_{\mathrm{N}}M}{\ell^2} \right)^2 (1 + e \cos \phi)
$$

$$
- \left(\frac{G_{\mathrm{N}}M}{\ell^2} \right)^2 (1 + e \cos \phi)^2,
$$

$$
\text{or,} \quad \left(\frac{G_{\mathrm{N}}M}{\ell^2} \right)^2 e^2 = \frac{2E}{\ell^2} + \left(\frac{G_{\mathrm{N}}M}{\ell^2} \right)^2,
$$

$$
\text{or,} \quad e^2 = 1 + \frac{2E\ell^2}{G_{\mathrm{N}}^2 M^2}. \tag{9.89}
$$

It is worth recalling here that for bound orbits, the total energy $E < 0$ so that the eccentricity of the orbit $e < 1$. ($e = 0$ leads to $r = \frac{1}{u} = \frac{\ell^2}{G_{\mathrm{N}}M}$ corresponding to circular motion.)

With this value of the constant, the radius takes values according to

$$u(\phi) = \frac{1}{r} = \frac{G_N M}{\ell^2} (1 + e \cos \phi).$$ (9.90)

The perihelion (closest point to the star) and the aphelion (farthest point from the star) on the orbit are determined from the condition that at such points

$$\frac{dr}{d\phi} = 0, \qquad \text{or,} \quad \frac{du}{d\phi} = 0,$$ (9.91)

which implies $\phi = 0, \pi$. This fixes the perihelion and the aphelion to be at $\phi = 0$ and $\phi = \pi$ respectively. The distances of these points from the star are given by

$$
\begin{aligned}
r_{\text{AP}} &= \frac{\ell^2}{G_N M} \frac{1}{1-e} = r_1, \\
r_{\text{PH}} &= \frac{\ell^2}{G_N M} \frac{1}{1+e} = r_2.
\end{aligned}
$$ (9.92)

Correspondingly we can write

$$
\begin{aligned}
u_{\text{AP}} &= \frac{G_N M}{\ell^2} (1 - e) = u_1, \\
u_{\text{PH}} &= \frac{G_N M}{\ell^2} (1 + e) = u_2.
\end{aligned}
$$ (9.93)

Parenthetically, we note here that u_1, u_2 are simply the two roots of the quadratic equation (9.88) with (9.91) and (9.89).

Let us now include the relativistic effects. We know that in the Schwarzschild background (in the relativistic case) the motion of the particle (planet) is described by (see (9.19))

$$
\begin{aligned}
\left(\frac{du}{d\phi}\right)^2 &= \frac{(k^2 - 1)}{\ell^2} + \frac{2G_N M}{\ell^2} u - u^2 + 2G_N M u^3 \\
&= \frac{2E}{\ell^2} + \frac{2G_N M}{\ell^2} u - u^2 + 2G_N M u^3.
\end{aligned}
$$ (9.94)

Here we have made the identification $2E = k^2 - 1$. Furthermore, we recognize that the last term in (9.94) represents the relativistic correction (compare with (9.88)) and noting that $2G_N M$ is a small quantity we can denote it by ϵ and write

$$\left(\frac{\mathrm{d}u}{\mathrm{d}\phi}\right)^2 = \frac{2E}{\ell^2} + \frac{2G_N M}{\ell^2}\, u - u^2 + \epsilon u^3. \qquad (9.95)$$

We can now ask how the perihelion and the aphelion ((9.92) or (9.93)) would modify when we include this relativistic correction. Using (9.95) these positions are determined from (see (9.91))

$$\frac{\mathrm{d}u}{\mathrm{d}\phi} = 0,$$

or, $$\epsilon u^3 - u^2 + \frac{2G_N M}{\ell^2}\, u + \frac{2E}{\ell^2} = 0,$$

or, $$u^3 - \frac{1}{\epsilon}\, u^2 + \frac{2G_N M}{\epsilon\ell^2}\, u + \frac{2E}{\epsilon\ell^2} = 0. \qquad (9.96)$$

This is a cubic equation with three solutions u_1, u_2, u_3 such that

$$u_1 + u_2 + u_3 = \frac{1}{\epsilon}, \qquad (9.97)$$

which is a large quantity since ϵ is small. On the other hand, we have seen that the aphelion and the perihelion given respectively by u_1 and u_2 are small (see (9.93)) compared to $\frac{1}{\epsilon}$ and hence it is the third solution which must be large and unphysical and we can assume the motion to be bounded by

$$u_1 \leq u \leq u_2. \qquad (9.98)$$

As a result, we can write (9.95) as

$$\left(\frac{du}{d\phi}\right)^2 = \epsilon\left[u^3 - \frac{u^2}{\epsilon} + \frac{2G_N M u}{\epsilon} + \frac{2E}{\epsilon\ell^2}\right]$$

$$= \epsilon(u - u_1)(u_2 - u)(u_3 - u)$$

$$= \epsilon(u - u_1)(u_2 - u)\left(\frac{1}{\epsilon} - u - u_1 - u_2\right)$$

$$= (u - u_1)(u_2 - u)(1 - \epsilon(u + u_1 + u_2)),$$

or, $$\frac{du}{d\phi} = [(u - u_1)(u_2 - u)]^{\frac{1}{2}}(1 - \epsilon(u + u_1 + u_2))^{\frac{1}{2}},$$

or, $$\frac{d\phi}{du} = \frac{(1 - \epsilon(u + u_1 + u_2))^{-\frac{1}{2}}}{[(u - u_1)(u_2 - u)]^{\frac{1}{2}}}$$

$$\approx \frac{1 + \frac{1}{2}\epsilon(u + u_1 + u_2)}{[(u - u_1)(u_2 - u)]^{\frac{1}{2}}}. \tag{9.99}$$

Let us define

$$\alpha = \frac{1}{2}(u_1 + u_2), \qquad \beta = \frac{1}{2}(u_2 - u_1), \tag{9.100}$$

so that

$$u_1 = \alpha - \beta, \qquad u_2 = \alpha + \beta, \tag{9.101}$$

and it follows that

$$(u - u_1)(u_2 - u) = (u - (\alpha - \beta))(\alpha + \beta - u)$$

$$= [\beta^2 - (u - \alpha)^2]. \tag{9.102}$$

Thus, we can write (9.99) in the form

$$\frac{d\phi}{du} = \frac{1 + \frac{1}{2}\epsilon u + \epsilon\alpha}{[\beta^2 - (u - \alpha)^2]^{\frac{1}{2}}} = \frac{\frac{1}{2}\epsilon(u - \alpha) + 1 + \frac{3}{2}\epsilon\alpha}{[\beta^2 - (u - \alpha)^2]^{\frac{1}{2}}}. \tag{9.103}$$

Integrating this, we can obtain the change in the angle between the perihelion and the aphelion as

$$
\begin{aligned}
|\delta\phi| &= \int_{u_1}^{u_2} du \, \frac{\left[\frac{1}{2}\epsilon(u-\alpha)+1+\frac{3}{2}\epsilon\alpha\right]}{[\beta^2-(u-\alpha)^2]^{\frac{1}{2}}} \\
&= \left[-\frac{1}{2}\epsilon(\beta^2-(u-\alpha)^2)^{\frac{1}{2}}+\left(1+\frac{3}{2}\epsilon\alpha\right)\sin^{-1}\frac{u-\alpha}{\beta}\right]_{u_1}^{u_2} \\
&= \left(1+\frac{3}{2}\epsilon\alpha\right)\left(\sin^{-1}(1)-\sin^{-1}(-1)\right) \\
&= \left(1+\frac{3}{2}\epsilon\alpha\right)\left(\frac{\pi}{2}+\frac{\pi}{2}\right) = \pi\left(1+\frac{3}{2}\epsilon\alpha\right). \qquad (9.104)
\end{aligned}
$$

Doubling this gives the angle between successive perihelia and shows that it does not return to its original position, rather it advances by an angle

$$
\begin{aligned}
\Delta\phi &= 2|\delta\phi|-2\pi = 3\pi\epsilon\alpha = 3\pi \times 2G_{\text{N}}M \times \frac{1}{2}(u_1+u_2) \\
&= 3\pi G_{\text{N}}M\left(\frac{1}{r_1}+\frac{1}{r_2}\right). \qquad (9.105)
\end{aligned}
$$

Clearly, the advance in the perihelion is a relativistic effect and is larger for planets closer to the sun. Therefore, Mercury is the natural candidate to study this phenomenon. Even for Mercury this angle is very small,

$$(\Delta\phi)/\text{revolution} \simeq 0.1038''. \qquad (9.106)$$

However, this effect is cumulative and per century the perihelion of Mercury advances by

$$(\Delta\phi)/\text{century} \simeq 43'', \qquad (9.107)$$

where we have used the fact that Mercury's revolution period is 88 days so that in a century there are about 415 revolutions. This advance of the perihelion (9.107) is appreciable and can be observed. In fact, the astronomer Leverrier had already discovered in 1845 that Mercury exhibited an anomalous precession of the perihelion of about $43''$ per century that could not be accounted for by known perturbations at the time. Thus the precession of the perihelion of Mercury constituted an early verification of general relativity. There are numerous other tests that show that general relativity conforms to astronomical observations quite accurately.

CHAPTER 10

Black holes

We have already seen that the Schwarzschild metric given by

$$
d\tau^2 = \left(1 - \frac{2G_{\mathrm{N}}M}{r}\right)dt^2 - \left(1 - \frac{2G_{\mathrm{N}}M}{r}\right)^{-1}dr^2 - r^2(d\theta^2 + \sin^2\theta d\phi^2),
$$

$$(10.1)$$

has manifest singularities at $r = 2G_{\mathrm{N}}M$ and at $r = 0$. The singularity at the origin, if the particle is a point particle, is, of course, a real singularity which is present even in the classical case. The singularity at the Schwarzschild radius $r = 2G_{\mathrm{N}}M$, on the other hand, has a different character. Note that although

$$
g_{tt}(r = 2G_{\mathrm{N}}M) = 0,
$$

$$(10.2)$$

the determinant of the metric is well behaved and that if we calculate the components of the curvature tensor, they are also well behaved at that point. Thus, we conclude that the point $r = 2G_{\mathrm{N}}M$ cannot represent a physical singularity of the space-time. However, we have also indicated earlier that for $r < 2G_{\mathrm{N}}M$ (see (8.61))

$$
g_{tt} < 0, \qquad g_{rr} > 0,
$$

$$(10.3)$$

so that the time and the radial components of the metric (and, therefore, the coordinates) seem to exchange their roles and hence there must be some interesting phenomena in that region. This is what we will investigate in this chapter.

10.1 Singularities of the metric

Whenever the metric components are badly behaved at a certain point in a coordinate system, one of two things may happen:

i) the space-time geometry is, in fact, singular at that point,

ii) the space-time geometry is nonsingular but the coordinates fail to cover a region of space-time containing that point properly.

Normally the first possibility can be demonstrated by calculating the scalar curvature and showing that it diverges at some point of space-time. Furthermore, this singularity must occur at a finite value of the affine parameter labelling the geodesic. That is, the singularity must not lie at infinity. However, even this is not fool proof because there are singularities of space-time where the scalar curvature is not necessarily singular. The second possibility is usually demonstrated by finding a coordinate system where the metric components are well behaved. Thus the study of singularities of a manifold is an extremely difficult one.

We would begin by illustrating with simple examples how a wrong choice of coordinates can make the metric components look ill behaved. Let us consider the two dimensional line element given by

$$d\tau^2 = \frac{1}{t^4}\,dt^2 - dx^2, \tag{10.4}$$

defined over $-\infty < x < \infty$ and $0 \leq t < \infty$. The metric in this case appears to have a singularity at $t = 0$. However, the true nature of the space-time geometry can be seen by making the coordinate transformation

$$t \to \frac{1}{t}, \tag{10.5}$$

so that the line element becomes

$$d\tau^2 = dt^2 - dx^2, \tag{10.6}$$

with $-\infty < x < \infty$ and $t \geq 0$. But this is nothing other than the upper half plane of the two dimensional Minkowski space-time.

Furthermore, the singularity has now disappeared and hence we can extend the form of the line element to the entire Minkowski space.

The second example is again of a two dimensional space known as the Rindler space and is closer in analogy to the Schwarzschild metric than the previous example. The line element, in this case, is given by

$$d\tau^2 = x^2 dt^2 - dx^2, \tag{10.7}$$

with $-\infty < t < \infty$ and $0 \leq x < \infty$. It is obvious from (10.7) that the metric is singular at $x = 0$. That is, at this point the determinant of the metric vanishes and consequently the contravariant metric components are undefined at such points. On the other hand, calculation of the scalar curvature shows no singularity as $x \to 0$ suggesting that the singularity may be a coordinate singularity. In fact, the curvature for the Rindler space vanishes identically suggesting that this space may be a part of the Minkowski space. We ignore this point for the present and continue with our analysis of the singularity.

To find a coordinate system that would get rid of the singularity, let us look at the null geodesics in this space. The null condition leads to (dots denote derivatives with respect to the affine parameter λ)

$$x^2 \dot{t}^2 - \dot{x}^2 = 0, \tag{10.8}$$

which yields

$$\left(\frac{dt}{dx}\right)^2 = \frac{1}{x^2},$$

or, $t = \pm \ln x + \text{constant}.$ \tag{10.9}

Thus the outgoing and the incoming null geodesics are naturally given in terms of the null coordinates

$$u = t - \ln x, \qquad v = t + \ln x. \tag{10.10}$$

Namely, an outgoing null geodesic is given by

$$u = \text{constant},\tag{10.11}$$

while an incoming null geodesic is given by

$$v = \text{constant}.\tag{10.12}$$

In terms of these coordinates, then, we can write

$$t = \frac{1}{2}(v + u), \qquad x = e^{\frac{1}{2}(v-u)},\tag{10.13}$$

so that the line element (10.7) takes the form

$$\mathrm{d}\tau^2 = e^{(v-u)}\,\mathrm{d}u\mathrm{d}v.\tag{10.14}$$

We are still not in a position to analyze the singularity at $x = 0$ since the coordinate range $-\infty < u < \infty$ and $-\infty < v < \infty$ still corresponds to $x > 0$ of the Rindler space. However, let us recall that for this space, t is a cyclic variable in the Lagrangian density, i.e., if we take (see, for example, (9.8))

$$\mathcal{L} = x^2\dot{t}^2 - \dot{x}^2,\tag{10.15}$$

where dots denote derivatives with respect to the affine parameter λ, it follows from the t equation that

$$\frac{\mathrm{d}}{\mathrm{d}\lambda}\left(\frac{\partial \mathcal{L}}{\partial \dot{t}}\right) = 0,$$

$$\text{or,} \qquad \frac{\partial \mathcal{L}}{\partial \dot{t}} = 2x^2\dot{t} = \text{constant},$$

$$\text{or,} \qquad x^2\,\frac{\mathrm{d}t}{\mathrm{d}\lambda} = E = \text{constant}.\tag{10.16}$$

Furthermore, we note from (10.13) that

$$u + v = 2t,\tag{10.17}$$

so that for a constant u (outgoing geodesic, see (10.11)),

$$\mathrm{d}v = 2\mathrm{d}t. \tag{10.18}$$

As a result, for the outgoing geodesic for which $u = $ constant, (10.16) can be integrated to give

$$
\begin{aligned}
\lambda_{\text{out}} &= \int \mathrm{d}\lambda = \frac{1}{E} \int \mathrm{d}t\ x^2 = \frac{1}{2E} \int \mathrm{d}v\ e^{(v-u)} \\
&= \text{constant} + \frac{e^{(v-u)}}{2E} = \text{constant} + \left(\frac{e^{-u}}{2E}\right) e^v.
\end{aligned} \tag{10.19}
$$

Similarly, for an incoming geodesic we can determine

$$
\begin{aligned}
\lambda_{\text{in}} &= \int \mathrm{d}\lambda = \frac{1}{E} \int \mathrm{d}t\ x^2 = \frac{1}{2E} \int \mathrm{d}u\ e^{(v-u)} \\
&= \text{constant} - \left(\frac{e^v}{2E}\right) e^{-u}.
\end{aligned} \tag{10.20}
$$

This shows that the affine parameters lead to incomplete null geodesics. This also suggests that if we now define new variables

$$U = -e^{-u}, \qquad V = e^v, \tag{10.21}$$

then, the line element (10.14) can be written as

$$\mathrm{d}\tau^2 = \mathrm{d}U\mathrm{d}V. \tag{10.22}$$

The ranges of the new variables are given by $U < 0$ and $V > 0$, but since there is no singularity in the line element any more, we can extend it to cover the entire space-time

$$-\infty < U < \infty, \qquad -\infty < V < \infty. \tag{10.23}$$

The final coordinate transformation

$$T = \frac{U+V}{2}, \qquad X = \frac{V-U}{2}, \tag{10.24}$$

renders the line element (10.22) into the form

$$d\tau^2 = dU dV = dT^2 - dX^2, \tag{10.25}$$

which we recognize to be the two dimensional Minkowski line element.

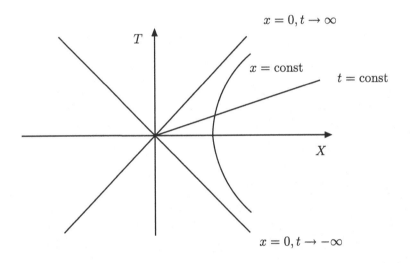

Figure 10.1: Rindler space-time.

The original coordinates (t, x) can be expressed in terms of the Minkowski coordinates (T, X) as (see (10.13), (10.21) and (10.24))

$$
\begin{aligned}
x &= e^{\frac{1}{2}(v-u)} = (-UV)^{\frac{1}{2}} = (-(T-X)(T+X))^{\frac{1}{2}} \\
&= (X^2 - T^2)^{\frac{1}{2}}, \\
t &= \frac{1}{2}(v+u) = \frac{1}{2}(\ln V - \ln(-U)) = \frac{1}{2}\ln\frac{1+\frac{T}{X}}{1-\frac{T}{X}} \\
&= \tanh^{-1}\left(\frac{T}{X}\right). \tag{10.26}
\end{aligned}
$$

We see from Fig. 10.1 that the original Rindler space corresponds to the wedge $X > |T|$ of the two dimensional Minkowski space-time. The original singularity has now disappeared and we can extend the metric to the entire Minkowski space-time.

10.2 Singularities of the Schwarzschild metric

The analysis for the Schwarzschild metric can be carried out in an analogous manner to that of the Rindler space. The main difference lies in the fact that the Schwarzschild solution is four dimensional in contrast to the two dimensional Rindler problem. However, we can effectively reduce the Schwarzschild problem to a two dimensional one by simply restricting the motion to $\theta = \frac{\pi}{2}$ (equatorial plane) with a fixed azimuthal angle ($\ell = 0$). In this case, the line element takes the form

$$d\tau^2 = \left(1 - \frac{2G_N M}{r}\right) dt^2 - \left(1 - \frac{2G_N M}{r}\right)^{-1} dr^2, \qquad (10.27)$$

which has a two dimensional character and the null condition in the present case is given by

$$\left(1 - \frac{2G_N M}{r}\right) \dot{t}^2 - \left(1 - \frac{2G_N M}{r}\right)^{-1} \dot{r}^2 = 0,$$

$$\text{or,} \quad \left(\frac{dt}{dr}\right)^2 = \left(\frac{1}{1 - \frac{2G_N M}{r}}\right)^2 = \left(\frac{\frac{r}{2G_N M}}{\frac{r}{2G_N M} - 1}\right)^2,$$

$$\text{or,} \quad \frac{dt}{dr} = \pm \left[1 + \frac{1}{\frac{r}{2G_N M} - 1}\right],$$

$$\text{or,} \quad t = \pm \left[r + 2G_N M \ln\left(\frac{r}{2G_N M} - 1\right)\right] + \text{constant}$$

$$= \pm \tilde{r} + \text{constant}, \qquad (10.28)$$

where $\tilde{r}$ defined as

$$\tilde{r} = r + 2G_N M \ln\left(\frac{r}{2G_N M} - 1\right), \qquad (10.29)$$

is known as the Wheeler-Regge coordinate. Therefore, the null coordinates in the Schwarzschild space are given by

$$u = t - \tilde{r}, \qquad v = t + \tilde{r}, \tag{10.30}$$

so that the outgoing and the incoming null geodesics can be written respectively as

$$u = \text{constant}, \qquad v = \text{constant}. \tag{10.31}$$

We note that the relations (10.30) can be inverted to give

$$t = \frac{1}{2}(v + u), \qquad \tilde{r} = \frac{1}{2}(v - u). \tag{10.32}$$

Furthermore, from (10.29) we obtain

$$\frac{\partial \tilde{r}}{\partial r} = 1 + \frac{1}{\frac{r}{2G_N M} - 1} = \left(1 - \frac{2G_N M}{r}\right)^{-1}, \tag{10.33}$$

and as a result, we can write the Schwarzschild line element (10.27) in terms of u and v coordinates as

$$d\tau^2 = \left(1 - \frac{2G_N M}{r}\right)(dt^2 - d\tilde{r}^2) = \left(1 - \frac{2G_N M}{r}\right) du dv. \tag{10.34}$$

Here we have used (10.32) and we are supposed to understand that r is expressed in terms of u and v through the relation

$$\tilde{r} = r + 2G_N M \ln\left(\frac{r}{2G_N M} - 1\right) = \frac{1}{2}(v - u),$$

$$\text{or,} \qquad \left(\frac{r}{2G_N M} - 1\right) e^{\frac{r}{2G_N M}} = e^{\frac{1}{4G_N M}(v-u)}, \tag{10.35}$$

which leads to

$$1 - \frac{2G_N M}{r} = 2G_N M \ \frac{e^{-\frac{r}{2G_N M}}}{r} \ e^{\frac{1}{4G_N M}(v-u)}. \tag{10.36}$$

We can, therefore, write the line element (10.34) as

$$d\tau^2 = 2G_N M \ \frac{e^{-\frac{r}{2G_N M}}}{r} \ e^{\frac{(v-u)}{4G_N M}} \ dudv. \tag{10.37}$$

This form of the parameterization is known as the Eddington-Finkelstein parameterization (coordinates). Furthermore, analogous to the earlier example of the Rindler space, we can now define the coordinates (see (10.21))

$$U = -e^{-\frac{u}{4G_N M}}, \qquad V = e^{\frac{v}{4G_N M}}, \tag{10.38}$$

so that the line element (10.37) takes the form

$$d\tau^2 = 32(G_N M)^3 \ \frac{e^{-\frac{r}{2G_N M}}}{r} \ dUdV. \tag{10.39}$$

The Schwarzschild singularity at $r = 2G_N M$ has now disappeared from the line element with only a singularity left at $r = 0$. We know that this is a physical singularity and so we can now extend this form of the line element to all space with $r > 0$. The final identification with the Kruskal coordinates is made by defining (see (10.24))

$$T = \frac{U+V}{2}, \qquad R = \frac{V-U}{2}, \tag{10.40}$$

so that the complete (four dimensional) Schwarzschild element is given by

$$d\tau^2 = 32(G_N M)^3 \ \frac{e^{-\frac{r}{2G_N M}}}{r} \ (dT^2 - dR^2) - r^2(d\theta^2 + \sin^2\theta d\phi^2). \tag{10.41}$$

Note that the original coordinates can be expressed in terms of T and R as (see (10.32), (10.37), (10.38) and (10.40))

$$\left(\frac{r}{2G_N M} - 1\right) e^{\frac{r}{2G_N M}} = -UV = R^2 - T^2,$$

$$\frac{t}{4G_N M} = \frac{1}{8G_N M}(v + u) = \frac{1}{2}(\ln V - \ln(-U))$$

$$= \frac{1}{2}(\ln(R+T) - \ln(R-T)) = \tanh^{-1}\left(\frac{T}{R}\right). \quad (10.42)$$

Thus, we can draw a space-time diagram similar to the Rindler space as shown in Fig. 10.2.

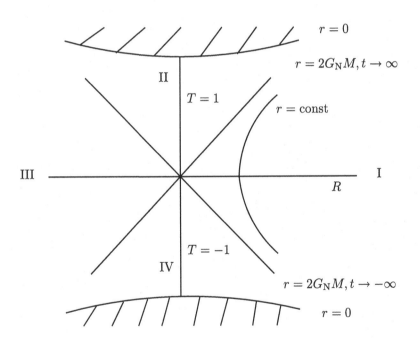

Figure 10.2: Schwarzschild space-time.

We see from Fig. 10.2 that the region of space outside the Schwarzschild radius $r = 2G_N M$ is given by the wedge

$$R \geq |T|. \quad (10.43)$$

The original Schwarzschild singularity at $r = 2G_N M$ corresponds to the null geodesics

$$R = |T|. \tag{10.44}$$

Clearly the apparent singularity was due to a wrong choice of coordinates. The space now divides into four regions. Region I is, of course, the physical space outside of the Schwarzschild radius. Furthermore, $r = 0$ corresponds to the hyperbola

$$R^2 - T^2 = -1. \tag{10.45}$$

It is denoted by the two shaded regions shown in Fig. 10.2. Note that in the two regions II and IV we have

$$R^2 - T^2 < 0, \tag{10.46}$$

so that they are space-like in nature (in terms of the original t, r coordinates) whereas region III is time-like and hence represents another asymptotically flat region. If a light signal falls radially from region I, then within a finite proper time it would enter region II. However, once it enters region II it will be trapped there forever. That is, since it is a space-like region it will not be able to escape. For an observer far away from the Schwarzschild radius, however, it would appear as if the light signal would take forever to reach the Schwarzschild radius (see discussion around (9.42)). The light ray would be red shifted so much that to an observer far away its intensity would drop to zero. Thus it is clear that an observer cannot detect anything that happens inside the radius $r = 2G_N M$. For this reason, the Schwarzschild radius is also known as the event horizon.

Note also from the above discussion that even though there are two asymptotically flat regions, observers in these two regions cannot communicate with one another since light signals sent by them would eventually be lost in the intermediate regions. Region II is known as a black hole, since according to classical ideas, it only absorbs and never emits any radiation. Region IV, on the other hand, is a time reversed case of region II and is known as a white hole. In this region, matter and radiation cannot be present. They would eventually be pushed out. Note also that the physical singularity at $r = 0$ occurs inside the regions II and IV and hence can never be detected by an observer outside the Schwarzschild radius.

10.3 Black holes

We may ask at this point whether the study of such regions is purely academic or whether there exist observational data to corroborate these. At the present time at least two black holes have been observed. The point is that although we cannot detect a black hole directly, its gravitational effect on other objects can be measured. There are several mechanisms to produce a black hole and one of the most popular ones is of the view that black holes are produced due to the gravitational collapse of a star.

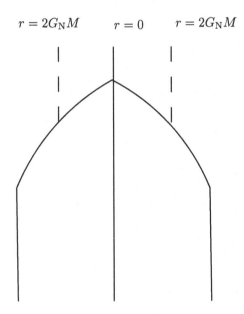

Figure 10.3: Gravitational collapse of a star.

For simplicity let us consider a spherically symmetric star undergoing spherical collapse because we can use the Schwarzschild solution which we have been studying so far. In this case the star would collapse under its own gravitational attraction as shown in

Fig. 10.3 and if the collapse is complete, then it would happen at the singularity $r = 0$. However, for gravitational purposes we would still feel as if a star is present with the same mass. If an observer is observing a star undergoing gravitational collapse, he would keep seeing it. The light ray from the star would be red shifted and hence the star would look steadily fainter until it completely fades out of the observer's view. The observer would merely think that the star is still there but is completely burnt out.

That gravitational collapse happens can be seen from the following simple argument. First of all collapse occurs in burnt out stars. This is because, when a star is burning, there is thermal motion which gives rise to the pressure needed to prevent collapse. If the star is burnt out, then the only pressure to balance collapse comes from the degeneracy pressure from the Pauli exclusion principle due to the presence of fermions. Thus for example, if the star contains n fermions of mass m per unit volume, then each fermion on an average would occupy a volume of space n^{-1}. Therefore, if we assume that

$$p \sim \Delta p, \tag{10.47}$$

then its momentum on an average can be estimated to be

$$p \sim \Delta p \sim \frac{\hbar}{\Delta x} = \hbar n^{\frac{1}{3}}. \tag{10.48}$$

As a result the average velocity of non-relativistic fermions can be obtained as

$$v \sim \frac{p}{m} = \hbar n^{\frac{1}{3}} m^{-1}. \tag{10.49}$$

On the other hand, if the fermions are relativistic then we can approximate their velocity by the speed of light ($c = 1$).

The degeneracy pressure due to fermions is given by

$$P = \text{momentum} \times \text{velocity} \times \text{number density}. \tag{10.50}$$

Therefore, if the fermions are nonrelativistic, we have

$$P_{\mathrm{NR}} \sim \hbar n^{\frac{1}{3}} \times \hbar n^{\frac{1}{3}} m^{-1} \times n = \hbar^2 n^{\frac{5}{3}} m^{-1}. \tag{10.51}$$

This shows that in the nonrelativistic case only the electrons would contribute significantly to the degeneracy pressure since they are the lightest. If the fermions are relativistic, on the other hand, the degeneracy pressure is given by

$$P_{\mathrm{R}} \sim \hbar n^{\frac{1}{3}} \times 1 \times n = \hbar n^{\frac{4}{3}}, \tag{10.52}$$

so that all fermion species contribute equally.

Let us now restrict to the case where the fermions are relativistic. Then, for stability, the degeneracy pressure has to balance the attractive gravitational pressure given by (we assume $G_{\mathrm{N}} = 1$ in this discussion)

$$P_{\mathrm{grav}} = \frac{M^2}{r^4}, \tag{10.53}$$

where the mass density can be identified with

$$\frac{M}{r^3} = nm,$$

$$\text{or,} \quad \frac{1}{r} = (nmM^{-1})^{\frac{1}{3}}. \tag{10.54}$$

Putting this back in (10.53), we see that the gravitational pressure is given by

$$P_{\mathrm{grav}} = M^2 n^{\frac{4}{3}} m^{\frac{4}{3}} M^{-\frac{4}{3}} = n^{\frac{4}{3}} m^{\frac{4}{3}} M^{\frac{2}{3}}. \tag{10.55}$$

The star would, therefore, be stable only if

$$P_{\mathrm{grav}} = P_{\mathrm{R}},$$

$$\text{or,} \quad n^{\frac{4}{3}} m^{\frac{4}{3}} M^{\frac{2}{3}} = \hbar n^{\frac{4}{3}}$$

$$\text{or,} \quad M = (\hbar m^{-\frac{4}{3}})^{\frac{3}{2}} = \hbar^{\frac{3}{2}} m^{-2}. \tag{10.56}$$

Taking baryons to be the most abundant fermions in stars we can calculate the value of this critical mass to be (we should restore all the dimensional constants such as G_N and the final number is more like $3M_{\text{sun}}$)

$$M = \hbar^{\frac{3}{2}} m^{-2} \simeq 2\, M_{\text{sun}}, \tag{10.57}$$

which shows that a star twice as massive as the sun or more would undergo gravitational collapse, if it is completely burnt out.

We have discussed only the simplest kind of a black hole, namely a Schwarzschild black hole. If the star is charged and/or rotating then we are also led to other kinds of black holes which we do not go into here.

Chapter 11

Cosmological models and the big bang theory

11.1 Homogeneity and isotropy

In building a model of our universe, a lot of experimental observations go into the basic assumptions. Quite important among them is the observation that our universe is quite homogeneous and isotropic on a length scale of 10^7 parsecs. (1 parsec is about 3.2 light years.) Thus, if we ignore the inhomogeneities and nonisotropic nature of the universe on smaller scales, the line element must reflect this. If we denote the line element as

$$\mathrm{d}\tau^2 = \mathrm{d}t^2 - \mathrm{d}s^2, \tag{11.1}$$

then the three dimensional line element $\mathrm{d}s^2$ must exhibit homogeneity and isotropy.

For a homogeneous and isotropic space, it is easy to show that the curvature tensor can be written in the form (see (8.76))

$$R_{\mu\nu\lambda\rho} = K\left(g_{\mu\lambda}g_{\nu\rho} - g_{\mu\rho}g_{\nu\lambda}\right). \tag{11.2}$$

This follows because if we choose a flat coordinate system at any point, then the metric will become the Minkowski metric. Furthermore, if we want isotropy, the curvature should be independent of rotations at that point. Only the Minkowski metric is invariant under rotations. Thus, the curvature tensor can be written in terms of the Minkowski metric combinations at that point. Furthermore, the requirement of the appropriate antisymmetry properties would then, select out the curvature tensor to have the above form. It also

311

follows from the definition of homogeneity that the quantity K, must be a constant since otherwise its value will distinguish between different points in this space. (This would also follow from isotropy. If K is not a constant, then its gradients can select out a preferred direction.)

Thus, we see that a homogeneous and isotropic space is a space of constant curvature. If we now apply this to our three dimensional space, we note that

$$R^{(3)}_{\mu\nu\lambda\rho} = K\left(g^{(3)}_{\mu\lambda}g^{(3)}_{\nu\rho} - g^{(3)}_{\mu\rho}g^{(3)}_{\nu\lambda}\right), \tag{11.3}$$

where the constant curvature K can depend on the time coordinate (since that is not part of the 3-dimensional space). It follows now that

$$R^{(3)}_{\mu\nu} = g^{(3)\lambda\rho}R^{(3)}_{\mu\lambda\nu\rho} = 2Kg^{(3)}_{\mu\nu}. \tag{11.4}$$

Choosing the 3-dimensional isotropic line element of the form

$$\mathrm{d}s^2 = e^{\lambda(r)}\mathrm{d}r^2 + r^2(\mathrm{d}\theta^2 + \sin^2\theta\mathrm{d}\phi^2), \tag{11.5}$$

we can calculate the nontrivial curvature components which yield

$$R^{(3)}_{11} = \frac{\lambda'(r)}{r},$$

$$R^{(3)}_{22} = \mathrm{cosec}^2\theta R^{(3)}_{33} = 1 + \frac{1}{2}\,r\lambda'(r)e^{-\lambda(r)} - e^{-\lambda(r)}. \tag{11.6}$$

Requiring

$$R^{(3)}_{\mu\nu} = 2Kg^{(3)}_{\mu\nu}, \tag{11.7}$$

we obtain

$$R^{(3)}_{11} = \frac{\lambda'(r)}{r} = 2Ke^{\lambda},$$

$$R^{(3)}_{22} = 1 + \frac{1}{2}\,r\lambda'(r)e^{-\lambda} - e^{-\lambda} = 2Kr^2. \tag{11.8}$$

This implies that

$$1 + \frac{1}{2} r\lambda'(r)e^{-\lambda} - e^{-\lambda} = 2Kr^2,$$

or, $\quad 1 + \frac{1}{2} \times 2Kr^2 - e^{-\lambda} = 2Kr^2,$

or, $\quad 1 - e^{-\lambda} = Kr^2,$

or, $\quad e^{-\lambda(r)} = (1 - Kr^2),$

or, $\quad e^{\lambda(r)} = (1 - Kr^2)^{-1}.$ $\qquad(11.9)$

Thus, the form of the three dimensional line element is determined to be (see also (8.88))

$$ds^2 = \frac{dr^2}{1 - Kr^2} + r^2(d\theta^2 + \sin^2\theta d\phi^2). \qquad(11.10)$$

If we now define a rescaled variable

$$r^* = \sqrt{|K|}\, r, \qquad\qquad k = \frac{K}{|K|}, \qquad(11.11)$$

we can write (the form of the line element below is true even when $K = 0$, $\frac{r^{*2}}{|K|} = r^2$, $\frac{dr^{*2}}{|K|} = dr^2$)

$$ds^2 = \frac{1}{|K|} \left(\frac{dr^{*2}}{1 - kr^{*2}} + r^{*2}(d\theta^2 + \sin^2\theta d\phi^2) \right), \qquad(11.12)$$

where $k = 1,\ -1,\ 0$ corresponding to a space of positive, negative and zero curvature respectively. Sometimes, this is also written as (ignoring the "*" on the radial coordinate in (11.12))

$$ds^2 = (R(t))^2 \left(\frac{dr^2}{1 - kr^2} + r^2(d\theta^2 + \sin^2\theta d\phi^2) \right). \qquad(11.13)$$

The four dimensional line element (11.1) correspondingly can be written as

$$d\tau^2 = dt^2 - ds^2$$

$$= dt^2 - R^2(t)\left(\frac{dr^2}{1 - kr^2} + r^2(d\theta^2 + \sin^2\theta d\phi^2)\right), \qquad (11.14)$$

and this is known as the Friedmann-Robertson-Walker line element. $R(t)$ is known as the scale factor for reasons that will become clear shortly.

11.2 Different models of the universe

There are now three distinct cases to consider depending on $k = \pm 1, 0$.

11.2.1 Close universe. Let us assume that $k = 1$. In this case, we can define coordinates

$$r = \sin\chi, \qquad dr = \cos\chi d\chi, \qquad (11.15)$$

such that the line element (11.13) has the form

$$\begin{aligned}
ds^2 &= R^2(t)\left(\frac{\cos^2\chi d\chi^2}{\cos^2\chi} + \sin^2\chi(d\theta^2 + \sin^2\theta d\phi^2)\right) \\
&= R^2(t)\left(d\chi^2 + \sin^2\chi(d\theta^2 + \sin^2\theta d\phi^2)\right). \qquad (11.16)
\end{aligned}$$

This indeed describes the three dimensional surface of a sphere of radius $R(t)$ which can be checked from the fact that if we define

$$\begin{aligned}
x_1 &= R\cos\chi, \\
x_2 &= R\sin\chi\sin\theta\cos\phi, \\
x_3 &= R\sin\chi\sin\theta\sin\phi, \\
x_4 &= R\sin\chi\cos\theta, \qquad (11.17)
\end{aligned}$$

we have

$$x_1^2 + x_2^2 + x_3^2 + x_4^2 = R^2, \tag{11.18}$$

and we can write

$$\mathrm{d}x_1^2 + \mathrm{d}x_2^2 + \mathrm{d}x_3^2 + \mathrm{d}x_4^2 = R^2 \left(\mathrm{d}\chi^2 + \sin^2\chi(\mathrm{d}\theta^2 + \sin^2\theta\mathrm{d}\phi^2)\right)$$

$$= \mathrm{d}s^2. \tag{11.19}$$

Thus, the case of constant positive curvature corresponds to a closed, spherical universe. It has a finite volume.

11.2.2 Flat universe. Let us next assume that $k = 0$. In this case, we can define the coordinates

$$
\begin{aligned}
x_1 &= Rr\sin\theta\cos\phi, \\
x_2 &= Rr\sin\theta\sin\phi, \\
x_3 &= Rr\cos\theta, \tag{11.20}
\end{aligned}
$$

which gives (see (11.13))

$$\mathrm{d}s^2 = R^2 \left(\mathrm{d}r^2 + r^2(\mathrm{d}\theta^2 + \sin^2\theta\mathrm{d}\phi^2)\right)$$

$$= \mathrm{d}x_1^2 + \mathrm{d}x_2^2 + \mathrm{d}x_3^2. \tag{11.21}$$

This is, of course, the three dimensional flat space which is infinite in volume and hence the case $k = 0$ is also said to correspond to an open universe.

11.2.3 Open universe. Finally, we consider the case $k = -1$. In this case, we can define the coordinates

$$r = \sinh\chi, \qquad \mathrm{d}r = \cosh\chi\mathrm{d}\chi, \tag{11.22}$$

so that the three dimensional line element (11.13) becomes

$$
\begin{aligned}
\mathrm{d}s^2 &= R^2 \left(\frac{\cosh^2 \chi \mathrm{d}\chi^2}{\cosh^2 \chi} + \sinh^2 \chi (\mathrm{d}\theta^2 + \sin^2 \theta \mathrm{d}\phi^2) \right) \\
&= R^2 \left(\mathrm{d}\chi^2 + \sinh^2 \chi (\mathrm{d}\theta^2 + \sin^2 \theta \mathrm{d}\phi^2) \right) .
\end{aligned}
\tag{11.23}
$$

It is easy to check that if we define

$$
\begin{aligned}
x_1 &= R \cosh \chi, \\
x_2 &= R \sinh \chi \sin \theta \cos \phi, \\
x_3 &= R \sinh \chi \sin \theta \sin \phi, \\
x_4 &= R \sinh \chi \cos \theta,
\end{aligned}
\tag{11.24}
$$

then,

$$
x_1{}^2 - x_2{}^2 - x_3{}^2 - x_4{}^2 = R^2(t),
\tag{11.25}
$$

and we can write

$$
\mathrm{d}s^2 = -\mathrm{d}x_1{}^2 + \mathrm{d}x_2{}^2 + \mathrm{d}x_3{}^2 + \mathrm{d}x_4{}^2.
\tag{11.26}
$$

This is what defines a three dimensional surface of a pseudosphere whose volume is infinite and, therefore, represents an open universe. However, this is different from the flat universe in that it has nontrivial curvature.

11.3 Hubble's law

Let us now restrict only to the cases $k = \pm 1$, the flat universe can be obtained in the limit $R \to \infty$. We can write the line element (11.21) as

$$
\begin{aligned}
k = 1 : \quad & \mathrm{d}\tau^2 = \mathrm{d}t^2 - R^2(t) \left(\mathrm{d}\chi^2 + \sin^2 \chi (\mathrm{d}\theta^2 + \sin^2 \theta \mathrm{d}\phi^2) \right), \\
k = -1 : \quad & \mathrm{d}\tau^2 = \mathrm{d}t^2 - R^2(t) \left(\mathrm{d}\chi^2 + \sinh^2 \chi (\mathrm{d}\theta^2 + \sin^2 \theta \mathrm{d}\phi^2) \right).
\end{aligned}
\tag{11.27}
$$

The coordinates χ, θ, ϕ are known as co-moving coordinates and are useful in the study of various cosmological phenomena. As the galaxies expand, they carry with them these coordinates and the distance between galaxies change with time not because the coordinates change with time – rather because the metric changes with time.

If we look at the radial 3-dimensional distance between two galaxies, then it has the form

$$\ell(t) = \int d\ell = \int d\chi\, R(t) = R(t) \int d\chi = R(t)\chi, \qquad (11.28)$$

where $R(t)$ is called the scale factor since it shows how distances scale with time. From this, we obtain

$$
\begin{aligned}
\frac{d\ell(t)}{dt} &= \frac{dR(t)}{dt}\chi = \frac{1}{R(t)}\frac{dR(t)}{dt}R(t)\chi \\[2mm]
&= \frac{\dot{R}(t)}{R(t)}\ell(t), \qquad\qquad\qquad\qquad (11.29)
\end{aligned}
$$

where a dot denotes a derivative with respect to t. This shows that if the galaxies are moving away from one another, then the velocity of recession is proportional to their distance, namely,

$$v(t) = \frac{d\ell(t)}{dt} = \frac{\dot{R}(t)}{R(t)}\ell(t) = H\ell(t). \qquad (11.30)$$

This is one form of Hubble's law and identifies

$$H(t) = \frac{\dot{R}(t)}{R(t)} = \text{Hubble's constant.} \qquad (11.31)$$

As we have noted, $R(t)$ is known as the scale factor. Hubble's constant (it is actually time dependent) is very crucial in cosmological studies and it is worth discussing how it is determined.

Let us consider the radial motion for either $k = \pm 1$. Then, from (11.27) we have

$$
\begin{aligned}
d\tau^2 &= dt^2 - R^2(t)d\chi^2 \\
&= R^2(\eta)(d\eta^2 - d\chi^2),
\end{aligned}
\tag{11.32}
$$

where we have defined

$$
dt = Rd\eta.
\tag{11.33}
$$

Propagation of light in this case, would correspond to

$$
d\eta^2 - d\chi^2 = 0.
\tag{11.34}
$$

The equation for the world line of the light signal in this case will be of the form

$$
\chi(\eta) = \eta - \eta_0.
\tag{11.35}
$$

If two light signals are emitted (from the same point) at times η_0 and $\eta_0 + \Delta\eta_0$, then the two world lines would be described by

$$
\begin{aligned}
\chi &= \eta - \eta_0, \\
\chi &= \eta - \eta_0 - \Delta\eta_0.
\end{aligned}
\tag{11.36}
$$

Thus, the two signals will arrive at a given coordinate χ with a time difference

$$
\Delta\eta(\chi) = \Delta\eta_0 = \text{constant}.
\tag{11.37}
$$

Namely, it will be a constant interval independent of the coordinate. The proper time interval, on the other hand, is given by (see (11.32))

$$
\Delta\tau = R\Delta\eta,
$$

or,
$$
\frac{R(\eta)}{\Delta\tau} = \frac{1}{\Delta\eta} = \text{constant},
$$

or,
$$
R(\eta)\nu(\chi) = \text{constant},
\tag{11.38}
$$

where ν is the frequency of the light wave at the coordinate χ.

In an expanding universe R increases and, therefore, ν must decrease inversely and this gives rise to a redshift in the signal. Since the product in (11.38) is a constant, we can write

$$R(\eta)\nu(\chi) = R(\eta_0)\nu(0)$$

or, $$\nu(\chi) = \nu(0)\frac{R(\eta_0)}{R(\eta)} = \nu(0)\frac{R(\eta - \chi)}{R(\eta)}. \tag{11.39}$$

If $\chi \ll 1$, that is, if the separation of the galaxies is not very large, we can Taylor expand the right hand side in (11.39) and write

$$\nu(\chi) \simeq \nu(0)\frac{1}{R(\eta)}\left(R(\eta) - \chi\frac{dR(\eta)}{d\eta}\right),$$

or, $$\frac{\nu(\chi) - \nu(0)}{\nu(0)} \simeq -\frac{1}{R(\eta)}\frac{dR(\eta)}{d\eta}\chi. \tag{11.40}$$

Using the defining relation (11.33)

$$dt = Rd\eta, \tag{11.41}$$

we have

$$\frac{d\eta}{dt} = R^{-1},$$

$$\frac{\nu(\chi) - \nu(0)}{\nu(0)} \simeq -\frac{d\eta}{dt}\frac{dR}{d\eta}\chi = -\dot{R}\chi. \tag{11.42}$$

In terms of the wave lengths, we can write this as

$$\frac{\lambda(\chi) - \lambda(0)}{\lambda(0)} \simeq \dot{R}\chi. \tag{11.43}$$

This, of course, gives the redshift and from the form of the Doppler shift formula (6.65) we note that

$$\frac{\lambda(\chi) - \lambda(0)}{\lambda(0)} = v = \dot{R}\chi = \frac{\dot{R}}{R}\,R\chi = H\ell, \tag{11.44}$$

giving us once again Hubble's law (11.30). This shows that Hubble's constant can be determined primarily from a study of the redshift of the galaxies and has the present value (from WMAP) of

$$H \simeq (7.04 \pm 0.15)\text{cm/sec/parsec}. \tag{11.45}$$

(There are further uncertainties associated with it. The error quoted here is only from random errors.) The inverse of the Hubble constant has the dimensions of time and defines a time scale which is normally identified with the age of the universe. The present value of the Hubble constant leads to an age of the universe

$$T_{\text{universe}} = H^{-1} \simeq 13.8 \times 10^9 \text{ yrs} \simeq 4.35 \times 10^{17} \text{sec}. \tag{11.46}$$

11.4 Evolution equation

Let us look at the Friedmann-Robertson-Walker line element (11.21) which has the form

$$d\tau^2 = dt^2 - R^2(t)\left(\frac{dr^2}{1 - kr^2} + r^2(d\theta^2 + \sin^2\theta d\phi^2)\right). \tag{11.47}$$

The Lagrangian for the geodesic can be obtained from this to have the form (see (5.127))

$$L = \frac{1}{2}\,\dot{t}^2 - \frac{1}{2}\,\frac{R^2}{1 - kr^2}\,\dot{r}^2 - \frac{1}{2}\,R^2 r^2 \dot{\theta}^2 - \frac{1}{2}\,R^2 r^2 \sin^2\theta\dot{\phi}^2, \tag{11.48}$$

where a dot denotes a derivative with respect to τ. We can now obtain the geodesic equations from the Euler-Lagrange equations (see (5.130) and (5.131)). The t equation gives

$$\ddot{t} + R \frac{dR}{dt} \left(\frac{1}{1 - kr^2} \dot{r}^2 + r^2 \dot{\theta}^2 + r^2 \sin^2 \theta \dot{\phi}^2 \right) = 0, \tag{11.49}$$

which determines the nontrivial components of $\Gamma^0_{\mu\nu}$ to be of the form (see (4.49) or (5.131))

$$\Gamma^0_{11} = -R \frac{dR}{dt} \frac{1}{1 - kr^2},$$

$$\Gamma^0_{22} = -R \frac{dR}{dt} r^2,$$

$$\Gamma^0_{33} = -R \frac{dR}{dt} r^2 \sin^2 \theta. \tag{11.50}$$

More compactly, we can write

$$\Gamma^0_{ij} = \frac{1}{R} \frac{dR}{dt} g_{ij}. \tag{11.51}$$

The r equation, leads to

$$-\frac{R^2}{1 - kr^2} \ddot{r} - 2R \frac{dR}{dt} \frac{1}{1 - kr^2} \dot{t}\dot{r} - \frac{2kR^2 r}{(1 - kr^2)^2} \dot{r}^2$$

$$+ \frac{dR^2 r}{(1 - kr^2)^2} \dot{r}^2 + R^2 r \dot{\theta}^2 + R^2 r \sin^2 \theta \dot{\phi}^2 = 0, \tag{11.52}$$

or, $\quad \ddot{r} + \frac{2dR}{Rdt} \dot{t}\dot{r} + \frac{kr}{1 - kr^2} \dot{r}^2 - r(1 - kr^2) \left(\dot{\theta}^2 - \sin^2 \theta \dot{\phi}^2 \right) = 0.$

This determines

$$\Gamma^1_{01} = \Gamma^1_{01} = -\frac{1}{R} \frac{dR}{dt},$$

$$\Gamma^1_{11} = -\frac{kr}{1 - kr^2},$$

$$\Gamma^1_{22} = r(1 - kr^2),$$

$$\Gamma^1_{33} = r(1 - kr^2) \sin^2 \theta. \tag{11.53}$$

Similarly, the θ equation gives

$$- R^2 r^2 \ddot{\theta} - 2Rr^2 \frac{\mathrm{d}R}{\mathrm{d}t} \dot{t}\dot{\theta} - 2R^2 r \dot{r}\dot{\theta} + R^2 r^2 \sin\cos\theta\dot{\phi}^2 = 0,$$

or, $\quad \ddot{\theta} + \dfrac{2}{R} \dfrac{\mathrm{d}R}{\mathrm{d}t} \dot{t}\dot{\theta} + \dfrac{2}{R} \dot{r}\dot{\theta} - \sin\theta\cos\theta\dot{\phi}^2 = 0, \qquad (11.54)$

which determines

$$
\begin{aligned}
\Gamma^2_{02} &= \Gamma^2_{20} = -\frac{1}{R} \frac{\mathrm{d}R}{\mathrm{d}t}, \\
\Gamma^2_{12} &= \Gamma^2_{21} = -\frac{1}{r}, \\
\Gamma^2_{33} &= \sin\theta\cos\theta. \qquad (11.55)
\end{aligned}
$$

Finally, the ϕ equation leads to

$$- R^2 r^2 \sin^2\theta\ddot{\phi} - 2Rr^2 \sin^2\theta \frac{\mathrm{d}R}{\mathrm{d}t} \dot{t}\dot{\phi} - 2R^2 \sin^2\theta r \dot{r}\dot{\phi}$$

$$- 2R^2 r^2 \sin\theta\cos\theta\dot{\theta}\dot{\phi} = 0,$$

or, $\quad \ddot{\phi} + \dfrac{2}{R} \dfrac{\mathrm{d}R}{\mathrm{d}t} \dot{t}\dot{\phi} + \dfrac{2}{r} \dot{r}\dot{\phi} + 2\cot\theta\dot{\theta}\dot{\phi} = 0, \qquad (11.56)$

and this determines

$$
\begin{aligned}
\Gamma^3_{03} &= \Gamma^3_{30} = -\frac{1}{R} \frac{\mathrm{d}R}{\mathrm{d}t}, \\
\Gamma^3_{13} &= \Gamma^3_{31} = -\frac{1}{r}, \\
\Gamma^3_{23} &= \Gamma^3_{32} = -\cot\theta. \qquad (11.57)
\end{aligned}
$$

These are all the nonvanishing components of the connection. Given these we can calculate the components of curvature tensor which are defined as

$$R_{\mu\nu} = \partial_\nu \Gamma^\lambda_{\lambda\mu} - \partial_\lambda \Gamma^\lambda_{\mu\nu} + \Gamma^\lambda_{\lambda\sigma}\Gamma^\sigma_{\mu\nu} - \Gamma^\lambda_{\nu\sigma}\Gamma^\sigma_{\mu\lambda}. \qquad (11.58)$$

This leads to

$$
\begin{aligned}
R_{00} &= \partial_0 \Gamma^\lambda_{\lambda 0} - \partial_\lambda \Gamma^\lambda_{00} + \Gamma^\lambda_{\lambda\sigma}\Gamma^\sigma_{00} - \Gamma^\lambda_{0\sigma}\Gamma^\sigma_{0\lambda} \\
&= \partial_0 \left(\Gamma^1_{10} + \Gamma^2_{20} + \Gamma^3_{30} \right) - \Gamma^1_{01}\Gamma^1_{01} - \Gamma^2_{02}\Gamma^2_{02} - \Gamma^3_{03}\Gamma^3_{03} \\
&= -3\, \frac{\mathrm{d}}{\mathrm{d}t}\left(\frac{1}{R}\frac{\mathrm{d}R}{\mathrm{d}t} \right) - 3\left(\frac{1}{R}\frac{\mathrm{d}R}{\mathrm{d}t} \right)^2 \\
&= -\frac{3}{R}\frac{\mathrm{d}^2 R}{\mathrm{d}t^2},
\end{aligned}
$$

$$
\begin{aligned}
R_{11} &= \partial_1 \Gamma^\lambda_{\lambda 1} - \partial_\lambda \Gamma^\lambda_{11} + \Gamma^\lambda_{\lambda\sigma}\Gamma^\sigma_{11} - \Gamma^\lambda_{1\sigma}\Gamma^\sigma_{1\lambda} \\
&= \partial_1 \left(\Gamma^1_{11} + \Gamma^2_{21} + \Gamma^3_{31} \right) - \partial_0 \Gamma^0_{11} - \partial_1 \Gamma^1_{11} \\
&\quad + \left(\Gamma^1_{10} + \Gamma^2_{20} + \Gamma^3_{30} \right)\Gamma^0_{11} + \left(\Gamma^1_{11} + \Gamma^2_{21} + \Gamma^3_{31} \right)\Gamma^1_{11} \\
&\quad - \Gamma^0_{11}\Gamma^1_{10} - \Gamma^1_{10} + \Gamma^0_{11} - \Gamma^1_{11}\Gamma^1_{11} - \Gamma^2_{12}\Gamma^2_{12} - \Gamma^3_{13}\Gamma^3_{13} \\
&= \frac{2}{r^2} + \frac{1}{1 - kr^2}\left[\frac{\mathrm{d}}{\mathrm{d}t}\left(R\frac{\mathrm{d}R}{\mathrm{d}t} \right) + \left(\frac{\mathrm{d}R}{\mathrm{d}t} \right)^2 + 2k \right] - \frac{2}{r^2} \\
&= \frac{1}{1 - kr^2}\left(2\left(\frac{\mathrm{d}R}{\mathrm{d}t} \right)^2 + R\frac{\mathrm{d}^2 R}{\mathrm{d}t^2} + 2k \right),
\end{aligned}
$$

$$
\begin{aligned}
R_{22} &= \partial_2 \Gamma^\lambda_{\lambda 2} - \partial_\lambda \Gamma^\lambda_{22} + \Gamma^\lambda_{\lambda\sigma}\Gamma^\sigma_{22} - \Gamma^\lambda_{2\sigma}\Gamma^\sigma_{2\lambda} \\
&= \partial_2 \Gamma^3_{32} - \partial_0 \Gamma^0_{22} - \partial_1 \Gamma^1_{22} + \left(\Gamma^1_{10} + \Gamma^2_{20} + \Gamma^3_{30} \right)\Gamma^0_{22} \\
&\quad + \left(\Gamma^1_{11} + \Gamma^2_{21} + \Gamma^3_{31} \right)\Gamma^1_{22} - \Gamma^0_{22}\Gamma^2_{20} - \Gamma^1_{22}\Gamma^2_{21} - \Gamma^2_{20}\Gamma^0_{22} \\
&\quad - \Gamma^2_{21}\Gamma^1_{22} - \Gamma^3_{23}\Gamma^3_{23} \\
&= \operatorname{cosec}^2\theta + r^2 \frac{\mathrm{d}}{\mathrm{d}t}\left(R\,\frac{\mathrm{d}R}{\mathrm{d}t} \right) - (1 - 3kr^2) + r^2\left(\frac{\mathrm{d}R}{\mathrm{d}t} \right)^2 \\
&\quad + r(1 - kr^2)\left(-\frac{kr}{1 - kr^2} - \frac{1}{r} \right) + \frac{1}{r}\, r(1 - kr^2) - \cot^2\theta \\
&= r^2\left(2\left(\frac{\mathrm{d}R}{\mathrm{d}t} \right)^2 + R\frac{\mathrm{d}^2 R}{\mathrm{d}t^2} + 2k \right),
\end{aligned}
$$

$$
R_{33} = \partial_3 \Gamma^\lambda_{\lambda 3} - \partial_\lambda \Gamma^\lambda_{33} + \Gamma^\lambda_{\lambda\sigma}\Gamma^\sigma_{33} - \Gamma^\lambda_{3\sigma}\Gamma^\sigma_{3\lambda}
$$

$$= - \partial_0 \Gamma^0_{33} - \partial_1 \Gamma^1_{33} - \partial_2 \Gamma^2_{33} + \left(\Gamma^1_{10} + \Gamma^2_{20} + \Gamma^3_{30} \right) \Gamma^0_{33}$$

$$+ \left(\Gamma^1_{11} + \Gamma^2_{21} + \Gamma^3_{31} \right) \Gamma^1_{33} + \Gamma^3_{32} \Gamma^2_{33} - \Gamma^0_{33} \Gamma^3_{30}$$

$$- \Gamma^1_{33} \Gamma^3_{31} - \Gamma^2_{33} \Gamma^3_{32} - \Gamma^3_{30} \Gamma^0_{33} - \Gamma^3_{31} \Gamma^1_{33} - \Gamma^3_{32} \Gamma^2_{33}$$

$$= r^2 \sin^2 \theta \frac{d}{dt} \left(R \frac{dR}{dt} \right) - \sin^2 \theta (1 - 3kr^2) - (\cos^2 \theta - \sin^2 \theta)$$

$$+ r^2 \sin^2 \theta \left[\left(\frac{dR}{dt} \right)^2 - k \right] + \cot \theta \sin \theta \cos \theta$$

$$= r^2 \sin^2 \theta \left(2 \left(\frac{dR}{dt} \right)^2 + R \frac{d^2 R}{dt^2} + 2k \right). \tag{11.59}$$

Thus, we see that we can write $(i, j = 1, 2, 3)$

$$R_{00} = -\frac{3}{R} \frac{d^2 R}{dt^2},$$

$$R_{ij} = -\frac{1}{R^2} \left(2 \left(\frac{dR}{dt} \right)^2 + R \frac{d^2 R}{dt^2} + 2k \right) g_{ij}. \tag{11.60}$$

We can now calculate the scalar curvature,

$$
\begin{aligned}
R &= g^{00} R_{00} + g^{ij} R_{ij} \\
&= -\frac{3}{R} \frac{d^2 R}{dt^2} - \frac{3}{R^2} \left(2 \left(\frac{dR}{dt} \right)^2 + R \frac{d^2 R}{dt^2} + 2k \right) \\
&= -\frac{6}{R^2} \left(\frac{dR}{dt} \right)^2 - \frac{6}{R} \frac{d^2 R}{dt^2} - \frac{6k}{R^2} \\
&= -6 \left(\frac{1}{R^2} \left(\frac{dR}{dt} \right)^2 + \frac{1}{R} \frac{d^2 R}{dt^2} + \frac{k}{R^2} \right).
\end{aligned} \tag{11.61}
$$

Therefore, we obtain the tt (00) component of Einstein's equation to be

$$G_{00} = R_{00} - \frac{1}{2} g_{00} R = 8\pi G T_{00},$$

$$\text{or,} \quad -\frac{3}{R}\frac{d^2R}{dt^2} - \frac{1}{2}(-6)\left(\frac{1}{R^2}\left(\frac{dR}{dt}\right)^2 + \frac{1}{R}\frac{d^2R}{dt^2} + \frac{k}{R^2}\right) = 8\pi G T_{00},$$

$$\text{or,} \quad \frac{1}{R^2}\left(\frac{dR}{dt}\right)^2 + \frac{k}{R^2} = \frac{8\pi G}{3} T_{00}. \tag{11.62}$$

This equation will determine the evolution of the scale factor depending on the value of k and the other components of Einstein's equation do not give any additional information.

To determine the evolution of $R(t)$, we need to assume the form of $T^{\mu\nu}$ for the matter distribution of the universe. We assume that the universe consists of a gas of particles (galaxies) and on a large scale the velocity of this gas is the same as the velocity of expansion. (Namely, there is no appreciable root mean square deviation of velocities.) Thus, it is appropriate to neglect the contribution of pressure to the stress tensor $T^{\mu\nu}$ of the system. In that case, we can write

$$T^{00} = T_{00} = \rho, \tag{11.63}$$

where ρ is the density of matter (energy) in our universe. Models where pressure is neglected (and also the cosmological constant is neglected) are known as the Friedmann models (universes) and the equation in this case becomes

$$\frac{1}{R^2}\left(\frac{dR}{dt}\right)^2 + \frac{k}{R^2} = \frac{8\pi G}{3} \rho. \tag{11.64}$$

Let us now study the three cases separately.

11.4.1 $k = 1$. The matter density varies inversely as the volume so that we can write (the unfamiliar factor in the volume comes from a nontrivial metric)

$$\rho = \frac{M}{2\pi^2 R^3}, \tag{11.65}$$

where, we can think of M as the total mass of the universe. In this case, equation (11.64) has the form (for $k = 1$)

$$\frac{1}{R^2}\left(\frac{dR}{dt}\right)^2 + \frac{1}{R^2} = \frac{4GM}{3\pi R^3}. \tag{11.66}$$

If we define, as before, (see (11.33))

$$dt = Rd\eta, \qquad \frac{d\eta}{dt} = \frac{1}{R}, \tag{11.67}$$

then equation (11.66) can be rewritten in the form

$$\left(\frac{1}{R}\frac{dR}{d\eta}\right)^2 + 1 = \frac{4GM}{3\pi}\frac{1}{R}. \tag{11.68}$$

The solution of this equation is quite simple.

$$
\begin{aligned}
R(\eta) &= R_*(1 - \cos\eta), \\
R_* &= \frac{2GM}{3\pi}, \\
t(\eta) &= R_*(\eta - \sin\eta).
\end{aligned}
\tag{11.69}
$$

As we had noted earlier (see subsection **11.2.1**), the case $k = 1$ corresponds to a closed universe and the evolution of the universe (scale factor) obtained in (11.69) is shown in Fig 11.1.

This is what is known as a periodic universe. Here the universe expands and then contracts eventually to a point and then starts all over again with a period of $2\pi R_*$.

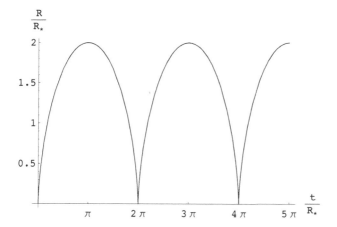

Figure 11.1: For $k = 1$ the universe is closed and the evolution periodic.

11.4.2 k = 0. In this case, equation (11.64) has the form

$$\frac{1}{R^2}\left(\frac{dR}{dt}\right)^2 = \frac{4GM}{3\pi R^3},$$

or, $$\frac{dR}{dt} = \sqrt{\frac{4GM}{3\pi}}\,\frac{1}{R^{\frac{1}{2}}} = \sqrt{2R_*}\,\frac{1}{\sqrt{R}},$$

or, $$R(t) = \left(\frac{9R_*}{2}\right)^{\frac{1}{3}} t^{\frac{2}{3}}. \tag{11.70}$$

These solutions can also be written in the parametric form

$$R(\eta) = \frac{R_*}{2}\,\eta^2,$$

$$t(\eta) = \frac{R_*}{6}\,\eta^3. \tag{11.71}$$

As we have seen, $k = 0$ corresponds to the flat universe and as is clear from (11.70), in this case, the universe expands forever as shown in Fig. 11.2 .

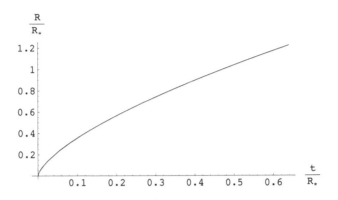

Figure 11.2: For $k = 0$ the universe is flat and open.

11.4.3 $k = -1$. In this case, the evolution equation (11.64) takes the form

$$\frac{1}{R^2}\left(\frac{dR}{dt}\right)^2 - \frac{1}{R^2} = \frac{4GM}{3\pi R^3}. \tag{11.72}$$

Here the volume of the space is infinite and hence we cannot identify M with the total mass of the universe. Rather, we should think of this as the total mass contained in a volume $2\pi^2 R^3$. The solution is again obtained simply by going to the η-variable (11.33)

$$dt = Rd\eta,$$

$$\frac{d\eta}{dt} = \frac{1}{R}. \tag{11.73}$$

The equation, in this variable, becomes

$$\left(\frac{1}{R}\frac{dR}{d\eta}\right)^2 - 1 = \frac{4GM}{3\pi R}, \tag{11.74}$$

whose solution is

$$R(\eta) \;=\; \frac{2GM}{3\pi}\,(\cosh\eta - 1) = R_*(\cosh\eta - 1),$$

$$t(\eta) \;=\; R_*(\sinh\eta - \eta). \tag{11.75}$$

Once again, here the universe expands forever and as $t \to \infty$, the universe becomes flat as shown in Fig. 11.3. We note that for smaller values of η (or time t), the solutions (11.75) reduce to (11.71). Namely, initially the open universe starts out as the flat universe. However, for larger times (η large) the evolution of the open universe is linear

$$R \sim t, \tag{11.76}$$

which is different from the flat universe.

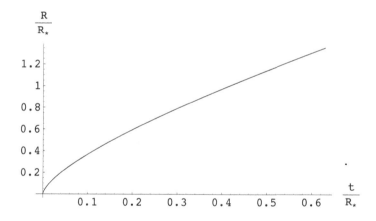

Figure 11.3: For $k = -1$ the universe is open.

11.5 Big bang theory and blackbody radiation

It is clear from the forms of the solutions in the previous section that our universe started out at some instant of time from a very small region and is expanding since then. Even in the case of the periodic universe, we are presently in an expanding epoch. One postulates that the univese started with a big bang from a small, densely packed, hot region and has been expanding ever since and the expansion is assumed to be adiabatic. This is known as the big bang theory due originally to Gamow.

If the big bang theory is true, then there must be some relic of this in the background photon radiation. Namely, throughout the evolution of the universe the background photons must have been in thermal equilibrium with the temperature of the background radiation falling with the expansion of the universe. A measurement of the temperature of the cosmic microwave background (CMB) radiation can, therefore, test the idea of the big bang theory if only the relic temperature can be calculated theoretically. Therefore, there was a lot of interest in such a calculation of the background radiation temperature quite early on. A careful calculation of this temperature involves a detailed analysis of the primordial nucleosynthesis and leads to a value of the background temperature to be approximately 3 K. However, there is also a much simpler (approximate) derivation of this temperature due to Gamow which only uses the information about the evolution of the universe (avoiding the technicalities associated with nucleosynthesis) and leads to a value of about 7 K for the temperature of the background radiation. The two numbers are in fact quite close and were theoretically derived much before the experimental observation of about 3 K for the temperature of the background radiation. The experimental observation of the appropriate relic temperature vindicates the big bang theory and in what follows we give the simpler derivation due to Gamow so as to avoid getting into the technicalities associated with the analysis of primordial nucleosynthesis. However, we would like to emphasize that a serious discussion of the background radiation temperature must involve the nucleosynthesis analysis.

In the context of the big bang theory, at very early times the temperature of the universe was quite high and consequently it was radiation dominated (matter was negligible). We recall from statistical mechanics that the energy density for radiation in equilibrium at a given temperature can be related to its temperature through the Stefan-Boltzmann law as

$$\rho_{\text{rad}} = \frac{4\sigma}{c^3} \, T^4, \tag{11.77}$$

where

$$\sigma = 5.67 \times 10^{-5} \text{gm/sec}^3\text{-K}^4, \tag{11.78}$$

represents the Stefan-Boltzmann constant (constant of proportionality). The entropy density associated with radiation is given by

$$\mathcal{S} \sim T^3. \tag{11.79}$$

Thus, the total entropy which remains constant in an adiabatic expansion has the form

$$S = \mathcal{S}R^3 \sim (RT)^3 = \text{constant},$$

or, $RT = \text{constant}.$ \hfill (11.80)

In his calculation Gamow assumed that our universe is open ($k = -1$). Therefore, using (11.77) as well as (11.80) in the Einstein equation (11.64) for an open universe ($k = -1$), we obtain for earlier times

$$\frac{1}{R^2}\left(\frac{\mathrm{d}R}{\mathrm{d}t}\right)^2 - \frac{1}{R^2} = \frac{8\pi G_\mathrm{N}}{3}\rho_\mathrm{rad},$$

or, $$\left(\frac{\mathrm{d}T}{\mathrm{d}t}\right)^2 - T^4 = \frac{8\pi G_\mathrm{N}}{3} \times \frac{4\sigma}{c^3} T^6,$$

or, $$\frac{\mathrm{d}T}{\mathrm{d}t} = -\sqrt{\frac{32\pi G_\mathrm{N}\sigma}{3c^3}}\, T^3,$$

or, $$T(t) = \left(\frac{3c^3}{128\pi G_\mathrm{N}\sigma}\right)^{\frac{1}{4}} t^{-\frac{1}{2}}. \tag{11.81}$$

Here G_N denotes the gravitational constant (Newton's constant) whose value is given by

$$G_\mathrm{N} = 6.6726 \times 10^{-8} \text{cm}^3/\text{gm} - \text{sec}^2, \tag{11.82}$$

and we have neglected the T^4 term since at high temperature (earlier times) the T^6 term on the right hand side dominates. (In this case, the constant in (11.80) also drops out of (11.81).) We have also chosen the negative sign for the square root to signify that the temperature of the universe decreases as time increases in an expanding universe. Equation (11.81) leads to the radiation density in the universe at any time as

$$
\begin{aligned}
\rho_{\text{rad}}(t) &= \frac{4\sigma}{c^3} T^4 = \frac{4\sigma}{c^3} \times \frac{3c^3}{128\pi G_{\text{N}}\sigma} t^{-2} \\
&= \frac{3}{32\pi G_{\text{N}}} t^{-2} \simeq 4.4722 \times 10^5 \, t^{-2} \text{gm-sec}^2/\text{cm}^3. \quad (11.83)
\end{aligned}
$$

This shows that the radiation density in the universe decreases with time as we would expect.

Our universe is presently matter dominated and in a matter dominated universe the matter density has the form

$$
\rho_{\text{matter}} \sim \frac{1}{R^3}. \tag{11.84}
$$

Consequently, the Einstein equation (11.64) for a matter dominated open universe $(k = -1)$

$$
\frac{1}{R^2} \left(\frac{dR}{dt}\right)^2 - \frac{1}{R^2} = \frac{8\pi G_{\text{N}}}{3} \rho \simeq \frac{8\pi G_{\text{N}}}{3} \frac{1}{R^3}, \tag{11.85}
$$

can be approximated, for large values of R, by

$$
\frac{1}{R^2} \left(\frac{dR}{dt}\right)^2 - \frac{1}{R^2} = 0,
$$

or, $\quad R(t) \sim t.$ \hfill (11.86)

Thus, in this case, we have

$$
\rho_{\text{matter}}(t) \sim t^{-3},
$$

or, $\quad \rho_{\text{matter}}(t)t^3 = \text{constant}.$ \hfill (11.87)

The constant in (11.87) can be determined as follows. Taking the present estimate of the matter density in the universe as well as the present age of the universe (see (11.46))

$$\rho_{\text{matter}}(t_{\text{p}}) \simeq 2.4 \times 10^{-31} \text{ gm/cm}^3,$$

$$t_{\text{p}} \simeq 13.73 \times 10^9 \text{ yrs} \simeq 4.35 \times 10^{17} \text{ sec,} \qquad (11.88)$$

we obtain

$$\rho_{\text{matter}}(t_{\text{p}})t_{\text{p}}^3 \simeq 2.4 \times (4.35)^3 \times 10^{20} \simeq 1.9755 \times 10^{22} \text{gm-sec}^3/\text{cm}^3,$$
$$(11.89)$$

which determines the matter density at any earlier time t to be given by

$$\rho_{\text{matter}}(t) \simeq 1.9755 \times 10^{22}\, t^{-3} \text{ gm-sec}^3/\text{cm}^3. \qquad (11.90)$$

Gamow's idea was to extrapolate the two asymptotic behaviors in (11.83) and (11.90) in the following way. (This is clearly very simple minded, nonetheless the result is quite impressive.) Namely, we know that in the very early times the universe was radiation dominated. As time increased the relative density of radiation and matter decreased and presently we are in a matter dominated universe. Therefore, at some instant of time, say t_*, during the evolution the radiation density must have equalled the matter density. That would determine

$$\rho_{\text{rad}}(t_*) = 4.4722 \times 10^5 t_*^{-2} = 1.9755 \times 10^{22} t_*^{-3} = \rho_{\text{matter}}(t_*),$$
$$(11.91)$$

which determines

$$t_* = \frac{1.9755 \times 10^{22}}{4.4722 \times 10^5} \text{ sec} \simeq 4.417 \times 10^{16} \text{ sec.} \qquad (11.92)$$

Using this in (11.81) we obtain the temperature of the universe at that instant to be

$$T(t_*) = \left(\frac{3c^3}{128\pi G_{\mathrm{N}}\sigma} \right)^{\frac{1}{4}} t_*^{-\frac{1}{2}} = \left(\frac{c^3}{4\sigma} \frac{3}{32\pi G_{\mathrm{N}}} \right)^{\frac{1}{4}} \times \frac{1}{\sqrt{t_*}}$$

$$= \left(\frac{27 \times 10^{30} \times 4.4722 \times 10^5}{4 \times 5.67 \times 10^{-5}} \right)^{\frac{1}{4}} \times \frac{1}{\sqrt{4.417} \times 10^8}$$

$$\simeq 72.27K. \tag{11.93}$$

Recalling that in the matter dominated epoch (see (11.86)),

$$R(t) \sim t, \tag{11.94}$$

and using (11.80) we obtain

$$\frac{T(t_{\mathrm{p}})}{T(t_*)} = \frac{t_*}{t_{\mathrm{p}}},$$

$$\text{or,} \quad T(t_{\mathrm{p}}) = T(t_*) \frac{t_*}{t_p} \simeq 72.27\text{K} \times \frac{4.417 \times 10^{16}}{4.35 \times 10^{17}} \simeq 7.3\text{K}. \tag{11.95}$$

This very crude calculation shows that the temperature of the background blackbody radiation in the universe at present must be about 7 K which is very close to the observed value of about 3 K. As we have emphasized earlier, a careful determination of the background radiation temperature involves a careful analysis of the primordial nucleosynthesis in the universe and leads to a value of about 3 K consistent with the observed value. This marks a significant confirmation of the big bang theory.

Index